STATISTICS IN RESEARCH AND DEVELOPMENT

Second edition

TEXTS IN STATISTICAL SCIENCE SERIES

Editors:

Dr Chris Chatfield
Reader in Statistics
School of Mathematical Sciences
University of Bath, UK

Professor Jim V. Zidek
Department of Statistics
University of British Columbia
Canada

OTHER TITLES IN THE SERIES INCLUDE

Practical Statistics for Medical Research
D.G. Altman

Interpreting Data
A.J.B. Anderson

Statistical Methods for SPC and TQM
D. Bissell

Statistics in Research and Development
Second edition
R. Caulcutt

The Analysis of Time Series
Fourth edition
C. Chatfield

Statistics in Engineering – A Practical Approach
A.V. Metcalfe

Statistics for Technology
Third edition
C. Chatfield

Introduction to Multivariate Analysis
C. Chatfield and A.J. Collins

Modelling Binary Data
D. Collett

Modelling Survival Data in Medical Research
D. Collett

Applied Statistics
D.R. Cox and E.J. Snell

Statistical Analysis of Reliability Data
M.J. Crowder, A.C. Kimber,
T.J. Sweeting and R.L. Smith

An Introduction to Generalized Linear Models
A.J. Dobson

Introduction to Optimization Methods and their Applications in Statistics
B.S. Everitt

Multivariate Studies – A Practical Approach
B. Flury and H. Riedwyl

Readings in Decision Analysis
S. French

Multivariate Analysis of Variance and Repeated Measures
D.J. Hand and C.C. Taylor

The Theory of Linear Models
B. Jorgensen

Statistical Theory
Fourth edition
B. Lindgren

Randomization and Monte Carlo Methods in Biology
B.F.J. Manly

Statistical Methods in Agriculture and Experimental Biology
Second edition
R. Mead, R.N. Curnow and
A.M. Hasted

Elements of Simulation
B.J.T. Morgan

Probability: Methods and Measurement
A. O'Hagan

Essential Statistics
Third edition
D.G. Rees

Large Sample Methods in Statistics
P.K. Sen and J.M. Singer

Decision Analysis: A Bayesian Approach
J.Q. Smith

Applied Nonparametric Statistical Methods
Second edition
P. Sprent

Elementary Applications of Probability Theory
Second edition
H.C. Tuckwell

Statistical Process Control: Theory and Practice
Third edition
G.B. Wetherill and D.W. Brown

Statistics for Accountants
S. Letchford

Full information on the complete range of Chapman & Hall statistics books is available from the publishers.

STATISTICS IN RESEARCH AND DEVELOPMENT

Second edition

Roland Caulcutt

BP Chemicals Lecturer
University of Bradford

CRC Press
Taylor & Francis Group
Boca Raton London New York

CRC Press is an imprint of the
Taylor & Francis Group, an **informa** business
A CHAPMAN & HALL BOOK

First published 1999 by Chapman & Hall
First edition 1983
Second edition 1991

Published 2019 by CRC Press
Taylor & Francis Group
6000 Broken Sound Parkway NW, Suite 300
Boca Raton, FL 33487-2742

First issued in paperback 2019

No claim to original U.S. Government works

ISBN-13: 978-0-367-45047-2 (pbk)
ISBN-13: 978-0-412-35890-6 (hbk)

Visit the Taylor & Francis Web site at
http://www.taylorandfrancis.com

and the CRC Press Web site at
http://www.crcpress.com

Library of Congress Cataloging-in-Publication Data

Catalog record is available from the Library of Congress.

Contents

Preface

The first edition of this book was written for chemists working in the chemical industry, with a suggestion in the preface that it might also be useful to engineers, scientists and managers. Correspondence I have received, however, indicates that the first edition found a wider readership, being well received outside the chemical industry. Thus I would like to suggest that this second edition could be of use to anyone concerned with the development or the management of complex processes in any industry.

Since the first edition was published in 1983 many readers will have experienced great change. The economic recession of the early 1980s caused many companies in the chemical and allied industries to make severe staff cuts. These cuts were not entirely beneficial. They severely interrupted, even reversed, the slow learning process by which we gain a better understanding of complex manufacturing processes. Thus we now have new staff attempting to solve old problems. The need for this book has, therefore, not diminished.

Further change has been thrust upon us by our customers. They demand quality. Furthermore, they demand that we demonstrate our ability to produce quality consistently. Customers have also realized the benefits to be gained by reducing the number of suppliers they deal with. Thus we are striving for never ending process improvement in order to stay in business. Clearly, the need for this book has increased.

To match the broadening perspectives of those dedicated to process improvement, this second edition has been enlarged by the addition of two new chapters on quality and by many references in other chapters to quality improvement and process capability. The new Chapter 8 on statistical process control and the new Chapter 16 on Taguchi methods are written in the non-mathematical and problem-centred style that was praised by many reviewers of the first edition. The main theme of the book remains unchanged, however, and it can be summarized as follows: Chapters 1 to 9 cover the basic techniques of data analysis which can be used to assess performance, to compare methods and to control quality. Every scientist and technologist in the process industries would benefit from a knowledge of these techniques. Chapters 10 to 16 cover experimental design and more advanced techniques of data analysis, including multiple regression analysis and analysis of variance. The range of experimental plans is restricted to factorial and fractional factorial designs, which have proved very useful in the process industries. The final chapter examines the collection of tools known as 'Taguchi methods' and shows how these are related to the techniques and concepts in earlier chapters. I certainly would not argue that this second part of the book should be studied by such a wide audience as the first. However, as one reader pointed out to me, 'You don't know what you are missing until you have read it all.'

I would like to take this opportunity to thank the hundreds of industrial scientists who have attended short courses based on the first edition. Their comments and questions have helped me to produce a second edition which should be more useful to other process developers. I would also like to thank the numerous colleagues who have assisted with the presentation of these courses and made many suggestions for their improvement. I am particularly grateful to Dr Les Porter of Bradford University and Dr Chris Pickles of Laporte plc.

Finally, I thank Barbara Shutt and Ann Fenoughty for wordprocessing my manuscript. They now understand, perhaps better than many readers, that it is difficult to produce a high quality product from low quality feedstock.

Any reader wanting details of short courses and consultancy in applied statistics or quality management should contact Roland Caulcutt, The Management Centre, University of Bradford, Emm Lane, Bradford BD9 4JL, Tel: (0274) 542299.

1

What is statistics?

The purpose of this chapter is to single out the prime elements of statistics so that you will not lose sight of what is most important as you work through the detail of subsequent chapters. Certain words will be introduced which will recur throughout the book; they are written in bold letters in this chapter when it first appears. Let us start with a simple definition: Statistics is a body of knowledge which can be of use to anyone who has taken a **sample**. It is appropriate that the first word which has been presented in bold letters is sample, since its importance cannot be over-emphasized. Many people in the chemical and allied industries occasionally find themselves in the unenviable position of needing to investigate a **population** but are constrained by the available resources to examine only a sample taken from the population. In essence a population is simply a large group. In some situations the population is a large group of people, in other situations it is a large group of inanimate objects, though it can in some cases be more abstract. Let us consider some examples of situations in which someone has taken a sample.

Example 1.1

A works manager wishes to **estimate** the average number of days of absence due to sickness for his employees during 1977. From the files which contain details of absences of the 2300 employees his secretary selects the record cards of 50 employees. She calculates that the average number of days of sickness absence of the 50 employees during 1977 is 3.61 days.

In this example the population consists of 2300 employees whilst the sample consists of 50 employees. After the sample was taken there was no doubt about the average sickness absence of the employees in the sample. It is the population average about which we are uncertain. The works manager intends to use the sample average (3.61 days) as an estimate of the population average. Common sense would suggest that there are dangers to be heeded. Perhaps it is worthwhile to spell out several points, some of which may be obvious:

(a) If the secretary had selected a different set of 50 employees to include in the sample the sample average would almost certainly have had a different value.

(b) The sample average (3.61) is very unlikely to be equal to the population average. (More formally we could say that the **probability** of the sample average being equal to the population average is very small.)

(c) Whether or not the sample average is close to the population average will depend on whether or not the sample is **representative** of the population.

(d) No one can guarantee that a sample will be representative of the population from which it came. We can only hope that, by following a reputable procedure for taking the sample, we will end up with a sample which is representative of the population. It is unfortunately true, as Confucius may have pointed out, that 'He who takes a sample takes a risk'.

(e) One reputable procedure for taking a sample is known as **random sampling**. The essential feature of this method is that every member of the population has the same chance (or probability) of being included in the sample. The end product of random sampling is known as a **random sample**, and all the statistical techniques which are introduced in this book are based upon the assumption that a random sample has been taken.

Example 1.2

Nicoprone is manufactured by a batch production process. The plant manager is worried about the percentage impurity, which appears to be higher in recent batches than it was in batches produced some months ago. He suspects that the impurity of the final product may depend on the presence of polystyline in the resin which is one of the raw materials of the process. The supplier of the resin has agreed that the polystyline content will not exceed 1% on average and that no single bag will contain more than 2.5%. Approximately 900 bags of resin are in the warehouse at the present time. The warehouse manager takes a sample of 18 bags by selecting every 50th bag on the shelf. From each of the selected bags a 20 gram sample* of resin is taken. The determinations of polystyline content are:

1.6%	0.5%	3.1%	0.7%	0.8%	1.7%	1.4%	0.8%	1.1%
0.9%	2.4%	0.6%	2.2%	2.9%	0.3%	0.5%	1.0%	1.3%

---

* Note the different usage of the word *sample* by the chemist and the statistician. In this example the chemist speaks of 18 samples whilst the statistician speaks of one sample containing 18 items.

We can easily calculate that the average polystyline content is equal to 1.32% and we notice that two of the determinations exceed 2.5%. So the sample average is certainly greater than the specified limit of 1.0%, but we need to consider what this average might have been if the warehouse manager had selected a different sample of bags, or perhaps we should ask ourselves what the average polystyline content would have been if he had sampled all the bags in the warehouse.

Questions like these will be answered in subsequent chapters. At this point we will probe the questions more deeply by translating into the language of the statistician. In this more abstract language we would ask 'What conclusions can we draw concerning the population, based upon the sample that has been examined?' This in turn prompts two further, very important questions, 'What exactly is the population about which we wish to draw conclusions?' and, perhaps surprisingly, 'What exactly is the sample on which the conclusions will be based?'

It is easier to answer these questions in reverse order. The sample can be looked upon as either:

(a) 18 bags of resin;
(b) 18 quantities of resin, each containing 20 g; or
(c) 18 measurements of polystyline content.

Whether it is better to take (a), (b) or (c) will depend upon such chemical/physical considerations as the dispersion of the polystyline within the resin and how the **variability** between bags compares with the variability within bags. Let us assume that each measurement gives a true indication of the polystyline content of the bag it represents and we will take (a) as our definition of the sample.

If our sample consists of 18 bags of resin then our population must also consist of bags of resin. The 18 bags were chosen from those in the warehouse, so it might seen reasonable to define our population as 'the 900 bags of resin in the warehouse', but does this cover *all* the resin about which the manager wishes to draw conclusions? He may wish to define the population in such a way as to include all bags of resin received in the past from this supplier and all bags to be received in the future. Before taking this bold step he would need to ask himself, 'Is the sample I have taken **representative** of the population I wish to define?'

It would obviously not be possible to take a random sample from such a population since the batches to be received in the future do not yet exist. Whenever we attempt to predict the future from our knowledge of the past we are talking about a population from which we cannot take a random sample. We may, nonetheless, be confident that our sample is representative. (The statistician prefers to discuss random samples rather than **representative samples** since the former are easier to define and are amenable to the tools of probability theory. The statistician does not, however, wish to see the statistical tail wagging the scientific/technical dog.)

Even if the plant manager defines his population as 'the 900 bags of resin in the warehouse' he still hasn't got a random sample. When the warehouse manager selected every 50th bag from the shelf he was practising what is known as **systematic sampling**. This is a procedure which is often used in the inspection of manufactured products and there is a good chance that systematic sampling will give a representative sample provided there are no hidden patterns in the population.

It has already been stated that the statistical techniques in this book are built upon the mathematical basis of random sampling, but this is only one of the many assumptions used by statisticians. An awareness of these assumptions is just as important to the scientist or technologist as the ability to select the appropriate statistical technique. For this reason a substantial part of Chapter 7 is devoted to the assumptions underlying the important techniques presented in Chapters 4 to 6.

2
Describing the sample

2.1 INTRODUCTION

In the previous chapter we focused attention on three words which are very important in statistics:

Sample
Population
Variability.

When a scientist or technologist is using statistical techniques he is probably attempting to make a generalization, based upon what he has found in one or more samples. In arguing from the particular to the general he will also be inferring from the sample to the population. Whilst doing so it is essential that he takes account of the variability within the sample(s).

It is the variability in the sample that alerts the scientist to the presence of random variation. Only by taking account of this random variation can we have an objective procedure for distinguishing between real and chance effects. Thus a prerequisite of using many statistical techniques is that we should be able to measure or describe the variability in a set of data.

In this chapter we will examine simple methods of describing variability and in doing so we will confine our attention to the sample.

2.2 VARIABILITY IN PLANT PERFORMANCE

Higson Industrial Chemicals manufactures a range of pigments for use in the textile industry. One particular pigment, digozo blue, is made by a well-established process in a plant which has recently been renovated. During the renovation various modifications were incorporated, one of which made the agitation system fully automatic. Though this programme of work was very successful in reducing the number of operators needed to run the plant, production of digozo blue has not been completely trouble-free since the work was completed. First, the anticipated increase in yield does not appear to have materialized, and secondly, several batches have been found to contain a disturbingly large percentage of a particular impurity.

The plant manager has suggested that the plant is unable to perform as well as expected because the agitator cannot cope with the increased capacity of the vessel. With the new control system the agitation speed is automatically reduced for a period of two minutes whenever the agitator drive becomes overloaded. The plant manager is of the opinion that it is during these slow periods that the reaction is unsatisfactory.

Whether his diagnosis is correct or not there can be no doubt that these overloads do occur because each incident is automatically recorded on a data logger. The number of overloads which occurred has been tabulated by the chief chemist (production) in Table 2.1 for the first 50 batches produced since the renovation was completed. Also tabulated is the yield and the percentage impurity of each batch.

Table 2.1   Yield, impurity and overloads in 50 batches of digozo blue pigment

Batch	Yield	Impurity	Number of overloads	Batch	Yield	Impurity	Number of overloads
1	69.0	1.63	0	26	69.5	1.78	1
2	71.2	5.64	4	27	72.6	2.34	1
3	74.2	1.03	2	28	66.9	0.83	0
4	68.1	0.56	2	29	74.2	1.26	0
5	72.1	1.66	1	30	72.9	1.78	1
6	64.3	1.90	4	31	69.6	4.92	3
7	71.2	7.24	0	32	76.2	1.58	0
8	71.0	1.62	1	33	70.4	5.13	5
9	74.0	2.64	2	34	68.3	3.02	1
10	72.4	2.10	2	35	65.9	0.19	3
11	67.6	0.42	0	36	71.5	2.00	1
12	76.7	1.07	1	37	69.9	1.15	0
13	69.0	1.52	1	38	70.4	0.66	0
14	70.8	11.31	2	39	74.1	3.24	3
15	78.0	2.19	0	40	73.5	2.24	2
16	73.6	3.63	2	41	68.3	2.51	2
17	67.3	1.07	1	42	70.1	2.82	1
18	72.5	3.89	0	43	75.0	0.37	0
19	70.7	2.19	2	44	78.1	2.63	1
20	69.3	0.71	3	45	67.2	4.17	2
21	75.8	0.83	1	46	71.5	1.38	0
22	63.8	1.38	0	47	73.0	3.63	4
23	72.6	2.45	3	48	71.8	0.76	2
24	70.5	2.34	0	49	68.8	3.09	3
25	76.0	8.51	5	50	70.1	1.29	0

The chief chemist hopes that the data in Table 2.1 will help him to answer several questions, including:

(a) If the present situation is allowed to continue what percentage of batches will contain more than 5% impurity?
(b) If the present situation is allowed to continue what percentage of batches will give a yield of less than 66%?
(c) Is the quality or the yield of a batch related to the number of overloads that occurred during the manufacture of the batch?

When we attempt to answer questions like those above, we make use of statistical techniques which come under the heading of inferential statistics. To answer the questions we would need to infer the condition of future batches from what we have found when examining the 50 batches which constitute our sample. We must postpone the use of inferential statistics until we have explored much simpler techniques which come under the alternative heading of descriptive statistics. The use of descriptive statistics will help us to summarize the data in Table 2.1 as a prelude to using it for inferential purposes.

2.3 FREQUENCY DISTRIBUTIONS

The yield values in Table 2.1 can be presented in a compact form if they are re-tabulated as in Table 2.2.

To produce Table 2.2 each of the 50 batches has been allocated to one of the nine groups listed in the upper row of the table. The number of batches in each group is indicated by the number below. The numbers in the bottom row of the table are known as frequencies and the whole table is often referred to as a frequency distribution. The frequencies do, of course, add up to 50 as there are 50 batches in the sample.

In Table 2.2, then, we can see at a glance exactly how many of the 50 batches have a yield between 66.0 and 67.9, say. By comparing the frequencies for the different groups we can easily see that many batches have been allocated to the three groups in the centre of the table with very few batches in the end groups.

Table 2.2 A frequency distribution for the yield of 50 batches

Yield (%)	62.0 to 63.9	64.0 to 65.9	66.0 to 67.9	68.0 to 69.9	70.0 to 71.9	72.0 to 73.9	74.0 to 75.9	76.0 to 77.9	78.0 to 79.9	Total
Number of batches	1	2	4	10	13	9	6	3	2	50

Table 2.3   Percentage frequency distribution – yield

Yield (%)	62.0 to 63.9	64.0 to 65.9	66.0 to 67.9	68.0 to 69.9	70.0 to 71.9	72.0 to 73.9	74.0 to 75.9	76.0 to 77.9	78.0 to 79.9	Total
Percentage of batches	2	4	8	20	26	18	12	6	4	100

Table 2.4   Proportional frequency distribution – yield

Yield (%)	62.0 to 63.9	64.0 to 65.9	66.0 to 67.9	68.0 to 69.9	70.0 to 71.9	72.0 to 73.9	74.0 to 75.9	76.0 to 77.9	78.0 to 79.9	Total
Proportion of batches	0.02	0.04	0.08	0.20	0.26	0.18	0.12	0.06	0.04	1.00

For some purposes it is convenient to express frequencies as percentages of the total frequency or as proportions of the total frequency. The percentage frequencies will add up to 100%, of course, and the proportions will add up to 1.0, as we see in Tables 2.3 and 2.4.

We could also produce frequency distributions for the other two columns

Table 2.5   Frequency distribution – number of overloads

Number of overloads	0	1	2	3	4	5	Total
Number of batches	15	13	11	6	3	2	50

Table 2.6   Frequency distribution – percentage impurity

Percentage impurity	Number of batches	Percentage impurity	Number of batches
0.00–0.99	9	6.00–6.99	0
1.00–1.99	16	7.00–7.99	1
2.00–2.99	12	8.00–8.99	1
3.00–3.99	6	9.00–9.99	0
4.00–4.99	2	10.00–10.99	0
5.00–5.99	2	11.00–11.99	1
		Total	50

of data in Table 2.1, as in Tables 2.5 and 2.6. There is no necessity to group the data for number of overloads as there are only six different values recorded for this variable.

(a) *The histogram and the bar chart*

The frequency distribution is a simple tabulation which gives us an overall appreciation of a set of data. It is particularly useful if the data set is large. On the other hand, it could be argued that we only get the maximum benefit from a frequency distribution if we display it in a pictorial form. Figures 2.1, 2.2 and 2.4 represent the frequency distributions of our three variables, yield, impurity and number of overloads.

Figure 2.1 and Figure 2.2 are known as histograms. In both diagrams the height of a bar (or block) tells us the number of batches which have yield (or impurity) in a particular range. We note that Figures 2.1 and 2.2 are very different in shape, with Figure 2.1 being virtually symmetrical whilst Figure 2.2 is very skewed. Perhaps we should have expected different shapes for the distributions of these two variables. In Figure 2.1 we see that a large number of batches are included in the middle three blocks, with yield between 68 and 74, and there are just a few batches in the tails of the distribution. In Figure 2.2 we see that most of the batches are included in the three blocks to the left of the picture, with impurity between 0% and 3%. It would not be possible for this distribution to 'tail-off' in both directions since we are already up against the lower limit (0%).

Whatever the reasons for the yield and the impurity of our 50 batches having very different frequency distributions, how does the difference in shape affect our confidence in future batches? It is probably fair to say that we can be confident that very few batches will have a yield outside the

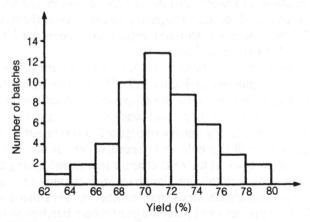

Figure 2.1   Yield of 50 batches of digozo blue

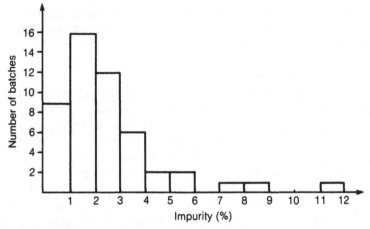

Figure 2.2   Impurity of 50 batches of digozo blue

range 62 to 80. Concerning the impurity of future batches we can be sure that none will have less than 0% impurity but we would surely be very reluctant to predict an upper limit. The right-hand tail of Figure 2.2 suggests that the next 50 batches might well contain one with impurity in excess of 12%.

The symmetrical pattern of Figure 2.1 is a great comfort to both chemist and statistician. A similar pattern is found in many sets of data and we will discuss this shape again in the next chapter. When confronted with a set of data which does not conform to this pattern, however, it is always possible to transform the data in order to change the shape of their distribution. A suitable transformation may result in a 'better' shape. One very popular transformation is to take logarithms of the individual observations. Taking logs (to the base 10) of the 50 impurity values gives us the 'log impurities' in Table 2.7 and these transformed values have been used to produce the frequency distribution in Table 2.8.

The frequency distribution of the log impurity values has been represented graphically in Figure 2.3. Clearly the transformation of the data has altered considerably the shape of the distribution. In Figure 2.3 we see the symmetrical shape that was noted in Figure 2.1.

Having admired the symmetry of Figure 2.3 you might be tempted to ask 'So what? How will this help us to speculate about the impurity of future batches?' We will see in the next chapter that transforming a set of data to get a distribution with this familiar shape may enable us to make use of one of the theoretical distributions that statisticians have made available. This can help us to predict what percentage of future batches will have impurity in excess of any particular value.

Table 2.7   Log impurity of 50 batches

Batch	Impurity	Log impurity	Batch	Impurity	Log impurity
1	1.63	0.21	26	1.78	0.25
2	5.64	0.75	27	2.34	0.37
3	1.03	0.01	28	0.83	−0.08
4	0.56	−0.25	29	1.26	0.10
5	1.66	0.22	30	1.78	0.25
6	1.90	0.28	31	4.92	0.69
7	7.24	0.86	32	1.58	0.20
8	1.62	0.21	33	5.13	0.71
9	2.64	0.42	34	3.02	0.48
10	2.10	0.32	35	0.19	−0.72
11	0.42	−0.38	36	2.00	0.30
12	1.07	0.03	37	1.15	0.06
13	1.52	0.18	38	0.66	−0.18
14	11.31	1.05	39	3.24	0.51
15	2.19	0.34	40	2.24	0.35
16	3.63	0.56	41	2.51	0.40
17	1.07	0.03	42	2.82	0.45
18	3.89	0.59	43	0.37	−0.43
19	2.19	0.34	44	2.63	0.42
20	0.71	−0.15	45	4.17	0.62
21	0.83	−0.08	46	1.38	0.14
22	1.38	0.14	47	3.63	0.56
23	2.45	0.39	48	0.76	−0.12
24	2.34	0.37	49	3.09	0.49
25	8.51	0.93	50	1.29	0.11

Table 2.8   Frequency distribution of log impurity

Log impurity	Number of batches	Log impurity	Number of batches
−0.80 to −0.61	1	0.20 to 0.39	15
−0.60 to −0.41	1	0.40 to 0.59	10
−0.40 to −0.21	2	0.60 to 0.79	4
−0.20 to −0.01	5	0.80 to 0.99	2
0.00 to 0.19	9	1.00 to 1.19	1
		Total	50

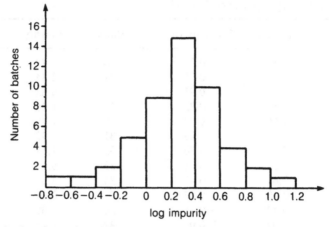

Figure 2.3   Log impurity of 50 batches of digozo blue

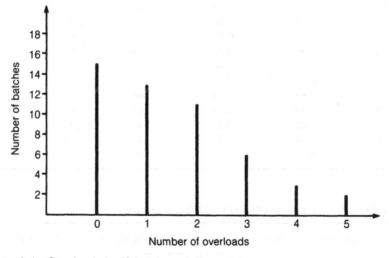

Figure 2.4   Overloads in 50 batches of digozo blue

The frequency distribution of our third variable (number of overloads) is displayed in Figure 2.4. This diagram is known as a bar chart. You will notice that it differs from the histograms that we have examined previously. In a histogram there are no gaps between blocks or bars whereas in a bar chart we have quite narrow bars with large gaps between them.

The two types of diagram are used to represent different types of variable. Yield and impurity are both continuous variables whereas number of overloads is a discrete variable. If we have two batches with yields of 70.14 and 70.15 it is possible to imagine a third batch which has yield between these two figures. Since all values (in a certain range) are possible we say that yield is a continuous variable. On the other hand we may have a batch which suffered one overload during its manufacture and we can have a second batch with two overloads but it would be meaningless to speak of a third batch with $1\frac{1}{2}$ overloads. As there are distinct gaps between the values which 'number of overloads' can have, we say that this is a discrete variable.

It is usual to represent the distribution of a continuous variable by a histogram and to represent the distribution of a discrete variable by a bar chart. The gaps between the bars in the bar chart emphasize the gaps between the values that the discrete variable can take.

2.4 MEASURES OF LOCATION AND SPREAD

A diagram such as a histogram or a bar chart can give a very clear impression of the distribution of a set of data and the use of diagrams is strongly advocated by statisticians. There is, nonetheless, a need to summarize a distribution in numerical form. This is usually achieved by calculating the mean and the standard deviation of the data.

The mean (or arithmetic mean, or average) is easily calculated by adding up the observations and dividing the total by the number of observations. The yields of the first six batches in our sample are:

69.0   71.2   74.2   68.1   72.1   64.3

Adding these six yields we get a total of 418.9 and dividing by 6 we get a mean yield of 69.816 666. Rounding this to two decimal places gives 69.82. (It is usual to round off the mean to one more decimal place than the original data.)

Clearly the calculation of the mean of a set of observations is very straightforward and there is no need to write down a formula for fear of forgetting how the calculation is carried out. We will nonetheless examine a formula as it will introduce a mathematical notation that will be used later.

$$\text{Sample mean} = (\Sigma x)/n \qquad (2.1)$$

$$\text{Population mean} = (\Sigma x)/N \qquad (2.2)$$

In these formulae n is the number of observations in the sample, N is the number of observations in the population and Σx means 'the sum of the observations'.

For those who are not familiar with the use of expressions such as Σx an explanation is offered in Appendix A, and a full list of the symbols used throughout this book is contained in Appendix B.

It is quite possible that you will use equation (2.1) on many occasions but it is doubtful whether you will ever use equation (2.2) since you are unlikely to be in the position of having observed every member of a population.

The mean is often referred to as a measure of location since it gives us an indication of where a histogram would be located on the horizontal axis. If all the observations in a set of data were increased by adding 10, say, to each then the mean would be increased by 10 and the location of the histogram would be shifted 10 units to the right along the axis.

An alternative measure of location is the median. We can find the sample median by listing the observations in ascending order; the median is then the 'middle one'. Listing the yields of the six batches in ascending order gives:

$$64.3 \quad 68.1 \quad 69.0 \quad 71.2 \quad 72.1 \quad 74.2$$

Having listed the six observations we realize that there isn't a 'middle one'. Clearly there will only be a middle observation if the set of data contains an odd number of observations. When a set of data contains an even number of observations we average the 'middle two' to find the median. Averaging 69.0 and 71.2 we get the median yield to be 70.10.

We have calculated the mean yield to be 69.82 and the median yield to be 70.10. Which is the better measure of location? The answer to this question depends upon what use we intend to make of the measure of location and upon the shape of the distribution. Certainly the mean is used more frequently than the median but with a very skewed distribution the median may be preferable. A survey of salaries of industrial chemists would be summarized by quoting a median salary, for example.

A measure of location is often more useful if it is accompanied by a measure of spread. In many situations the spread or variability of a set of data is of prime importance. This is particularly true when we are discussing the precision and bias of a test method. The variability or spread of a set of measurements can be used as an indication of the lack of precision of the method or the operator.

The simplest measure of spread is the range. The range of a set of data is calculated by subtracting the smallest observation from the largest. For our six batches the range of the yield measurements is 74.2 minus 64.3 which is 9.9.

Though the range is very easy to calculate, other measures of spread are often preferred because the range is very strongly influenced by two particular observations (the largest and smallest). Furthermore, the range of a large sample is quite likely to be larger than the range of a small sample taken from the same population. The range, therefore, is not a useful measure of spread if we wish to compare samples of different size. This deficiency is not shared by the two most useful measures of variability:

(a) the standard deviation;
(b) the variance.

We will discuss both at this point since they are very closely related. The standard deviation of a set of data is the square root of the variance. Conversely the variance can be found by squaring the standard deviation. Many electronic calculators have a facility for calculating a standard deviation very easily. If you have such a calculator the easiest way to obtain the variance of a set of data is first to calculate the standard deviation and then square it. On the other hand if you have only a very simple calculator you will find it easier to calculate a variance directly from the formula:

$$\text{Sample variance} = \Sigma(x - \bar{x})^2/(n - 1) \qquad (2.3)$$

in which $\bar{x}$ is the sample mean.

The use of this formula will be illustrated by calculating the variance of the yield values for the first six batches. The calculation is set out in Table 2.9.

Rounding the results of Table 2.9 to three decimal places gives the

Table 2.9 Calculation of sample variance by equation (2.3)

Yield (%) x	Yield − mean $(x - \bar{x})$	(Yield − mean)^2 $(x - \bar{x})^2$
69.0	−0.82	0.6724
71.2	1.38	1.9044
74.2	4.38	19.1844
68.1	−1.72	2.9584
72.1	2.28	5.1984
64.3	−5.52	30.4704
Total 418.9	−0.02	60.3884
Mean 69.82		

$$\text{Sample variance} = \Sigma(x - \bar{x})^2/(n - 1)$$
$$= 60.3884/5$$
$$= 12.077\,68$$
$$\text{Sample standard deviation} = \sqrt{(\text{sample variance})}$$
$$= \sqrt{12.077\,68}$$
$$= 3.475\,295\,6$$

sample variance of 12.078 and the sample standard deviation of 3.475. It is a common practice to round off a standard deviation to two decimal places more than the original data.

With regard to the calculation in Table 2.9 several points are worthy of mention:

(a) The total of the second column (-0.02) would have been exactly zero if the mean had not been rounded.

(b) It was said earlier that it is a common practice to round off a sample mean to one decimal place more than the data. There is no guarantee that a mean rounded in this way will be sufficiently accurate for calculating a standard deviation.

(c) It is also common practice to round off a sample standard deviation to two decimal places more than the data. Rounding in this way would give a sample standard deviation of 3.475.

(d) The entries in the third column of Table 2.9 are all positive; therefore the sum of the column [i.e. $\Sigma(x - \bar{x})^2$] must be positive, therefore the sample variance must be positive and the sample standard deviation must be positive.

(e) In equation (2.3) we divide $\Sigma(x - \bar{x})^2$ by $(n - 1)$. The question is often asked 'Why do we divide by $(n - 1)$ and not by n?' The blunt answer is that dividing by $(n - 1)$ is likely to give us a better estimate of the population variance. If we adopt the practice of dividing by n our sample variance will tend to underestimate the population variance.

(f) $\Sigma(x - \bar{x})^2$ is often referred to as the 'sum of squares' whilst $(n - 1)$ is referred to as the 'degrees of freedom'. To obtain a variance we divide the sum of squares by its degrees of freedom. Using this terminology we could translate the question in (e) into 'Why does the sum of squares have $(n - 1)$ degrees of freedom?' A simple answer can be seen if we refer to the $(x - \bar{x})$ column of Table 2.9. Though this column contains six entries it contains only five independent entries. Because the column total must be zero (ignoring rounding errors) we can predict the sixth number if we know any five numbers in the column. If we had n rows in Table 2.9, the centre column would contain $(n - 1)$ independent entries and $\Sigma(x - \bar{x})^2$ would have $(n - 1)$ degrees of freedom.

Equation (2.3) which we have used to calculate the sample variance is just one of several that we might have used. Equation (2.4) is an alternative which is easier to use especially if a simple calculator is available.

$$\text{Sample variance} = (\Sigma x^2 - n\bar{x}^2)/(n - 1) \qquad (2.4)$$

The use of this alternative method will be illustrated by repeating the calculation of the variance of the six yield values which is set out in Table 2.10.

Table 2.10   Calculation of variance by equation (2.4)

	Yield (%) x	(Yield)^2 x^2
	69.0	4 761.00
	71.2	5 069.44
	74.2	5 505.64
	68.1	4 637.61
	72.1	5 198.41
	64.3	4 134.49
Total	418.9	29 306.59
Mean	69.82	

$$
\begin{aligned}
\text{Sample variance} &= (\Sigma x^2 - n\bar{x}^2)/(n - 1) \\
&= [29\,306.59 - 6(69.82)^2]/5 \\
&= (29\,306.59 - 29\,248.994)/5 \\
&= 57.596/5 \\
&= 11.5192
\end{aligned}
$$

$$
\begin{aligned}
\text{Sample standard deviation} &= \sqrt{}(\text{sample variance}) \\
&= \sqrt{}11.5192 \\
&= 3.394
\end{aligned}
$$

You will note that the value of standard deviation calculated in Table 2.10 (3.394) is not in close agreement with the value calculated earlier (3.475) when using equation (2.3). We noted earlier that the first calculation was in error because we had rounded the mean to two decimal places. This same rounding of the mean has led to a much greater error when using equation (2.4). Had we used a sample mean of 69.816 666 when using equation (2.4) we would have obtained a standard deviation equal to 3.475 294 2 (working with eight decimal places). This agrees favourably with our first result and with the value of 3.475 293 753 calculated on a larger electronic calculator. Perhaps you find these discrepancies alarming. Be warned! Carry as many significant figures as your calculator will hold when performing statistical calculations. The sample standard deviation and the sample variance are closely related and it is very easy to calculate one if we know the other. The important difference between the two measures of spread is the units in which they are measured. The sample standard deviation, like the sample mean, is expressed in the same units as the original data. If, for example, the weight of an object has been determined five times then the mean and the standard deviation of the five determinations will be in grams if the five measurements were in grams.

Table 2.11 Fifty batches of digozo blue pigment

Variable	Yield	Impurity	Log impurity	Number of overloads
Mean	71.23	2.486	0.266	1.5
Standard deviation	3.251	2.1204	0.3488	1.53
Coefficient of variation (%)	4.6	85.3	131	102

The variance of the five determinations would be in grams squared. In some situations we use the sample standard deviation whilst in others we use the sample variance. Whilst the standard deviation can be more easily interpreted than the variance, we are forced to use the latter when we wish to combine measures of variability from several samples, as you will see in a later chapter.

The coefficient of variation is a measure of spread which is dimensionless. To calculate a coefficient of variation we express a standard deviation as a percentage of the mean.

$$\text{Coefficient of variation} = (\text{standard deviation/mean}) \times 100 \quad (2.5)$$

Equation (2.5) has been used to calculate the coefficients of variation for each of the three variables recorded in Table 2.1 and for the log impurities from Table 2.7. The mean, standard deviation and coefficient of variation of each of the four variables are given in Table 2.11.

Comparing the standard deviations in Table 2.11 we see that the standard deviation in the yield column (3.251) is the greatest of the four. You could say that yield is the most variable of the four variables. However, as the four standard deviations are in different units, this is a dubious comparison. Comparing the coefficients of variation we see that the smallest of the four (4.6%) is in the yield column. In *relative* terms, then, yield is the least variable of the four sets of measurements.

The comparisons we have just made in Table 2.11 are rather artificial since it is of no practical importance to be aware that impurity is more variable than yield or vice versa. In this particular situation we are more concerned with comparing the yield and/or impurity of individual batches with limits laid down in a specification. There are many other situations, however, in which the comparison of standard deviations and/or coefficients of variation are of fundamental importance. In the analytical laboratory, for example, we might find that the errors of measurement in a particular method of test tend to be greater when a larger value is being measured. If we obtain repeat determinations at several levels of concentration we would not be surprised to find results like those in Table 2.12.

Table 2.12   Precision of a test method

Level of concentration	1	2	3	4
Determinations	15.1, 15.2, 15.2	29.9, 30.1, 30.0	40.4, 40.3, 40.6	57.3, 57.1, 57.5
Mean determination	15.17	30.00	40.43	57.30
Standard deviation	0.0577	0.1000	0.1528	0.2000
Coefficient of variation (%)	0.38	0.33	0.38	0.35

We can see in Table 2.12 that the standard deviation increases as the level of concentration increases. When, however, the standard deviation is expressed as a percentage of the mean to give the coefficient of variation, we find that this measure of relative precision is substantially constant.

Note that the use of coefficients of variation is restricted to measurements which have a true zero (i.e. ratio scale measurements). When measurements have an arbitrary zero (interval scale), then the coefficient of variation may be meaningless. Suppose, for example, we have three repeat determinations of temperature which could be expressed in degrees centigrade or degrees Fahrenheit, as in Table 2.13. The three Fahrenheit temperatures are exact equivalents of the centigrade temperatures, with no rounding errors being introduced. We see that the coefficients of variation differ and the difference arises because of the 32 degrees shift in the zero as we change from one scale to the other.

2.5   CUMULATIVE FREQUENCY DISTRIBUTIONS

Earlier we tabulated four frequency distributions to describe the batch to batch variability in yield, impurity, log impurity and number of overloads. From any of these tables we can quickly see how many batches fall into a particular group. In Table 2.2, for example, we see that 10 batches had a yield between 68.0 and 69.9. In Table 2.6 we see that 12 batches had percentage impurity between 2.00 and 2.99%.

Table 2.13   Measurements with an arbitrary zero

Temperature scale	Determinations	Mean	SD	C of V (%)
Centigrade	38.0, 40.0, 40.5	39.50	1.323	3.35
Fahrenheit	100.4, 104.0, 104.9	103.1	2.381	2.31

Table 2.14 Cumulative frequency distribution – yield

Yield (%)	62.0	64.0	66.0	68.0	70.0	72.0	74.0	76.0	78.0	80.0
Number of batches having yield less than the above value	0	1	3	7	17	30	39	45	48	50

For some purposes we need to know how many batches had a yield less than a specified value or how many batches had impurity less than a certain percentage. In order to furnish such information in a convenient form we tabulate what is known as a cumulative frequency distribution. This is produced by adding (or cumulating) the frequencies in a frequency distribution table. To produce a cumulative frequency distribution for the yield of the 50 batches we read Table 2.2 from left to right, adding the frequencies as we progress. Thus we see that:

$$1 \text{ batch had yield less than } 64.0$$

$$(1 + 2) = 3 \text{ batches had yield less than } 66.0$$

$$(1 + 2 + 4) = 7 \text{ batches had yield less than } 68.0$$

Putting this information into a tabular form gives us the cumulative frequency distribution in Table 2.14.

The numbers in the bottom row of Table 2.14 are known as cumulative frequencies. Note how these cumulative frequencies increase from 0 at the left of the table to 50 at the right. This is very different to the pattern in the bottom row of Table 2.2. If we put Table 2.14 into graphical form we would expect it to give a picture very different from the histogram of Figure 2.1. This expectation is confirmed by Figure 2.5.

Figure 2.5 is known as an *ogive* or a *cumulative frequency curve*. The latter name could be misleading as it is more usual to join the points by straight lines rather than by a continuous curve. The distinctive S shape of Figure 2.5 is associated with the symmetrical histogram that we saw in Figure 2.1 which represents the frequency distribution of yield. If we draw a cumulative frequency curve for the impurity determinations we would have obtained a rather different shape which is a consequence of the skewness in the impurity distribution.

2.6   THE SEVEN BASIC TOOLS

In subsequent chapters we will discuss a whole range of statistical techniques. Some will be more complex than others, but it would be unwise to assume that the most complex techniques are the most useful. In

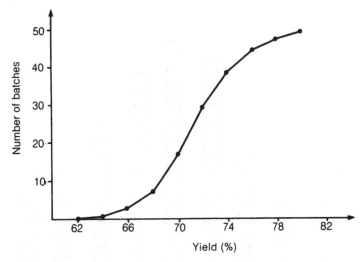

Figure 2.5   Cumulative frequency curve

fact it can be argued that there is more to be gained from widespread use of the simple methods than from the use of more advanced methods by specialist personnel. This is a view that would have been shared by the late Japanese quality control guru Kaoru Ishikawa. In his book *What is Total Quality Control?* he lists the 'seven indispensable tools for quality improvement':

(a) tally chart or check sheet
(b) histogram or dot plot
(c) Pareto chart
(d) cause and effect diagram
(e) scatter diagram
(f) control chart
(g) stratification.

Ishikawa states quite clearly that all of these simple tools should be used by everyone in manufacturing industry. It may well be true that widespread use of the 'seven tools' is one reason why Japanese companies have performed so well in recent years. I am sure that the profitability of many Western industrial companies could be improved if these techniques were understood at all levels from process operator to managing director.

Some of the seven basic tools will be covered in later chapters. Control charts in Chapters 8 and 9, for example, and scatter diagrams in Chapter 10. For details of the other tools you should refer to a text on statistical process control, such as Oakland (1990), as a full coverage would not be appropriate in this book. I would, however, like to illustrate the power of

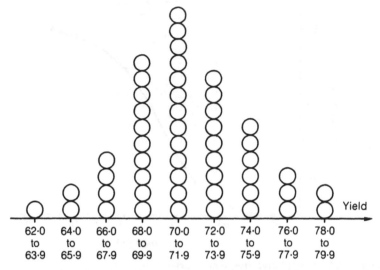

Figure 2.6   A dot plot of yield from 50 batches of digozo blue

stratification as this can be used to great effect in combination with other techniques.

You can see in the list of the seven basic tools that the dot plot is an alternative to the histogram. The dot plot in Figure 2.6 conveys the same information as the histogram in Figure 2.1. Which do you prefer? I think the dot plot is particularly useful when we have very few data. Suppose the 50 batches of digozo blue had been manufactured over a period of time by three shift teams. We could identify which batches were produced by each team and plot three separate dot plots, or stratify the dot plot, in Ishikawa's words.

Stratification is simply the splitting of a set of data into meaningful subgroups. By splitting the 50 yields into three groups we can compare the performance of the three shift teams. There is strong evidence in Figure 2.7 that shift team B is obtaining a higher average yield than team A. Could we reasonably claim that team B was superior to C? This type of question will be dealt with in Chapter 5. The important point to note at this time is that stratification can be used with histograms, dot plots and other diagrams to reveal important features of a set of data.

2.7 SUMMARY

In this chapter we have examined some of the calculations, tabulations and diagrams which can be used to condense a set of data into a more digestible

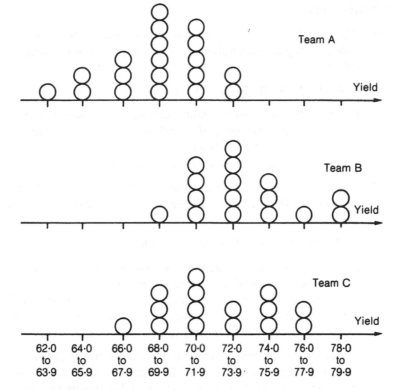

Figure 2.7 A stratified dot plot

form. Diagrams and tables are invaluable when reporting the results of empirical research. There is some truth in the saying 'A picture says more than a thousand words' but we must not underestimate the summary power of simple calculations.

There are several measures of location and spread but the most commonly used are the mean and standard deviation. These two will certainly be used repeatedly throughout this book. In later chapters, when we raise our eyes to look beyond the sample towards the population from whence the sample came, we will use the sample mean and standard deviation to estimate the location and spread of the population. These estimates can assist us to draw conclusions about the important features of the population so that we can predict what is likely to occur in the future or select from alternative courses of action.

Before we can devote ourselves exclusively to this main task of 'testing and estimation' we will spend the next chapter discussing ways of describing a population. Unfortunately that is not quite so easy as describing a sample.

PROBLEMS

2.1 The number of occasions on which production at a complex production plant was halted each day over a period of ten days are as follows:

4　2　7　3　0　3　1　13　4　3

(a) Find the mean and median number of production halts each day.
(b) Find the variance and standard deviation of the number of production halts each day.
(c) Is the variable measured above discrete or continuous?

2.2 An experiment was conducted to find the breaking strength of skeins of cotton. The breaking strengths (in pounds) for a small initial sample were:

90　99　97　89　108　99　82　96

(a) Find the mean, median, variance, standard deviation and coefficient of variation of this sample.
(b) Is the variable measured above discrete or continuous?
(c) Use the automatic routine on a calculator to confirm your answers for the mean and standard deviation.

2.3 Crosswell Chemicals has obtained a long-term order for triphenolite from a paint manufacturer who has specified that the plasticity of the triphenolite must be between 240.0 and 250.0. The first 50 batches of triphenolite produced by Crosswell give the plasticities listed below.
  If a batch is above specification it has to be scrapped at a cost of £1000. If it is below specification it can be blended to give a consignment within specification. The cost of blending is £300 for each

		Batch number		
1–10	11–20	21–30	31–40	41–50
240.36	245.71	248.21	242.16	247.52
245.21	247.32	243.69	250.24	244.71
246.09	244.67	246.42	241.74	246.09
242.39	244.56	242.73	247.71	249.31
249.31	245.61	249.31	245.61	243.28
247.32	241.51	245.32	244.91	247.11
243.04	246.21	244.11	245.72	242.66
245.72	242.01	250.23	246.31	245.91
244.22	246.51	247.01	243.54	246.17
251.51	243.76	240.71	244.62	248.26

batch below specification. The process can easily be adjusted to give a different mean level of plasticity and any such adjustment would not be expected to change the batch to batch variability of the production process.

The plasticity of the first 50 batches is summarized in the following table and Figure 2.8:

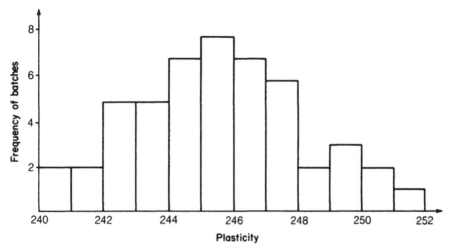

Figure 2.8   Histogram of batch plasticities

Sample size	50	Range	11.15
Mean	245.49	Variance	6.543
Standard deviation	2.561	Coefficient of variation	1.04%
Median	245.66		

Grouped frequency table

Plasticity	Number of batches
240.0 and under 241.0	2
241.0 and under 242.0	2
242.0 and under 243.0	5
243.0 and under 244.0	5
244.0 and under 245.0	7
245.0 and under 246.0	8
246.0 and under 247.0	7
247.0 and under 248.0	6
248.0 and under 249.0	2
249.0 and under 250.0	3
250.0 and under 251.0	2
251.0 and under 252.0	1

Using the analysis of data given above:

(a) To what value should the mean be set to minimize losses due to 'out-of-specification' batches?
(b) Do you consider that the sample has been drawn from a population with a symmetrical distribution?
(c) What is the population referred to in part (b)?

3

Describing the population

3.1 INTRODUCTION

In Chapter 2 we examined various ways of describing a sample. The methods we employed were basically very simple but they are, nonetheless, very powerful as a means of condensing a set of data into a more digestible form. You will recall that we made use of:

(a) frequency distributions;
(b) diagrams (such as histograms and bar charts);
(c) measures of location and spread (such as means and standard deviations).

We can adopt a similar approach when we wish to describe a population. There are, however, one or two complications which make the description of a population rather more difficult than the description of a sample. Firstly, the population will be much larger and may in some cases be infinitely large. Secondly, we will very rarely be able to measure every member of the population and this will necessitate that we infer some of the characteristics of the population from what we have found in the sample.

Statistical inference is so important that it will dominate the later chapters of this book. In this chapter we will confine our attention to several widely used population descriptions, which are known as probability distributions. These probability distributions are necessarily more abstract than the frequency distributions of the previous chapter and you may feel that you are being led astray if your main interest is data analysis. Let it be stated very clearly at the outset, therefore, why it is necessary to study probability distributions as a means of describing populations. There are two main reasons:

(a) We can use a probability distribution to predict unusual events. (When doing so we will be using the probability distribution as a model. Clearly it is important to choose a good model.)
(b) The methods of statistical inference that are described in later chapters

Table 3.1 Frequency distribution for 50 batches of pigment

Number of overloads	0	1	2	3	4	5	
							Total
Number of batches	15	13	11	6	3	2	50

are based on probability distributions. It is necessary to have some appreciation of the assumptions underlying these methods and this is only possible if you have some knowledge of the well-known distributions.

3.2 PROBABILITY DISTRIBUTIONS

When we transfer our attention from the sample to the population we must abandon the frequency distribution and adopt a probability distribution. There are similarities between the two, however, especially if we are describing a discrete variable. It will ease the transition from frequencies to probabilities therefore if we reconsider the frequency distribution in Table 2.5 which is reproduced as Table 3.1.

Table 3.1 describes the distribution of number of overloads for the 50 batches in the sample. What would this distribution have looked like if we had examined the whole population? Unfortunately we cannot answer this question unless we know exactly what we mean by 'the population'. Clearly the population consists of batches of digozo blue pigment, but which batches and how many batches?

We are free to define the population in any way we like. Perhaps the population should consist of those batches about which we wish to draw a conclusion. It would be wise, then, to include future batches in our population since the chief chemist's prime concern is to use what has happened in the past to predict what is likely to happen in the future. We will, therefore, define our population to be 'all batches of digozo blue pigment including past, present and future production'.

Having included future batches we cannot know exactly how many there will be. When speaking of the whole population, then, we will abandon frequencies and use proportions. We could, of course, have used proportions when describing the sample and this is illustrated by Table 3.2.

Table 3.2 Proportional frequency distribution for 50 batches

Number of overloads	0	1	2	3	4	5	
							Total
Proportion of batches	0.30	0.26	0.22	0.12	0.06	0.04	1.00

Table 3.3 Probability distribution for the population of batches

Number of overloads	0	1	2	3	4	5	
							Total
Probability							1.00

In Table 3.2 the numbers in the bottom row are proportions. We can see that the proportion of batches which were subject to two overloads was 0.22, for example. We note that the proportions add up to 1.00, which would be true no matter how many batches we had in the sample. Perhaps, then, something similar to Table 3.2 would be suitable for describing the whole population of batches. We do in fact use a table known as a probability distribution. This is based on probabilities, which in many respects are very similar to proportions. A formal definition of 'probability' will not be given in this book but the reader can safely regard a probability as a population proportion. A set of probabilities can be estimated by calculating a set of sample proportions like those in Table 3.2.

The probabilities which are missing from Table 3.3 will each be numbers between 0.00 and 1.00, like the proportions in Table 3.2. Furthermore the probabilities will add up to 1.00 which is another similarity between Tables 3.2 and 3.3.

How are we to obtain the probabilities which will enable us to complete Table 3.3? One possibility is to estimate them from the sample. This could be done by simply transferring the proportions from Table 3.2 to Table 3.3 and calling them probabilities. This would, of course, give misleading results if the sample were not representative of the population. Another possibility is to make use of one of the well-known probability distributions bequeathed to us by mathematical statistics. One of these theoretical distributions which has been found useful in many situations is known as the Poisson distribution.

3.3 THE POISSON DISTRIBUTION

Throughout the population the 'number of overloads' will vary from batch to batch. If the 'number of overloads' has a Poisson distribution then we can calculate the required probabilities using:

$$\text{Probability of } r \text{ overloads} = \mu^r e^{-\mu}/r! \tag{3.1}$$

$r!$ is read as 'r factorial' and is defined as follows:

$$0! = 1, \quad 1! = 1, \quad 2! = 2 \times 1, \quad 3! = 3 \times 2 \times 1, \quad 4! = 4 \times 3 \times 2 \times 1, \text{etc.}$$

The calculation of Poisson probabilities is quite easy on a pocket calculator once we have a value for μ and a value for r. μ is the mean number of overloads for all the batches in the population. Clearly we do not know the value of μ but we can estimate it from the sample. As the mean number of overloads for the 50 batches in the sample is 1.50 we will let $\mu = 1.5$. Letting $r = 0$ we can now calculate:

$$\text{Probability of 0 overloads} = (1.5)^0 e^{-1.5}/0!$$

$$= 0.2231$$

(Note that $(1.5)^0 = 1$)

This result is telling us that, if we select a batch at random from the population, there is a probability of 0.2231 that 0 overloads will occur during its production. Most people prefer percentages to probabilities so we could say, alternatively, that 22.3% of batches in the population would be subjected to 0 overloads during manufacture.

Letting $r = 1$, then letting $r = 2$, etc. in equation (3.1), we can calculate:

$$\text{Probability of 1 overload} = (1.5)^1 e^{-1.5}/1! = 0.3347$$

$$\text{Probability of 2 overloads} = (1.5)^2 e^{-1.5}/2! = 0.2510$$

$$\text{Probability of 3 overloads} = (1.5)^3 e^{-1.5}/3! = 0.1255$$

$$\text{Probability of 4 overloads} = (1.5)^4 e^{-1.5}/4! = 0.0471$$

$$\text{Probability of 5 overloads} = (1.5)^5 e^{-1.5}/5! = 0.0141$$

These probabilities could now be inserted in Table 3.3. Unfortunately the completed table would be a little ambiguous since the probabilities would not add up to 1.0. The total of the calculated probabilities is only 0.9955.

The missing 0.0045 is explained by the fact that the Poisson distribution does not stop at $r = 5$. We can insert higher values of r into equation (3.1) to obtain more probabilities. Doing so gives:

$$\text{Probability of 6 overloads} = 0.0035$$

$$\text{Probability of 7 overloads} = 0.0008$$

$$\text{Probability of 8 overloads} = 0.0001$$

The calculations have been discontinued at $r = 8$ because the probabilities have become very small though in theory the Poisson distribution continues forever. Tabulating the probabilities gives us the probability distribution in Table 3.4.

Table 3.4 Poisson distribution with $\mu = 1.5$

Number of overloads	0	1	2	3	4	5	6	7 or more
Probability	0.223	0.335	0.251	0.126	0.047	0.014	0.004	0.001

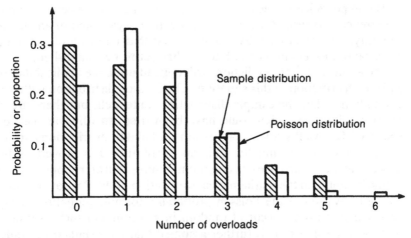

Figure 3.1 Multiple bar chart to compare two distributions

How do the probabilities in Table 3.4 compare with the proportions in Table 3.2? Perhaps the best way to compare the two distributions is by means of a graph such as Figure 3.1.

It is quite clear that the two distributions portrayed in Figure 3.1 are not identical. Perhaps it is unreasonable to expect them to be exactly the same. Whenever we take a sample from a population we must not expect that the scatter of values in the sample will have a distribution that is a perfect replica of the population distribution. Just how much variability we can expect from sample to sample will be discussed in a later chapter; at this point we will simply note that a sample is most unlikely to resemble in every respect the population from which it was taken.

So the two distributions in Figure 3.1 are different, but not very different. If someone put forward the suggestion that the population of batches had a Poisson distribution with $\mu = 1.5$ would you reject this suggestion on the evidence of the 50 batches in the sample? It is very doubtful!

Suppose, then, we accept that the number of overloads has a Poisson distribution with $\mu = 1.5$, what conclusions can we draw concerning future batches? We can use the probabilities in Table 3.4 to predict what percentage of future batches will be subjected to any particular number of overloads. We could predict, for example, that 1.4% of batches (i.e. 0.014×100) will have five overloads and that 0.4% of batches will have six overloads. By adding probabilities from Table 3.4 we can predict that 1.9% of batches [i.e. $(0.014 + 0.004 + 0.001) \times 100$] will have been subjected to at least five overloads during manufacture.

The importance of such predictions will depend on whether or not the occurrence of overloads is causing a reduction in yield and/or an increase in impurity. The chief chemist has suggested that interruptions in agitation have caused excess impurity, but this has yet to be established.

The accuracy of the predictions will depend upon the applicability of the Poisson distribution to this situation. The Poisson distribution with $\mu = 1.5$ certainly matches the sample distribution quite well, but there are many other distributions which could have been tried, so why was the Poisson selected? It is well known that certain probability distributions are useful in certain situations and the Poisson distribution has proved useful for describing the occurrence of equipment failures. Further support for its use would be found in the mathematical foundation of the Poisson distribution. In the derivation of the Poisson probability formula [equation (3.1)] it is assumed that events occur at random in a continuum such as time, or a line, or a space, etc. It is further assumed that the events occur independently of each other and at a fixed rate per unit time or per unit length, etc. In any succession of intervals the number of events observed will vary in a random manner from interval to interval with the probability of r events occurring in any particular interval being given by equation (3.1). The Poisson distribution has been found by the textile technologist to account for the scatter of faults along a continuous length of yarn and has been found useful by geographers when attempting to explain the scatter of certain natural phenomena in a continuous area of land. The decision to use the Poisson distribution, rather than an alternative, could have serious consequences, of course. Only if we select a suitable model can we hope to make accurate predictions of unlikely events such as 'eight overloads occurring during the manufacture of a batch'. You would need to know much more about several other distributions, before you were able to select the 'best' distribution for this task.

It is not our intention to discuss the finer points of probability distributions in this book. The reader who wishes to study further should consult Johnson and Leone, Vol. 1 (1964), Freund (1962) or Moroney (1966). We must now leave the Poisson distribution and examine what is undoubtedly the most important of all probability distributions.

3.4 THE NORMAL DISTRIBUTION

We have referred to 'number of overloads' as a discrete variable. Clearly the number of overloads occurring during the manufacture of any batch can only be a whole number (i.e. 0, 1, 2, 3, etc.) and when using equation (3.1) to calculate Poisson probabilities we substituted whole number values for r. The other two variables that were measured on each of the 50

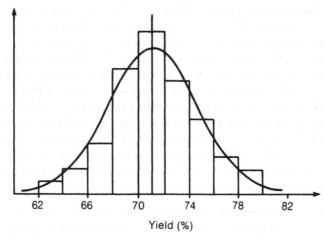

Figure 3.2 Comparing a histogram with a normal distribution curve

batches, yield and impurity, are not discrete variables. It would be most inappropriate to attempt to use the Poisson distribution to describe these continuous variables.

You will recall that we used histograms to describe the sample distributions for yield and impurity in contrast with the bar chart used to describe the distribution of 'number of overloads'. We noted at the time that one of the histograms (yield) was roughly symmetrical whilst the other was rather skewed. The symmetrical histogram is reproduced in Figure 3.2.

In Figure 3.2 a smooth curve has been superimposed on the histogram. Clearly the curve and the histogram are similar but not identical. The histogram represents the distribution of the yield values for the 50 batches in the *sample*. It is suggested that the curve might represent the distribution of the yield values of all the batches in the population.

The curve in Figure 3.2 represents a probability distribution which is known as the normal distribution. To be more precise it is a normal distribution which has mean equal to 71.23 and standard deviation equal to 3.251, these values having been chosen to match the mean and standard deviation of the 50 yield values in the sample. You would probably agree that it is reasonable to use the sample mean and standard deviation as estimates of the population mean and standard deviation, but you may wonder why we should introduce this particular curve. The main reasons why we turn to the normal distribution to describe this population are:

(a) The normal distribution has been found useful in many other situations.

(b) Tables are readily available which allow us to make use of the normal distribution without performing difficult calculations.
(c) The normal distribution can be shown to apply to any variable which is subjected to a large number of 'errors' which are additive in their effects. (If any one source of 'error' becomes dominant we may well get a skewed distribution, as we found with the impurity determinations.)
(d) It is reasonable to expect a single peak in the distribution curve since the operating conditions of the plant were chosen to give maximum yield. Any 'error' in the setting of these conditions would result in a loss of yield and this loss might be expected to be roughly equal for a departure in either direction. We would, therefore, expect a roughly symmetrical distribution curve.

The normal distribution has been studied by mathematicians for nearly 200 years and the equation of the normal curve also arises in the study of other physical phenomena such as heat flow. It has been found to be very useful for describing the distribution of errors of measurement and many natural measurements are found to have distributions which can be closely approximated by the normal curve. The heights of adult males and females, for example, are said to have normal distributions as illustrated in Figure 3.3. The two distributions have different means (69 inches and 63 inches) but the same standard deviation (3 inches).

Scores on intelligence tests are also said to have normal distributions. Tests are standardized so that the mean score for all adults is 100. The mean score for males and the mean score for females are found to be equal but the two distributions are not identical as the standard deviation for

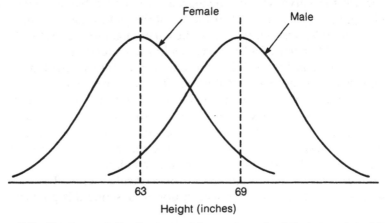

Figure 3.3 Two normal distributions with equal standard deviations but different means

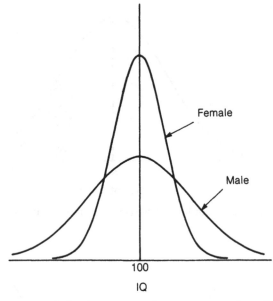

Figure 3.4 Two normal distributions with equal means but different standard deviations

males (17) is greater than that for females (13). The two distributions are illustrated in Figure 3.4.

You will notice that in Figure 3.4 the female scores are not so widely spread as the male scores. Note also that the narrower of the two distributions is also taller. This must be so because the areas under the two curves are equal. In fact the area under each curve is equal to 1.0.

Areas are very important when we are dealing with the probability distributions of continuous variables. By calculating the area under a distribution curve we automatically obtain a probability as we see in Figure 3.5. The probability that an adult female in Britain will have a height between 65 inches and 68 inches is given by the shaded area in Figure 3.5.

To find the shaded area in Figure 3.5 we could resort to mathematical integration but this is completely unnecessary if we make use of Statistical Table A. To obtain the probability of a height lying between 65 and 68 inches we proceed in two stages as follows. In stage 1 we concentrate on those females who are taller than 68 inches. The probability of a female having a height greater than 68 inches is represented by the shaded area in Figure 3.6(a).

To find this shaded area we must convert the height (x) into a standardized height (z) using:

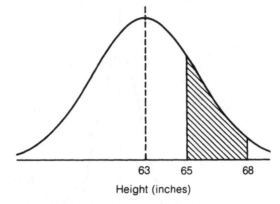

Figure 3.5   Females with heights between 65 and 68 inches

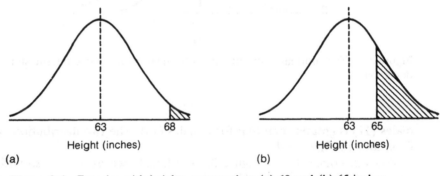

(a)   (b)

Figure 3.6   Females with heights greater than (a) 68 and (b) 65 inches

$$z = \frac{x - \mu}{\sigma}$$

where μ is the population mean and σ is the population standard deviation.

$$z = \frac{68 - 63}{3}$$

$$= 1.67$$

Turning to Table A we find that a z value of 1.67 gives us a probability of 0.0475. Thus, if we select an adult female at random the probability that her height will exceed 68 inches is 0.0475. Alternatively we might say that 4.75% of adult females in Britain are taller than 68 inches.

In stage 2 we concentrate on those females who are taller than 65 inches. The probability of a female having a height greater than 65 inches is represented by the shaded area in Figure 3.6(b). To find this probability we must once more calculate a standardized height:

$$z = \frac{x - \mu}{\sigma} = \frac{65 - 63}{3} = 0.67$$

Referring to Table A we obtain the probability 0.2515. Multiplying this by 100 leads us to the conclusion that 25.15% of adult females are taller than 65 inches.

Returning to our original problem, it is the shaded area in Figure 3.5 that we wish to find. This area represents the probability of an adult female having a height between 65 and 68 inches. We can now obtain this result by finding the difference between the two probabilities represented in Figure 3.6(a) and (b):

$$0.2515 - 0.0475 = 0.2040$$

If, therefore, an adult female is selected at random from the population there is a probability of 0.204 that her height will be between 65 and 68 inches.

Using Table A we could obtain other interesting probabilities concerning the heights of females. We could also make use of Table A to draw conclusions about the heights of males or the intelligence scores of males or females. Table A represents a particular normal distribution which has a mean of 0, and a standard deviation of 1. It is often referred to as the 'standard normal distribution' and it can be used to draw conclusions about any normal distribution. When we calculate a standardized value (z) we are in fact jumping to the standardized normal distribution from the normal distribution in which we are interested.

By using Table A we can draw general conclusions which are applicable to all variables which have a normal distribution. For example, we see that a z value of 1.96 in Table A corresponds to a probability of 0.025, which is $2\frac{1}{2}$% if expressed as a percentage. This information is displayed in Figure 3.7(a) which also shows that $2\frac{1}{2}$% of the area lies to the left of -1.96. Thus we can conclude that 95% of z values lie between -1.96 and $+1.96$. Transferring our attention to Figure 3.7(b), which represents a normal distribution with mean equal to μ and standard deviation equal to σ, we can say that 95% of values will lie between ($\mu - 1.96\sigma$) and ($\mu + 1.96\sigma$). Without using mathematical symbols we can conclude that for any variable which has a normal distribution, 95% of values will be within 1.96 standard deviations of the mean. Now that we appreciate the wide applicability of Table A and of the standard normal distribution let us return to our batches of digozo blue pigment. After inspecting Figure 3.2 you may be prepared to accept that the whole population of batches could have a normal distribution. If you are also prepared to accept that the mean of

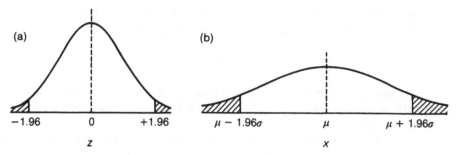

Figure 3.7   Central 95% of a normal distribution

this distribution is approximately 71.2 and the standard deviation approximately 3.25 we can predict the yield we are likely to get in future batches. (These estimates of the population mean and standard deviation are based on the sample mean and standard deviation. Estimation will be discussed fully in later chapters.) Using $\mu = 71.2$ and $\sigma = 3.25$ we calculate:

$$\mu + 1.96\sigma = 71.2 + 1.96(3.25) = 77.4$$

$$\mu - 1.96\sigma = 71.2 - 1.96(3.25) = 65.0$$

We can therefore expect that 95% of future batches will have yield between 65.0 and 77.4.

If we wish to quote an interval that will embrace 99% of batches we can make use of the following result: for any variable which has a normal distribution 99% of values will lie within 2.58 standard deviations of the mean.

$$\mu + 2.58\sigma = 71.2 + 2.58(3.25) = 79.6$$

$$\mu - 2.58\sigma = 71.2 - 2.58(3.25) = 62.8$$

Thus we can expect that 99% of future batches will have yield between 62.8 and 80.2.

You may recall that the chief chemist was particularly concerned about those batches with low yield and he wished to estimate the percentage of future batches which will have yield less than 66.0. If we assume that batch yields came from a normal distribution which has a mean of 71.2 and a standard deviation of 3.25 then the estimate we need is given by the shaded area in Figure 3.8.

Following the usual procedure we will first calculate a standardized yield and then refer to Table A.

$$z = \frac{x - \mu}{\sigma} = (66.0 - 71.2)/3.25 = -1.60$$

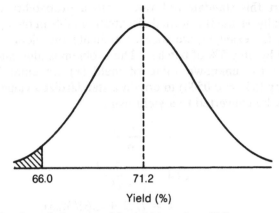

66.0   71.2

Yield (%)

Figure 3.8 Batches of pigment with yield less than 66.0

The standardized value is negative because we have standardized a yield value (66.0) which is below the mean. Table A does not contain any negative z values because it is a tabulation of only the right-hand side of the standard normal distribution. Because of the symmetry of the distribution there is no need to tabulate both sides. When we calculate a z value which is negative we simply ignore the minus sign whilst using Table A. Following this practice we get a probability of 0.0548 from Table A. Thus we conclude that 5.48% of future batches will have yield less than 66.0 if the present situation continues.

In our use of the normal distribution table we have first converted a value into a standardized value and then referred to Table A in order

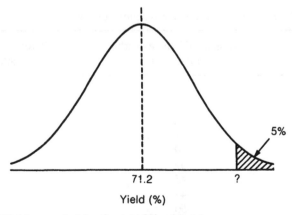

71.2   ?

Yield (%)

Figure 3.9 Yield exceeded by the top 5% of batches

to convert this standardized value into a probability or a percentage. Occasionally we use the normal distribution table in the opposite direction. Suppose, for example, we wish to estimate the yield value that will be exceeded by only 5% of batches. The problem is illustrated in Figure 3.9.

To find the unknown value of yield (x) we enter Table A with a probability (5% = 0.0500) to obtain a standardized value ($z = 1.64$). This must now be converted to a yield using:

$$z = \frac{x - \mu}{\sigma}$$

$$1.64 = \frac{x - 71.2}{3.25}$$

$$x = 71.2 + 1.64(3.25)$$

$$x = 76.53$$

Thus we can expect that only 5% of future batches will have yield greater than 76.53.

3.5 NORMAL PROBABILITY GRAPH PAPER

Our decision to use the normal distribution as a model was based entirely on Figure 3.2. The histogram, which describes the 50 batches in the sample, closely resembles the normal curve so we conclude that the normal curve might well describe the population of batches. The normal distribution in Figure 3.2 has the same mean and standard deviation as the histogram.

Superimposing a normal curve onto a histogram is more difficult than you might imagine. In fact the task is so difficult that in practice an

Table 3.5 Percentage cumulative frequencies – yield

Yield (x)	Number of batches with yield less than x	Percentage of batches with yield less than x
64.0	1	2
66.0	3	6
68.0	7	14
70.0	17	34
72.0	30	60
74.0	39	78
76.0	45	90
78.0	48	96
80.0	50	100

alternative approach is used. This approach involves plotting the sample data onto special graph paper known as 'normal probability paper'. In order to use this special paper we must return to the percentage cumulative frequency distribution that was tabulated in Chapter 2 and is reproduced as Table 3.5.

In Chapter 2 we used the pairs of numbers in the above table to plot a cumulative frequency curve. At the time we commented on the distinctive ogive shape of the graph. If we now plot the same points on normal probability paper we get a very different shape, as seen in Figure 3.10. (Note that the percentages in Table 3.5 have each been reduced by 1% before plotting. For further details of normal probability plots see Chatfield, 1978.)

Plotting the cumulative frequency distribution on normal probability paper has straightened out the ogive shape so that the points lie almost on a straight line. Drawing the 'best straight line' on Figure 3.10 corresponds to drawing the normal distribution curve in Figure 3.2; clearly such a line would fit the data very well.

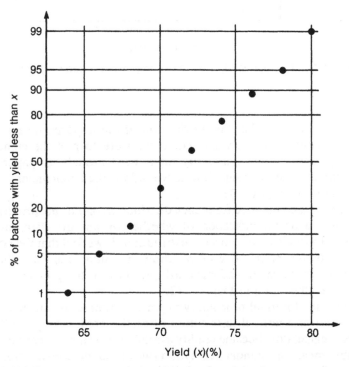

Figure 3.10 Percentage cumulative frequencies plotted on normal probability paper

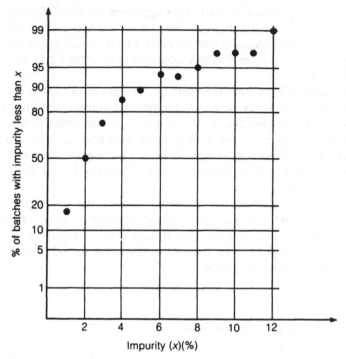

Figure 3.11   Cumulative distribution of impurity on normal probability paper

You may recall that the distribution of the impurity determinations of the 50 batches was very skewed. If we were to plot the cumulative frequency distribution of impurity on normal probability paper we would not expect to get a straight line as we did for the batch yields. The point is illustrated by Figure 3.11.

Plotting a cumulative frequency distribution on normal probability paper can give a useful indication of whether or not the data came from a population which has a normal distribution. If we feel that the points would fit closely to a straight line we can draw in a suitable line and then proceed to estimate the mean and standard deviation of the population. The procedure is illustrated in Figure 3.12.

The use of normal probability paper to estimate the mean and standard deviation of a normal distribution offered a great saving in effort before pocket calculators became readily available. It is now so easy to calculate a sample mean and standard deviation (which can be used as estimates of the population mean and standard deviation) that this use of normal probability paper is not important.

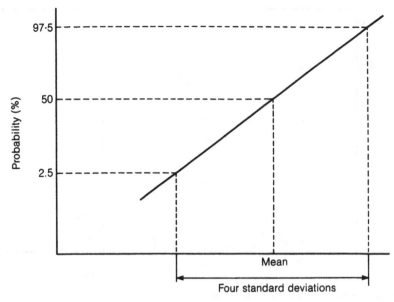

Figure 3.12 Estimation of mean and standard deviation from normal probability paper

3.6 PROCESS CAPABILITY

In Chapters 4 to 6 we will discuss several very useful statistical techniques. Then, in Chapter 7, we will explore the assumptions underlying these techniques, so that you will be able to analyse data with safety. When you have studied Chapter 7 you will understand why the normal distribution is so important in statistics. More immediately let us turn to a problem in which the normal distribution is directly applicable.

How can we determine whether or not a production process is capable of manufacturing a product which will satisfy the customer? First, we need to express the needs of the customer in terms of a product specification. This will list the important variables and specify a range of acceptable values for each. Then, for each of these variables we need to assess the variability in the product as it leaves the process. If this variation is so great that it cannot be contained within the specification, then we cannot claim to have a capable process.

Let us consider, for example, the manufacture of RGX 200 at R & G Chemicals Ltd. A specification for this product has been agreed with the customer. One clause in this specification states that the viscosity of the product should be between 180 and 220 centistokes (cSt). To see if the process is capable of producing RGX 200 with a viscosity between these

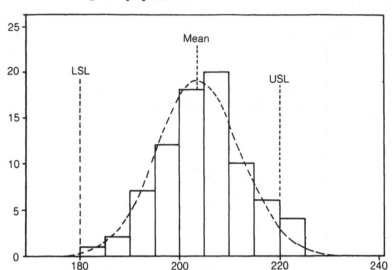

Figure 3.13   Process variability and product specification

limits we turn to the plant records. Data for the last 80 batches give a mean viscosity of 203.86 cSt and a standard deviation of 8.336 cSt. The scatter of the viscosity about the mean is illustrated by the histogram in Figure 3.13.

We also see in Figure 3.13 a normal distribution curve. This has the same mean and standard deviation as the data that gave us the histogram. What can we deduce from the normal curve and the specification limits (LSL and USL) about the capability of the process? Well, the histogram shows that 76 of the 80 batches of RGX 200 had viscosity within the specification. The normal curve suggests that 97.2% of future batches will satisfy the specification. This percentage is based on the assumption that future production will have a normal distribution with a mean of 203.86 and a standard deviation of 8.336 cSt. Is this a reasonable assumption to make? Figure 3.13 does indicate that the histogram based on the last 80 batches is similar in shape to a normal curve. However, it seems very unlikely that the mean viscosity of future batches will be exactly 203.86 cSt. Perhaps it would be more reasonable to assume a mean viscosity of 200 for future batches, as this is the target value, midway between the specification limits. With a mean of 200 and a standard deviation of 8.336 we would get only 1.64% of batches outside the specification. Thus, we might be tempted to suggest to our customer that 98.36% of our batches of RGX 200 will have viscosity within specification. If we did so our customer might reasonably question the basis of this prediction. For example, he or she might ask:

(a) How will you attempt to ensure that the mean viscosity is equal to 200 cSt?
(b) Is the 8.336 a good estimate of the process standard deviation that would exist in the short term, or in the long term?
(c) How stable is the process variability and how will you monitor it?

These questions will be discussed in Chapter 8 when we explore statistical process control. For the moment let me point out that our estimate, 98.36% of batches within specification, may not be very realistic. Furthermore, it is more common, in the dialogue between supplier and customer, to speak of capability indices rather than percentages. The most common of these indices is given the symbol Cp:

$$Cp = 2T/6\sigma$$

where $2T$ is the width of the specification interval and σ is the standard deviation. The 6σ in the formula represents the approximate width of the normal distribution curve and the $2T$ represents the width of the specification. When these two are equal we get a Cp of 1. This situation is illustrated by Figure 3.14. With a Cp of 1 virtually all the product would be

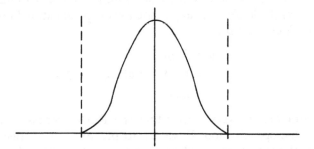

Figure 3.14 $Cp = 1$

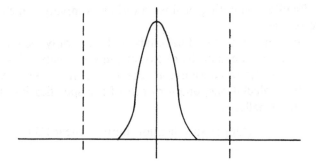

Figure 3.15 $Cp = 2$

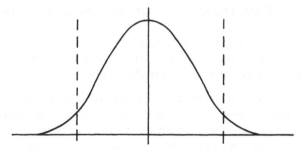

Figure 3.16 $Cp = 0.8$

within specification, provided the population mean were equal to the target value. In practice it will not be possible to control the mean with perfect precision, so it is desirable to have a Cp greater than 1. Indeed, some industrial customers hope to reach the happy state in which all their suppliers can demonstrate a Cp of 2, as in Figure 3.15. Clearly a Cp value which is less than 1 gives a strong indication that out-of-spec product is inevitable, as we see in Figure 3.16.

The data represented by the histogram in Figure 3.13 have a standard deviation of 8.336 cSt. The upper and lower specification limits are 220 and 180 cSt. Thus the Cp is

$$Cp = 2T/6\sigma$$
$$= (220 - 180)/(6 \times 8.336)$$
$$= 0.80$$

This result tells us very clearly that the RGX process is not capable of meeting the specification. If we are to prevent out-of-spec product from reaching the customer we will need to test each batch and screen out those which fail. It may, of course, be possible to rework these faulty batches or to blend them with others. This would introduce extra cost and we would have difficulty competing against another company which had a more capable process.

The capability index that I have labelled Cp is only meaningful if we are able to target the mean at the centre of the specification interval. If this is not possible or desirable we use an alternative index, known as Cpk. This involves two calculations, one for each of the specification limits.

Cpk is the smaller of:

$$(\text{Upper specification limit} - \text{mean})/3\sigma$$

and

$$(\text{Mean} - \text{lower specification limit})/3\sigma$$

Suppose, for example, we decided to aim for a mean viscosity of 195 cSt. It would then be more meaningful to quote the Cpk rather than the Cp and its value would be the smaller of:

$$(220 - 195)/(3 \times 8.336) = 1.00$$

and

$$(195 - 180)/(3 \times 8.336) = 0.60$$

Clearly, by running the process off-centre we get a lower value for the capability index. With a mean of 195 the Cpk is only 0.60. If we had maintained the mean at 200 the Cpk would have been 0.80, which is equal to the Cp calculated earlier. The reduction in the capability index as we move the mean away from the centre of the specification is accompanied by an increase in out-of-spec batches, of course.

Capability indices, such as Cp, Cpk and others, are often quoted in the dialogue between customers and suppliers. Unfortunately, these indices are all too often abused, and care should be exercised. If a Cp or Cpk is calculated from very few data or from unrepresentative data it can be misleading.

3.7 SUMMARY

It is doubtful that you will appreciate the usefulness of probability distributions until you have read later chapters. We have used them as models to describe populations and hence to make predictions of unlikely events.

Many other probability distributions are described in standard statistical texts such as Johnson and Leone (1964). If the author had wished to devote more space to probability distributions, then he would certainly have discussed the log-normal distribution and the exponential distribution. Both are applicable to continuous variables. The log-normal distribution might well give an adequate description of the batch to batch variation in impurity. The exponential distribution is used extensively in reliability assessment since it is applicable to the time intervals between failures when the number of failures in an interval has a Poisson distribution. A probability distribution known as the binomial distribution will be introduced in Chapter 6. This also has certain connections with the Poisson distribution.

An extremely important use of probability distributions is in the production of the statistical tables which we will use in significance testing and estimation. Though you may never produce a statistical table yourself it is most important that you have an awareness of the assumptions underlying these tables and the statistical techniques which they support.

PROBLEMS

3.1   A manufacturer knows from experience that the resistance of resistors he produces is normally distributed with mean $\mu = 80$ ohms and standard deviation $\sigma = 2$ ohms.

(a) What percentage of resistors will have resistance between 78 and 82 ohms?
(b) What percentage of resistors will have resistance of more than 83 ohms?
(c) What is the probability that a resistor selected at random will have resistance of less than 79 ohms?
(d) Between what symmetrical limits will the resistance of 99% of resistors lie?

3.2   The skein strength of a certain type of cotton yarn varies from piece to piece. Records show that the skein strengths have a normal distribution, with a mean of 96 lb and a standard deviation of 8 lb.

(a) What percentage of skeins will have a strength in excess of 100 lb?
(b) What is the probability that a randomly chosen skein has a strength of between 90 and 100 lb?
(c) What strength will be exceeded by 20% of the skeins?
(d) (More difficult.) What is the probability that two randomly chosen skeins will both have a strength of less than 100 lb? (You will need to make use of the 'multiplication rule for probabilities': probability of A and B = (probability of A) × (probability of B) where A and B are independent events.)

3.3   A manufacturer of paper supplies a particular type of paper in 1000 metre rolls. It has been agreed with a regular customer that a particular defect known as 'spotting' should not occur more than once per 10 metres on average. The manufacturer feels confident that they can meet this specification since the records show that spotting occurs at a rate of once per 15 metres on average when the production process is working normally.

(a) A 15 metre length of paper is taken from a roll produced under normal conditions. What is the probability that this length of paper will contain no occurrences of spotting?
(b) A 30 metre length of paper is taken from a roll produced under normal conditions.

   (i) What is the probability that this length of paper will contain no occurrences of spotting?

(ii) What is the probability that this length of paper will contain more than four occurrences of spotting?

(c) (More difficult.) The regular customer decides to check the quality of each roll by inspecting two 30 metre lengths selected at random from the 1000 metres on the roll. If fewer than five faults are found in each of the two pieces the roll is accepted but if five or more faults are found in either or both the roll will be subjected to full inspection.

    (i) What is the probability that a roll produced under normal conditions will be subjected to full inspection? (You will need the multiplication rule again.)

    (ii) What is the probability that a roll with a fault rate of one occurrence of spotting per 5 metres of paper will avoid full inspection?

3.4 The Weights and Measures Act of 1963 specifies that prepacked goods which have a weight marked on the package should conform to certain restrictions. The Act specifies that no prepack should have a weight which is less than the nominal weight (i.e. the weight marked on the package). Smith and Co. manufactures cornflour which is sold in cartons having a nominal weight of 500 grams. The packaging is carried out by sophisticated automatic machines which, despite their great cost, are variable in performance. The mean weight of cornflour per packet can be adjusted, of course, but the variability from packet to packet cannot be eliminated and this variability gives rise to a standard deviation of 10 grams regardless of the mean weight at which the machine is set.

(a) If the weights dispensed into individual packages have a normal distribution with a standard deviation of 10 grams what setting would you recommend for the mean weight dispensed in order to comply with the 1963 Act?

The *Guide to Good Practice in Weights and Measures in the Factory*, published by the Confederation of British Industry in 1969, is regarded by manufacturers and by local authority inspectors as an amendment to the 1963 Act. This document relaxes the restrictions in the Act by means of two concessions:

Concession 1 permits a 1 in 1000 chance of a package being grossly deficient (i.e. having a net weight which is less than 96% of the nominal weight).
Concession 2 permits a 1 in 40 chance of a package being deficient (i.e. having a net weight less than the nominal weight).

(b) What is the minimum setting for the process mean weight which will satisfy Concession 1?

(c) What is the minimum setting for the process mean weight which will satisfy Concession 2?

(d) What setting would you recommend for the mean weight in order to satisfy both concessions?

On 1 January 1980 new regulations came into force which replace the minimum system of the 1963 Act with an average system which conforms to EC recommendations. This new system is perhaps best described by quoting the three packers' rules on which it is based:

Rule 1: The contents must not be less on average than the nominal weight.
Rule 2: The content of the great majority of prepacks (97.5%) must lie above the tolerance limit derived from the directive. (Tolerance limit is defined as nominal weight minus tolerance and the tolerance specified for a package with weight around 500 grams is 7.5 grams.)
Rule 3: The contents of very prepack must lie above the absolute tolerance limit. (Absolute tolerance limit is defined as nominal weight minus twice the tolerance.)

(e) If Smith and Co. continues to pack with the process mean set at the figure recommended in (d) will the company comply with Rules 1, 2 and 3?

(f) Assuming that the phrase 'every prepack' in Rule 3 should read '99.99% of prepacks' would the process mean setting recommended in (d) lead to violation of Rules 1, 2 and 3?

(g) Assuming the modification to Rule 3 suggested in (f), what is the lowest value of process mean that will allow all three rules to be satisfied?

4
Testing and estimation: one sample

4.1 INTRODUCTION

Perhaps the previous chapter was a digression from the main purpose of this book. No matter how interesting you found the content of Chapter 3 you may have felt that it was a diversion from the main theme of using data from a sample to draw conclusions about a population. Even after studying Chapter 3 it is probably not entirely clear how an understanding of the normal distribution can help us to answer some of the questions posed in Chapter 2, questions to which we must now return.

You will recall that the manufacture of digozo blue pigment had been interrupted whilst modifications were made to the plant. The chief chemist is anxious to know whether the mean yield of the plant has increased as a result of these modifications. He is also concerned to check that any increase in yield which might have taken place has not been accompanied by an increase in impurity.

In this chapter, then, we are going to talk about the performance of the plant on average. We are not going to concern ourselves unduly with individual batches of pigment. We will in fact focus our attention on three questions:

(a) Has the mean yield of the plant increased since the modifications were carried out?
(b) If the mean yield has increased, what is its new value?
(c) When we want to detect an increase in yield, how many batches do we need to inspect?

4.2 HAS THE YIELD INCREASED?

Before the modifications were carried out the mean yield was 70.30. This figure was obtained by averaging the yield from many hundreds of batches produced during a two-year period when the production process was operating satisfactorily. Not all the batches produced during this period were included when calculating the mean. A small number of batches were excluded because they were thought to be special cases. In each of the excluded batches the yield was very low and/or the impurity was very high; we will discuss these rogue batches more fully when we talk about outliers in a later chapter.

So the mean yield was 70.30 before the modifications were carried out and the mean yield is 71.23 for the batches produced after the modifications. Can we conclude then that the incorporation of the modifications has been effective in increasing the yield? You may be tempted to answer simply, 'yes'. Before doing so, however, it would be wise to take account of the fact that the question refers to a population of batches whilst the mean yield of 71.23 was obtained from only a sample.

When the chief chemist asks the question 'Has the mean yield increased?' he is surely thinking beyond the 50 batches produced since the modification, to the batches which might be produced in the future. In attempting to answer this question, then, we must regard the 50 batches as being a sample from a population which consists of all batches made after the modification, including future batches. The question can then be re-worded to read: is the population mean greater than 70.30?

The only evidence available to help us answer this question is the yield values of the 50 batches in the sample. You might imagine therefore that the mean yield of future batches is simply a matter of opinion and that one person's opinion is as good as any other. If this were so the evaluation of the effectiveness of a modification would be very subjective indeed. Fortunately, however, statisticians have devised a technique known as significance testing (or hypothesis testing) which is purely objective and can be used to answer many of the questions we have asked. Before we carry out a significance test let us decide what factors will need to be taken into account. We will do this by examining two hypothetical samples of batches.

Suppose that the chief chemist was anxious to make an early decision concerning the effectiveness of the modifications and he decided to draw a conclusion about the mean yield of the modified plant after producing only six batches. His conclusion would depend upon what he found when the yield of each batch was measured.

Suppose the six yield values were:

Sample A: 72.1   69.0   74.3   71.2   70.1   70.7

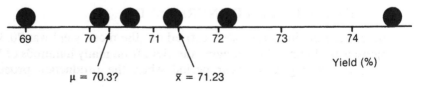

Can we conclude with confidence that the population mean yield is greater than 70.3? It is true that the sample mean yield (71.23) is greater than this figure but one doubts if the chief chemist would risk his reputation by proclaiming categorically that the mean yield had increased since the modifications were incorporated. Perhaps he would feel that it would be wise to examine further batches before reaching a decision.

Suppose, on the other hand, the first six batches had given the following yield measurements:

Sample B: 71.3   70.7   71.6   72.0   71.1   70.7

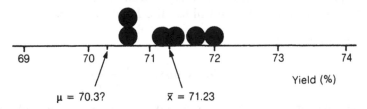

With these results the chief chemist might feel quite strongly that the mean yield had increased. He could not be certain, of course, but he might have sufficient confidence to announce that the modifications had effectively increased the yield.

In both of these hypothetical sets of data the mean is 71.23. Thus in both cases the sample mean differs from the old mean (70.30) by 0.93, which is a difference of 1.3%. Yet, with sample B we conclude that the mean yield has increased whereas with sample A we prefer to reserve judgement. What is the essential difference between these two samples which leads to the different conclusions? As we can see by comparing the two blob charts, the batch yields are much more widely scattered in sample A than in sample B. This point is confirmed if we compare the standard deviations of samples A and B given in Table 4.1.

When trying to decide whether or not the mean yield has increased it is not sufficient to calculate the difference between the sample mean and the old mean; we must compare this difference with some measure of the batch to batch variation such as the sample standard deviation.

Table 4.1   Variation in yield in three samples

Sample	Number of batches n	Sample mean $\bar{x}$	Sample SD s
Hypothetical A	6	71.23	1.828
Hypothetical B	6	71.23	0.5125
Taken by chief chemist	50	71.23	3.251

We will now carry out a significance test using the yield data from the 50 batches inspected by the chief chemist. We will make use of the sample mean ($\bar{x} = 71.23$), the old population mean ($\mu = 70.30$) and the sample standard deviation ($s = 3.251$). As this standard deviation is even larger than those from samples A and B (see Table 4.1) you might expect that we will be unable to conclude that the mean has increased. We will, however, be taking into account the sample size ($n = 50$) and this will have a considerable influence on our decision.

One important step in carrying out a significance test is to calculate what is known as a *test statistic*. For this particular test, which is often referred to as a one-sample *t*-test, we will use the following formula.

One-sample-test

$$\text{Test statistic} = \frac{|\bar{x} - \mu|}{s/\sqrt{n}}$$

The 'one-sample *t*-test' is carried out using a six-step procedure which we will follow in many other types of significance tests.

Step 1: Null hypothesis – the mean yield of all batches after the modification is equal to 70.30 ($\mu = 70.30$).

Step 2: Alternative hypothesis – the mean yield of all batches after the modification is greater than 70.30 ($\mu > 70.30$).

Step 3: Test statistic $= \dfrac{|\bar{x} - \mu|}{s/\sqrt{n}}$

$= \dfrac{|71.23 - 70.30|}{3.251/\sqrt{50}}$

$= 2.022$

Step 4: Critical values – from the *t*-table (Statistical Table B) using 49 degrees of freedom for a one-sided test:

1.68 at the 5% significance level
2.41 at the 1% significance level.

Step 5: Decision – as the test statistic lies between the two critical values we reject the null hypothesis at the 5% significance level.

Step 6: Conclusion – we conclude that the mean yield is greater than 70.30.

As a result of following this six-step procedure the chief chemist draws the conclusion that the mean yield has increased since the modifications were incorporated. He hopes that he has made the right decision in rejecting the null hypothesis but he realizes that he is running a risk of making a wrong decision every time he carries out a significance test. Whatever procedure he follows the risk will be there. It is unfortunately true that whenever we draw a conclusion about a population using data from a sample we run a risk of making an error. One advantage of the six-step procedure is that the risk can be quantified even though it cannot be eliminated. In fact the probability of wrongly rejecting the null hypothesis is referred to as the 'significance level'.

It might be interesting to return to the two sets of hypothetical data and apply the significance testing procedure. Would we reach the same conclusions that were suggested earlier, when we were simply using common sense? Carrying out a one-sample t-test on the six yield values in sample A gives:

Null hypothesis $\qquad \mu = 70.30$

Alternative hypothesis $\quad \mu > 70.30$

$$\text{Test statistic} = \frac{|\bar{x} - \mu|}{s/\sqrt{n}}$$

$$= \frac{|71.23 - 70.30|}{1.828/\sqrt{6}}$$

$$= 1.25$$

Critical values – from the t-table using 5 degrees of freedom for a one-sided test:

2.02 at the 5% significance level.
3.36 at the 1% significance level.

Decision – as the test statistic is less than both critical values we cannot reject the null hypothesis.

Conclusion – we are unable to conclude that the mean yield is greater than 70.30.

Thus the chief chemist would not be able to conclude that the modifications had been effective in increasing the yield of the plant. This conclusion is in agreement with that suggested earlier, by a visual inspection of the data.

Turning to the data from sample B, which previously led us to a very different conclusion, we carry out a one-sample t-test as follows:

Null hypothesis $\qquad \mu = 70.30$

Alternative hypothesis $\quad \mu > 70.30$

$$\text{Test statistic} = \frac{|\bar{x} - \mu|}{s/\sqrt{n}}$$

$$= \frac{|71.23 - 70.30|}{0.5125/\sqrt{6}}$$

$$= 4.44$$

Critical values – from the t-table using 5 degrees of freedom for a one-sided test:

2.02 at the 5% significance level.
3.36 at the 1% significance level.

Decision – as the test statistic is greater than both critical values we reject the null hypothesis at the 1% significance level.

Conclusion – we conclude that the mean yield is greater than 70.30.

This conclusion is again in agreement with that suggested by a visual examination of the data. In many cases the significance testing procedure will lead us to the same conclusion that would have been reached by any reasonable person. It is in the borderline case that the significance test is most useful.

4.3   THE SIGNIFICANCE TESTING PROCEDURE

Carrying out a one-sample t-test is quite straightforward if we stick to the recommended procedure. Understanding the theory underlying a one-sample t-test is much more demanding, however, and few chemists who carry out significance tests have a thorough understanding of the relevant statistical theory. Fortunately, a really deep understanding is not necessary. We will discuss significance testing theory to some extent in the later chapters and a number of comments are offered at this point in the hope that they will make the procedure seem at least reasonable.

1. The *null hypothesis* is a simple statement about the population. The purpose of the significance test is to decide whether the null hypothesis is true or false. The word 'null' became established in significance testing because the null hypothesis often specifies no change, or no difference. However, in some significance tests the word null appears to be most inappropriate, and a growing number of statisticians prefer the phrase initial hypothesis. We will continue to use null hypothesis throughout this book.
2. The *alternative hypothesis* also refers to the population. The two hypotheses, between them, cover all the possibilities we wish to con-

sider. Note that the null hypothesis is the one which contains the equals sign.

3. In order to calculate the test statistic we need the data from the sample (to give us $\bar{x}$, s and n) and the null hypothesis (to give us a hypothesized value for μ). If the null hypothesis were true we would expect the value of the test statistic to be small. Conversely, the larger the value of the test statistic the more confidently we can reject the null hypothesis.

4. Critical values are obtained from statistical tables. For a t-test we use the t-table, but many other tables are available and some will be used later. Users of significance tests do not calculate their own critical values; they use published tables, the construction of which will be discussed in a later chapter. The critical values are used as yardsticks against which we measure the test statistic in order to reach a decision concerning the null hypothesis. The decision process is illustrated by Figure 4.1.

5. In step 5 of the significance testing procedure we have quoted two critical values, at the 5% and the 1% significance levels. Many scientists use only one significance level, whereas others use two or even three. In different branches of science different conventions are adopted. If only one significance level is used this should be chosen before the data are analysed. Furthermore, the choice should not be left to a statistician. The scientist selects his level of significance to correspond with the risk he is prepared to take of rejecting a null hypothesis which is true. If the chief chemist wishes to run a 5% risk of concluding that the modification has been effective, when in fact no increase in yield has occurred, then he should use a 5% significance level. Many social scientists and many biologists prefer to use several significance levels and then report their findings as indicated in Figure 4.1. The chief chemist, for example, could report a significant increase in yield was found ($p < 0.05$).

6. Significance tests can be classed as one-sided tests or two-sided tests. Each of the three t-tests we have carried out was a one-sided test

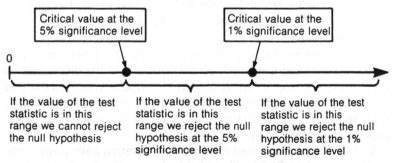

Figure 4.1 Comparing the test statistic with the critical values

because we were looking for a change in one direction. This one-sided search is indicated by the 'greater than' in the alternative hypothesis ($\mu > 70.30$). With a two-sided test the alternative hypothesis would state: The mean yield of all batches after the modification is not equal to 70.30, (i.e. $\mu \neq 70.30$). We will return to this point in Chapter 5.

7. In order to obtain critical values from the t-table we must use the appropriate number of degrees of freedom. For a one-sample t-test this will be 1 less than the number of observations used in the calculation of the standard deviation. When we carried out a significance test using the yield values from 50 batches we had 49 degrees of freedom and when we used the yield values from only 6 batches we had 5 degrees of freedom.

8. In almost all significance tests (and certainly in all tests in this book) a large test statistic is an indication that either:

(a) the null hypothesis is not true, or
(b) the sample is not representative of the population referred to in the null hypothesis.

If we have taken a random sample there is a good chance that it will be representative of the population, in that the sample mean ($\bar{x}$) will be close to the population mean (μ). When we get a large test statistic, therefore, we reject the null hypothesis.

Note that a small test statistic does not inspire us to accept the null hypothesis. At the decision step we either reject the null hypothesis or we fail to reject the null hypothesis. It would be illogical to accept the null hypothesis since the theory underlying the significance test is based on the assumption that the null hypothesis is true.

4.4 WHAT IS THE NEW MEAN YIELD?

The chief chemist has concluded that the mean yield is now greater than 70.30. Having rejected the null hypothesis at the 5% level of significance he is quite confident that his conclusion will be verified as further batches are produced. He feels quite sure that the mean yield of future batches will be greater than 70.30.

In reporting his findings, however, the chief chemist feels that a more positive conclusion is needed. When informed of the increase in yield, the production director is sure to ask: If the modification has been effective, what is the new mean level of yield? The chief chemist would like, therefore, to include in his report an estimate of the mean yield that could be expected from future batches.

The 50 batches produced since the modification have given a mean yield of 71.23 (i.e. $\bar{x} = 71.23$). Should the chemist simply pronounce then that the mean yield of future batches will be 71.23 (i.e. $\mu = 71.23$)? If he does so he will almost certainly be wrong. Common sense suggests that if he had taken a different sample he would have got a different mean. If, for example, he had taken only the first 20 batches after the modification his sample mean would have been 71.17. The chief chemist surmises that if the sample mean varies from sample to sample it is quite possible that none of these sample means will be exactly equal to the population mean. To state that $\mu = 71.23$ simply because his sample has given $\bar{x} = 71.23$, would be unwise. (For a more detailed discussion of how the sample mean can be expected to vary from sample to sample see Appendix C.)

One solution to this problem is to quote a *range of values* in which we think the population mean must lie. Thus the chief chemist could report that the mean yield of future batches will lie in the interval 71.23 ± 0.15 (i.e. 71.08 to 71.38), for example. To quote an interval of values in which we expect the population mean to lie does seem to be an improvement on the single value estimate. It is still open to question, however, and the chief chemist might well be asked 'How certain are you that the mean yield of future batches will be between 71.08 and 71.38?'

To overcome this difficulty we quote an interval and a confidence level in what is known as a *confidence interval*. To calculate such an interval we use the formula: a confidence interval for the population mean (μ) is given by $\bar{x} \pm ts/\sqrt{n}$. To perform this calculation we obtain $\bar{x}$, s and n from the sample data; t is a critical value from the t-table for a two-sided test using the appropriate degrees of freedom. Using 49 degrees of freedom the critical values are 2.01 at the 5% significance level and 2.68 at the 1% level. We can use these critical values to calculate a 95% confidence interval and a 99% confidence interval as follows.

A 95% confidence interval for μ

$$= 71.23 \pm 2.01(3.251)/\sqrt{50}$$
$$= 71.23 \pm 0.92$$
$$= 70.31 \text{ to } 72.15$$

A 99% confidence interval for μ

$$= 71.23 \pm 2.68(3.251)/\sqrt{50}$$
$$= 71.23 \pm 1.23$$
$$= 70.00 \text{ to } 72.46$$

Having carried out these calculations it is tempting to conclude, as many books do, that we can be 95% confident that the population mean lies between 70.31 and 72.15. On reflection, however, the statement appears to have little meaning. The author does not judge himself competent to explain what is meant by the phrase, 'we can be 95% confident', and he would certainly not wish to dictate to the reader how confident he or she should feel. However, several statements can be made which have a sound statistical basis. For example, if the experiment, with 50 batches, were repeated many, many times and a 95% confidence interval calculated on each occasion, then 95% of the confidence intervals would contain the population mean. If the phrase: we can be 95% confident is simply an abbreviation of the above statement then it is reasonable to suggest that we can be 95% confident that the mean yield of future batches will lie between 70.31 and 72.15. If we wish to be even more confident that our interval includes the population mean then we must quote a wider interval and we find that the 99% confidence interval is indeed wider, from 70.00 to 72.46.

When quoting a confidence interval we are making use of statistical inference just as we did when carrying out a significance test. We are again arguing from the sample to the population. A different sample would almost certainly give us a different confidence interval. If we had monitored only the first ten batches produced after the modifications we would have found $\bar{x} = 70.75$ and $s = 2.966$. Using $n = 10$ and $t = 2.26$ (9 degrees of freedom) we obtain a 95% confidence interval which extends from 68.63 to 72.87. This is very much wider than the interval obtained when we used the data from all 50 batches.

A larger sample can be expected to give us more information and we would expect a narrower confidence interval from a larger sample. This is reflected in the $\sqrt{n}$ divisor in the formula. It is interesting to compare the confidence intervals we would have obtained if we had examined only the first five batches or the first ten batches, etc., after the modifications had been incorporated. These 95% confidence intervals for the mean yield of future batches are given in Table 4.2 and are illustrated in Figure 4.2.

Concerning the confidence intervals in Table 4.2 and Figure 4.2, several points are worthy of note:

Table 4.2   95% confidence intervals for mean yield of future batches

Number of batches n	Sample mean $\bar{x}$	Sample SD s	t	Confidence interval	
5	70.92	2.443	2.78	70.92 ± 3.04	67.88 to 73.96
10	70.75	2.966	2.26	70.75 ± 2.12	68.63 to 72.87
20	71.15	3.261	2.10	71.15 ± 1.53	69.62 to 72.68
30	71.26	3.417	2.04	71.26 ± 1.27	69.99 to 72.53
40	71.19	3.280	2.02	71.19 ± 1.05	70.14 to 72.24
50	71.23	3.251	2.01	71.23 ± 0.92	70.31 to 72.15

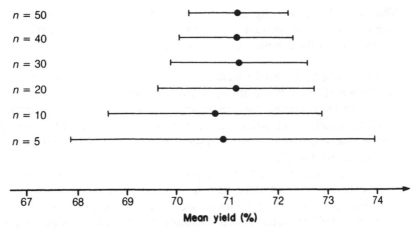

Figure 4.2 95% confidence intervals for mean yield of future batches

1. There is some indication that the sample mean and the sample standard deviation increase as the sample size (n) increases. This is simply due to chance and is not an indication of a general rule.
2. The t value decreases as the sample size (n) increases. This is certainly not due to chance but is due to the greater reliability of the estimate of population standard deviation as the sample size increases. It would lead us to expect a narrower interval from a larger sample. (The ratio of the largest t value (2.78) to the smallest (2.01) is not very large and will not have a great effect on the width of the confidence interval. If, however, we calculated a 95% confidence interval using only the first two batches the t value would be 12.71 and the interval would extend from 56.12 to 84.08.)
3. The width of the confidence intervals gets narrower as the sample size gets larger. This is largely due to the $\sqrt{n}$ divisor in the equation. Because of the square root this will not be a linear relationship and we would need, for example, to quadruple the sample size in order to halve the width of the confidence interval.

4.5 IS SUCH A LARGE SAMPLE REALLY NECESSARY?

Perhaps you get the impression from Figure 4.2 that there is little to be gained by increasing the sample size above 30. It can be predicted in advance that a sample size of 120 would be needed to halve the width of the confidence interval that would be expected with a sample size of 30. Is it worth the extra effort and the increased cost to get the better estimate?

In carrying out an investigation we are, after all, buying information.

The cost of the investigation may well be directly proportional to the sample size but the amount of information may only be proportional to the square root of the sample size. This would imply that there is a best sample size.

Alternatively, we might wish to carry out an investigation in order to decide whether or not a process improvement has been achieved. The larger the improvement we are looking for the easier it should be to find, if it has actually been achieved. It is desirable that we should be able to decide in advance how large a sample is needed in order to detect a change of a specified magnitude.

Perhaps it would have been convenient for the chief chemist if he could have calculated in advance just how many batches he would need to examine in order to detect a mean yield increase of 1 unit (i.e. from 70.30 to 71.30). Unfortunately, such calculations are only possible if we have some information about the variability of the process. It is, of course, the batch to batch variability which prevents us from detecting immediately any change in mean yield which might occur.

Now that the chief chemist has some data he is in a good position to calculate the number of batches he would need in order to estimate the mean yield with any required precision. From his data we have calculated the standard deviation to be 3.251, with 49 degrees of freedom. This standard deviation can be substituted into the formula: The sample size needed to estimate a population mean to within $\pm c$, using a confidence interval, is given by

$$n = (ts/c)^2$$

where s is a suitable standard deviation and t is taken from Table B with appropriate degrees of freedom. Suppose the chief chemist wished to estimate the mean yield to within ± 1 unit (i.e. $c = 1$) we could use the standard deviation ($s = 3.251$) and a t value with 49 degrees of freedom ($t = 2.01$) to calculate the required sample size.

$$n = (ts/c)^2 = [2.01(3.251)/1]^2 = 42.7$$

Thus the chief chemist would need to make at least 43 batches (rounding up 42.7) to be 95% confident of estimating the mean yield to within ± 1 unit. To get a more precise estimate an even larger sample would be needed. On the other hand, to estimate the mean yield to within ± 2 units would required only 11 batches.

For greater confidence a larger sample is required. If the chief chemist wished to estimate the mean yield to within ± 1 unit with *99% confidence* the t value required for the calculation would be 2.68.

$$n = (ts/c)^2 = [2.68(3.251)/1]^2 = 75.9$$

4.6 STATISTICAL SIGNIFICANCE AND PRACTICAL IMPORTANCE

Perhaps the time is ripe to summarize the statistical activity of the chief chemist. Firstly, he carried out a significance test using the yield values of the first 50 batches produced after the modifications had been incorporated. He concluded that the mean yield had increased and was now greater than the old mean of 70.3. Secondly, he calculated a confidence interval for the mean yield of the modified plant. He was able to report with 95% confidence that the mean yield is now between 70.31 and 72.15. Finally, he calculated the size of sample that would be needed to estimate the mean yield with a specified precision. Using the standard deviation of yield values from the 50 batches he was able to show that a sample of 43 batches would be needed to estimate the mean yield to within ± 1 unit, with 95% confidence.

What benefits has the chief chemist gained by his use of statistical techniques? A significance test has enabled him to demonstrate that the apparent increase in yield indicated by the first 50 batches is very likely a reflection of a real and permanent increase. That does not imply that this increase is of practical importance or has great financial value. The confidence interval would indicate that the increase in mean yield is around 1 unit but it could be as much as 2 units. The confidence interval, like the significance test, is indicating statistical significance and not practical importance.

The statistician has no desire to tell the chemist what constitutes an important change. The statistician would certainly not wish the chemist to equate statistical significance with practical importance. Only the chemist can decide what is important and he must take into account many factors including the possible consequences of his decision.

For example, Indichem manufactures two pigments which are very similar in chemical composition and method of production. In terms of profitability to the company, however, they are very different. Pigment A is in great demand and is manufactured in an uninterrupted stream of batches in one particular vessel. Pigment B, on the other hand, is in low demand and only one batch is manufactured each year. Production costs of pigment A are dominated by the cost of raw materials and a 1% increase in yield would be worthwhile. Production costs of pigment B are much more dependent on the costs of the labour-intensive cleaning and setting-up, so that a 1% increase in yield would be considered trivial.

The chemist must decide, then, whether or not x% change is important. He must not, however, equate percentage change with significant change. When assessing the results of an investigation it is important to ask two questions:

(a) Could the observed change be simply due to chance?
(b) If the change is not due to chance, is it large enough to be important?

Significance testing helps us to answer the first question, but we must bear in mind that an observed change of $x\%$ may be significant if it is based on a large sample but not significant if based on a small sample. A significance test automatically takes into account the sample size and the inherent variability which is present in the data, whereas a simple statement of percentage increase does not.

4.7 SUMMARY

This is a very important chapter. In it we have introduced the fundamentals of significance testing and estimation. These ideas will serve as a foundation for later chapters and should be consolidated by working through the problems before proceeding further. When attempting the problems you would be well advised to follow the six-step procedure which will be used again and again in later chapters.

There are some difficult concepts underlying significance testing and if you worry about details which have not been explained you may lose sight of why a chemist or technologist would wish to do a significance test. Try to remember that the purpose of a significance test is to reach a decision about a population using data obtained from a sample.

PROBLEMS

4.1   A polymer manufacturer specifies that his produce has a mean intrinsic viscosity greater than 6.2. A customer samples a consignment and obtains the following values:

6.21   6.32   6.18   6.34   6.26   6.31

Do the results confirm the manufacturer's claim?

4.2   A new analytical method has been developed by a laboratory for the determination of chloride in river samples. To evaluate the accuracy of the method eight determinations were made on a standard sample which contained 50 mg/litre. The determinations were:

49.2   51.1   49.6   48.7   50.6   49.9   48.1   49.6

(a) Carry out a significance test to determine whether the method is biased.
(b) Calculate a 95% confidence interval for the population mean chloride content. Using this interval what is the maximum bias of the method?
(c) What is the population under investigation?

4.3   A recent modification has been introduced into a process to lower the
      impurity. Previously the impurity level had a mean of 5.4. The first
      ten batches with the modification gave the following results:

      4.0   4.6   4.4   2.5   4.8   5.9   3.0   6.1   5.3   4.4

      (a) Has there been a decrease in mean impurity level?
      (b) Obtain a 95% confidence interval for the mean impurity level.
      (c) How many results are needed in order to be 95% confident of
          estimating the mean impurity to within ±0.15?

5

Testing and estimation: two samples

5.1 INTRODUCTION

The previous chapter was centred around the *one-sample t-test*. We made use of this particular type of significance test to decide whether or not the mean yield of a production process had increased as a result of carrying out a modification. We showed that the mean yield of a sample of 50 batches was significantly greater than a standard value (70.3) which had previously been established.

In this chapter we will progress to the *two-sample t-test*. Once again we will be using the six-step procedure to reach a decision about a null hypothesis so the *t*-test itself should present no problems. There are, however, two complications which cannot be ignored:

(a) There are two types of two-sample *t*-test and we must be careful to select the appropriate type.
(b) Before we can use a *t*-test to compare two sample means we must first compare the sample variances using a different type of significance test which is known as the *F*-test.

Because the *F*-test is a prerequisite of the two-sample *t*-test we will examine the former before the latter. First of all, therefore, we will explore a situation in which the *F*-test is very meaningful in its own right. This will necessitate leaving the production process and moving into the analytical laboratory. There we will make use of the *F*-test to compare the precision of two alternative test methods.

5.2 COMPARING THE PRECISION OF TWO TEST METHODS

In previous chapters we examined the impurity determinations made on 50 batches of digozo blue. We found a skewed distribution and considerable

Table 5.1 Determination of impurity by two analytical methods

	Mean	Standard deviation
New method	1.432	0.2190
Old method	1.290	0.3341

variation from batch to batch. The standard deviation of the 50 determinations was 2.12. The plant manager has often asked how much of the apparent variation in impurity is real, and how much is due to analytical errors. He is assured by the chief chemist that repeat determinations with this method give a standard deviation of only 0.3, but the plant supervisor suspects that this is an underestimate.

Suspecting that this issue could undermine the good relations between the analytical and production departments, the chief chemist discusses with the analytical development team the possibility of improving the precision of the method. A modification is suggested. Unfortunately, the introduction of this modification would increase the time taken to produce an impurity determination. Thus, the modification will only be adopted if an improvement in precision can be demonstrated. A sample of digozo blue from batch number 50 is split into eleven subsamples. Five are analysed using the existing method and six are analysed using the new method. The determinations are:

Old method: 0.87   1.12   1.66   1.20   1.60
New method: 1.29   1.72   1.20   1.25   1.65   1.48

How do the six measurements compare with those made using the old analytical method? The mean and standard deviation of both sets of measurements are given in Table 5.1, whilst Figure 5.1 offers a pictorial comparison which is perhaps more useful.

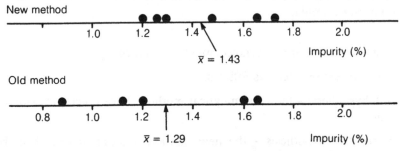

Figure 5.1 Determination of impurity by two analytical methods

Table 5.2 Symbols used in the F-test and the two-sample t-test

	Sample		Population	
	Mean	SD	Mean	SD
New method	$\bar{x}_1$	s_1	μ_1	σ_1
Old method	$\bar{x}_2$	s_2	μ_2	σ_2

Summarized in Table 5.1 and Figure 5.1 are two samples of measurements. By the old method we have one sample of five results and by the new method we have one sample of six results. If we continued testing indefinitely we would obtain two populations of results. Obviously it is not possible to continue testing forever but it is important to question the nature of the two populations when comparing the two analytical methods. To simplify the discussion and to distinguish clearly between populations and samples we will use the internationally accepted symbols given in Table 5.2.

Clearly the sample means are not equal ($\bar{x}_1 = 1.43$ and $\bar{x}_2 = 1.29$) but is it possible that the population means are equal? Later we will use a two-sample t-test to answer this question and in this test the null hypothesis will be $\mu_1 = \mu_2$. Before we can carry out the t-test, however, we must concern ourselves with the variability of the impurity measurements.

It is equally clear that the two sample standard deviations are not equal ($s_1 = 0.2190$ and $s_2 = 0.3341$) but is it possible that the population standard deviations are equal? To answer this question we can use an F-test in which the null hypothesis will be $\sigma_1^2 = \sigma_2^2$. (The F-test is based on variances not standard deviations, but the two are very closely related, of course.) If this null hypothesis is true the two analytical methods are equally variable; in other words they have equal precision. We have found that the sample standard deviations are not equal but it is possible that the two methods would be revealed to be equally variable if testing had been continued. When carrying out an F-test the test statistic is calculated by means of the following formula:

F-test
Test statistic = larger variance/smaller variance

The F-test proceeds as follows:

Null hypothesis – the two analytical methods have equal precision ($\sigma_1^2 = \sigma_2^2$).

Alternative hypothesis – the new method is more precise than the old method ($\sigma_1^2 < \sigma_2^2$).

Test statistic $= s_2^2/s_1^2$

$\qquad\qquad = (0.3341)^2/(0.2190)^2$

$\qquad\qquad = 2.33$

Critical values – from the F-table (Statistical Table C), using 4 degrees of freedom for the larger variance and 5 degrees of freedom for the smaller variance (one-sided test), the critical values are:

5.19 at the 5% significance level
11.39 at the 1% significance level.

Decision – as the test statistic is less than both critical values we cannot reject the null hypothesis.

Conclusion – we are unable to conclude that the new method is more precise than the old method.

It appears, then, that the apparently smaller variability of the new method of test could simply be due to chance, and it would be unwise to conclude that the new method has superior precision. If the chief chemist is not satisfied with this conclusion he can, of course, carry out a more extensive investigation. He could embark upon a full-scale precision experiment involving several laboratories, many operators and more than one batch. Such interlaboratory trials are discussed in a separate volume, *Statistics for Analytical Chemists* (Caulcutt and Boddy, 1983).

The F-test we have just carried out was a *one-sided* significance test. The one-sided test was appropriate as we were looking for a change in one direction, i.e. an improvement. The chief chemist does not seek a new method which is different, he seeks a method which is better. This point is emphasized by the presence of $<$ in the alternative hypothesis, rather than $\neq$ which would have indicated the need for a two-sided test.

Note that we would not have calculated a test statistic if the sample variance of the new method (s_1^2) had been greater than the sample variance of the old method (s_2^2). Obviously the discovery that $s_1^2 > s_2^2$ does not constitute support for the alternative hypothesis that $\sigma_1^2 < \sigma_2^2$. When carrying out a one-sided significance test we only calculate a test statistic if the sample data appear to support the alternative hypothesis. If the data appear to contradict the alternative hypothesis then we certainly cannot reject the null hypothesis. Consider, for example, the one-sided, one-sample t-test in Chapter 4. When carrying out this test it would have been very foolish to conclude that the mean yield had increased (i.e. $\mu > 70.30$) if the mean yield of the sample had been less than the old mean yield (i.e. $\bar{x} < 70.30$).

5.3   COMPARING THE BIAS OF TWO TEST METHODS

When comparing two analytical test methods we must never forget that the more precise method is not necessarily the more accurate. Precision is concerned solely with consistency or self-agreement. Whilst precision is highly desirable it does not guarantee accuracy. It is possible, for example, that a third method of measuring impurity might have given the following measurements when applied repeatedly to batch number 50:

  0.81   0.81   0.81   0.81

This third method appears to have very high precision but it may not be accurate. If the true impurity is 1.30, say, then the first method is superior even though it is less precise. We would say that method three was biased whereas the first method was unbiased. If we are to declare an analytical method to be accurate then we must be satisfied that it has adequate precision and that it is unbiased.

In practice it may be easy to see that a test method lacks precision but to detect bias we need to know the true value of that which is being measured. For the chief chemist to establish that either test method was unbiased he would need to know the true impurity of batch number 50. Unfortunately he does not have this information. We can state categorically, then, that with the impurity measurements he has obtained so far, he is in no position to conclude that:

(a) either the old method is unbiased;
(b) or the new method is unbiased.

Despite this limitation, which is imposed by his not knowing the true impurity, the chief chemist can compare the average performance of the two methods. If the two sample means ($\bar{x}_1 = 1.43$ and $\bar{x}_2 = 1.29$) are significantly different this might give some indication that one or other of the methods is biased. To compare the two means we would use a significance test which is known as a 't-test for two independent samples' and the test statistic is calculated using the following formula:

t-test for two independent samples

$$\text{Test statistic} = \frac{|\bar{x}_1 - \bar{x}_2|}{s\sqrt{\left(\dfrac{1}{n_1} + \dfrac{1}{n_2}\right)}}$$

where

$$s = \sqrt{\{[(n_1 - 1)s_1^2 + (n_2 - 1)s_2^2]/[(n_1 - 1) + (n_2 - 1)]\}}$$

and s_1^2 does not differ significantly from s_2^2.

The t-test for two independent samples makes use of our six-step procedure as follows:

Null hypothesis – the two analytical methods would give the same mean measurement in the long run (i.e. $\mu_1 = \mu_2$).

Alternative hypothesis – the two analytical methods would not give the same mean measurement in the long run (i.e. $\mu_1 \neq \mu_2$).

$$s = \sqrt{\{[(n_1 - 1)s_1^2 + (n_2 - 1)s_2^2]/[(n_1 - 1) + (n_2 - 1)]\}}$$
$$= \sqrt{\{[(6 - 1)(0.2190)^2 + (5 - 1)(0.3341)^2]/[(6 - 1) + (5 - 1)]\}}$$
$$= 0.2761$$

$$\text{Test statistic} = \frac{|\bar{x}_1 - \bar{x}_2|}{s\sqrt{\left(\dfrac{1}{n_1} + \dfrac{1}{n_2}\right)}}$$

$$= \frac{|1.43 - 1.29|}{0.2761\sqrt{\left(\dfrac{1}{6} + \dfrac{1}{5}\right)}}$$

$$= 0.83$$

Critical values – from the t-table (Statistical Table B) using 9 degrees of freedom [i.e. $(n_1 - 1) + (n_2 - 1)$] for a two-sided test:

2.26 at the 5% significance level
3.25 at the 1% significance level.

Decision – as the test statistic is less than both critical values we cannot reject the null hypothesis.

Conclusion – we are unable to conclude that there is any significant difference between the means of the results from the two analytical methods.

It could be, then, that in the long run neither of the two methods will be superior to the other in terms of precision or bias. As the new method is more costly to use, it would be unwise to introduce it on the strength of the very minimal evidence gathered to date.

You will have noticed that in carrying out the two-sample t-test we chose to do a *two-sided* test. This was appropriate because we were attempting to answer the question: do the two methods give significantly different results on average? We were looking for a difference in either direction. Previously when carrying out the F-test we chose to do a one-sided test because we were looking for a change in one particular direction. We were seeking a decrease in variability of the new method with respect to the old method.

The *t*-test for two independent samples is probably the most widely used of all significance tests. Because of its importance we will explore a second application of this test before we examine the second type of two-sample *t*-test.

5.4 DOES THE PRODUCT DETERIORATE IN STORAGE?

The chief chemist suspects that the brightness of the digozo blue pigment decreases whilst the product is in storage. His opinion on this matter has been influenced by recent complaints from Asian customers about the dullness of pigment which was manufactured some years ago. In order to investigate the occurrence of this deterioration the chief chemist has measured the brightness of the pigment in ten drums of digozo blue which have been in the warehouse for many months. It is known that these drums are not all from the same batch and it is thought that they are the residue from at least three different batches. The brightness determinations are:

2   1   3   0   1   3   1   1   1   0

As a basis for comparison the chief chemist measures the brightness of the pigment produced more recently. This is taken from ten drums of digozo blue, five drums from each of the last two batches. The brightness determinations are:

2   2   4   1   2   3   3   2   1   2

You may well be struck by the coarseness of the measurements. Brightness is measured on a scale which ranges from -7 to $+7$ with more positive numbers corresponding to greater brightness. Only five points of the scale (0, 1, 2, 3 and 4) are covered by the data, however. If each measurement had a tolerance of ± 1 there would appear to be little information on which to base a decision. You may have further reservations concerning the unit of measurement. Is the difference between and 1 and 2 on the brightness scale really equal to the difference between 3 and 4? The colour physicist who established the scale is confident that the answer is yes, and that a genuine unit of measurement does exist. If this were not so, the data would be unsuitable for the use of an *F*-test or a *t*-test. Many statisticians would express concern about the use of these tests because of the coarseness of the data. Unfortunately the alternative significance tests which one might adopt in this situation suffer from two defects:

(a) they are less powerful than the *t*-test or *F*-test;
(b) they are not within the repertoire of the majority of scientists (nor are they included in this book).

Table 5.3 Brightness of old and new pigment

	Number of drums	Brightness	
		Mean	SD
Old pigment	$n_1 = 10$	$\bar{x}_1 = 1.3$	$s_1 = 1.059$
New pigment	$n_2 = 10$	$\bar{x}_2 = 2.2$	$s_2 = 0.919$

It is certainly true to say that the ubiquitous t-test is occasionally used in situations where it is not strictly valid and we will now make use of the t-test to compare the brightness of the two samples.

The two sets of measurements are summarized in Table 5.3. We see in this table that the mean brightness of the old pigment (1.3) is less than the mean brightness of the new pigment (2.2). To see if the difference between the sample means is significant we will carry out a t-test for two independent samples. Before we do so, however, we must check that brightness measurements are equally variable whether made on old pigment or new pigment. For this purpose we carry out an F-test as follows:

Null hypothesis: $\sigma_1^2 = \sigma_2^2$ (i.e. variability of brightness is the same for new and old pigment).

Alternative hypothesis: $\sigma_1^2 \neq \sigma_2^2$.

$$\text{Test statistic} = (1.059)^2/(0.919)^2$$
$$= 1.33$$

Critical values – from the F-table using 9 and 9 degrees of freedom for a two-sided test:

4.03 at the 5% significance level
6.54 at the 1% significance level.

Decision – we cannot reject the null hypothesis.

Conclusion – we cannot conclude that the two population variances are unequal. Thus we provisionally assume that variability of brightness is the same for old and new pigment.

If we had rejected the null hypothesis when carrying out the F-test we would have been unable to proceed with the t-test for two independent samples. As we did not reject the null hypothesis we can proceed with the t-test on the assumption that the population variances are equal. The t-test proceeds as follows:

Null hypothesis: $\mu_1 = \mu_2$ (i.e. the mean brightness of old pigment is equal to the mean brightness of new pigment).

Alternative hypothesis: $\mu_1 < \mu_2$ (i.e. the mean brightness of old pigment is less than the mean brightness of new pigment)

Combined standard deviation:

$$
\begin{aligned}
s &= \sqrt{\{[(n_1 - 1)s_1{}^2 + (n_2 - 1)s_2{}^2]/[(n_1 - 1) + (n_2 - 1)]\}} \\
&= \sqrt{\{[9(1.059)^2 + 9(0.919)^2]/[9 + 9]\}} \\
&= 0.992
\end{aligned}
$$

$$
\text{Test statistic} = \frac{|\bar{x}_1 - \bar{x}_2|}{s\sqrt{\left(\dfrac{1}{n_1} + \dfrac{1}{n_2}\right)}} = \frac{|1.3 - 2.2|}{0.992\sqrt{\left(\dfrac{1}{10} + \dfrac{1}{10}\right)}}
$$

$$
= 2.02
$$

Critical values – from the t-table using 18 degrees of freedom for a one-sided test:

1.73 at the 5% significance level
2.55 at the 1% significance level.

Decision – we reject the null hypothesis at the 5% significance level.

Conclusion – we conclude that the mean brightness of old pigment is less than the mean brightness of new pigment.

The evidence furnished by the 20 drums of pigment leads us to the conclusion that the sample means are significantly different. It is reasonable therefore to ask: How large is the difference between the population means? This question can only be answered by using the sample means as estimates of the population means and as usual we express the uncertainty of our estimates in a confidence interval. A confidence interval for the difference between two population means is given by:

$$
|\bar{x}_1 - \bar{x}_2| \pm ts\sqrt{\left(\frac{1}{n_1} + \frac{1}{n_2}\right)}
$$

where s is the combined standard deviation:

$$
s = \sqrt{\{[(n_1 - 1)s_1{}^2 + (n_2 - 1)s_2{}^2]/[(n_1 - 1) + (n_2 - 1)]\}}
$$

To calculate a 95% confidence interval we would use a two-sided t value and a 5% significance level. With ten drums in each sample we have 18 degrees of freedom and the t value is 2.10, giving the following confidence interval:

$$|1.3 - 2.2| \pm 2.10(0.992)\sqrt{\left(\frac{1}{10} + \frac{1}{10}\right)}$$

$$= 0.9 \pm 0.94$$

$$= -0.04 \text{ to } 1.84$$

This result is telling us that we can be 95% certain that the reduction in brightness during storage will be less than 1.84 and greater than -0.04 (i.e. a slight gain in brightness). We might be surprised that this confidence interval should include zero when the t-test has already told us that the difference between the two population means is greater than zero. The reason why the confidence interval and the t-test are not in complete agreement is because in one case we used a two-sided t value whilst in the other we used a one-sided t value. It is perfectly reasonable that we should carry out a one-sided t-test because we were looking for a reduction in brightness which is a change in one direction. (Had we done a two-sided test, incidentally, we would not have rejected the null hypothesis.) On the other hand it is common practice to use a two-sided t value when calculating a confidence interval. The confidence intervals we have calculated in this and earlier chapters should, strictly speaking, be called two-sided confidence intervals to distinguish them from the much less common one-sided confidence intervals. The latter would be calculated using a one-sided t value.

In our particular problem, then, we could use the one-sided t value (1.73) that we used in the t-test, to calculate either a lower confidence limit or an upper confidence limit as follows:

Lower limit for the difference between the population means

$$= |\bar{x}_1 - \bar{x}_2| - ts\sqrt{\left(\frac{1}{n_1} + \frac{1}{n_2}\right)}$$

$$= 0.9 - 1.73(0.992)\sqrt{\left(\frac{1}{10} + \frac{1}{10}\right)}$$

$$= 0.13$$

Upper limit for the difference between the population means

$$= |\bar{x}_1 - \bar{x}_2| + ts\sqrt{\left(\frac{1}{n_1} + \frac{1}{n_2}\right)}$$

$$= 0.9 + 1.73(0.992)\sqrt{\left(\frac{1}{10} + \frac{1}{10}\right)}$$

$$= 1.67$$

Thus we can state that we are 95% confident that the reduction in brightness is greater than 0.13 or we can state that we are 95% confident that the reduction in brightness is less than 1.67 but we cannot make the two statements simultaneously. We could, however, quote 0.13 and 1.67 as the limits of a 90% *confidence interval*.

Whether we are calculating a one-sided confidence interval or the much more usual two-sided confidence interval we would expect the width of the interval to depend upon the sample size. In the last chapter we used the formula:

$$n = \left(\frac{ts}{c}\right)^2$$

to calculate the sample size needed to estimate the population mean to within $\pm c$, with a certain level of confidence. That was in the simple situation where we intended to take one sample from one population.

In the more complex situation where we wish to estimate a difference between two population means by taking a sample from each population, we can make use of the formula. The smallest sample sizes needed to estimate the difference between two population means to within $\pm c$ are given by:

$$n_1 = n_2 = 2\left(\frac{ts}{c}\right)^2$$

Thus we can calculate that sample sizes of 6 would be needed to be 95% confident of estimating a decrease in brightness to within ± 1 unit. To obtain this sample size we let $t = 1.73$ (one-sided with 18 degrees of freedom), $s = 0.992$ and $c = 1$.

5.5 A MUCH BETTER EXPERIMENT

The investigation carried out by the chief chemist is far from ideal. You will recall that he wished to determine whether or not the brightness of digozo blue pigment decreased during an extended period of storage. He compared the mean brightness of ten drums of pigment that had been recently manufactured, with the mean brightness of ten drums which were produced many months ago. By means of a two-sample t-test he has demonstrated that the two sample means are significantly different. Before the chief chemist concludes that brightness does decrease during storage there are several questions he should ask:

(a) Were the ten drums of new pigment representative of all drums of digozo blue produced recently?
(b) Were the ten drums of old pigment representative of all drums produced at that time?

(c) Was the mean brightness of all old drums the same, at the time they were produced, as the present mean?

It is very doubtful if the chief chemist is in a position to answer all of these questions and we must, therefore, regard his investigation with some suspicion. Perhaps he would be wise to put his results on one side and carry out an experiment which is more suited to detecting a change in brightness with age. The chief chemist would be able to speak with more authority if he carried out a *longitudinal* investigation. This would involve observing the same drums of pigment on two (or more) occasions. He could, for example, measure the brightness of ten drums of pigment selected at random from recently produced batches, and then repeat these measurements on the same ten drums in two years' time. This experiment would be in contrast with the cross-sectional investigation that he has already carried out, in which two different samples are measured at the same time.

Suppose that the chief chemist had chosen to carry out a longitudinal experiment and had obtained the results of Table 5.4. We see in this table that the mean brightness has decreased from 2.2 to 1.3 during the period of storage. These figures may ring a bell because they occurred earlier. In fact the brightness measurements in Table 5.4 are exactly the same as those used in the *t*-test for two independent samples when we were analysing the cross-sectional experiment.

Though the longitudinal experiment has yielded the same results it is not appropriate to use the same method of analysis. Instead of the *t*-test for two independent samples we will carry out a *t*-test for two matched samples. The latter test is often referred to as a paired comparison test

Table 5.4 Brightness of pigment, measured on two occasions

Drum	First occasion	Second occasion
A	2	2
B	2	1
C	4	3
D	1	0
E	2	1
F	3	3
G	3	1
H	2	1
I	1	1
J	2	0
Mean	2.2	1.3

Table 5.5 Decrease in brightness

Drum	First occasion	Second occasion	Decrease d
A	2	2	0
B	2	1	1
C	4	3	1
D	1	0	1
E	2	1	1
F	3	3	0
G	3	1	2
H	2	1	1
I	1	1	0
J	2	0	2
Mean	—	—	$d = 0.9$
SD	—	—	$s_d = 0.738$

which is perhaps a better name as the word *paired* indicates an important feature of the data in Table 5.4. Each brightness determination in the right-hand column is paired with one of the determinations in the centre column. This pairing of the observations occurs, of course, because we have two measurements on each drum, a feature which could not be made use of if we carried out the independent samples test on the data in Table 5.4.

The paired comparison test is actually quite straightforward. We simply calculate the decrease in brightness for each of the ten drums, as in Table 5.5, and then carry out a one-sample *t*-test on these calculated values.

Null hypothesis – mean decrease in brightness for all drums would be equal to zero (i.e. $\mu_d = 0$).

Alternative hypothesis – mean decrease in brightness for all drums would be greater than zero ($\mu_d > 0$).

$$\text{Test statistic} = \frac{|\bar{d} - \mu_d|}{s_d/\sqrt{n}}$$

$$= \frac{|0.9 - 0.0|}{0.738/\sqrt{10}}$$

$$= 3.86$$

Critical values – from the *t*-table with 9 degrees of freedom for a one-sided test:

1.83 at the 5% significance level
2.82 at the 1% significance level.

Decision – we reject the null hypothesis at the 1% level of significance.

Conclusion – we conclude that the brightness of the pigment has decreased during the storage period.

Note that the test statistic (3.86) is much larger than that calculated when doing the t-test for two independent samples (2.02) using the same data. This arises because when we calculate differences in the paired comparison test we automatically eliminate the drum to drum variability. This becomes clear if we compare the standard deviations used in the calculation of the test statistic.

Paired comparison test: $\qquad\qquad s_d = 0.738$

t-test for two independent samples: $\quad s_1 = 1.059$ and $s_2 = 0.919$

The power of the paired comparison test will be illustrated more dramatically in the problems which follow.

5.6   SUMMARY

In this chapter we have introduced three more significance tests to complement the one-sample t-test dealt with earlier. All four tests have followed the same six-step procedure which should now be very familiar.

We used the F-test to compare the precision of two test methods and also as a prerequisite to the t-test for two independent samples. The latter was used to check the statistical significance of the difference between two sample means, in order to decide whether or not the brightness of a pigment deteriorated during storage. Finally we used the paired comparison test for the same purpose, using data from a longitudinal experiment.

As your repertoire of significance tests grows you may become increasingly anxious that we are neglecting the assumptions which underlie these tests. Discussion of these assumptions will be reserved for the following chapters after we have examined more significance tests which are rather different from those studied so far.

PROBLEMS

5.1   Two methods of assessing the moisture content of cement are available. On the same sample of cement the same operator makes a number of repeat determinations and obtains the following results (mg/cc):

Method A   60   68   65   69   63   67   63   64   66   61
Method B   65   64   62   62   64   64   65   66   64

Is one method more repeatable than the other? (A method that is more repeatable will have less variability in repeat determinations.)

5.2 Tertiary Oils have developed a new additive, PX 235, for their engine oil which they believe will decrease petrol consumption. To confirm this belief they carry out an experiment in which miles per gallon (m.p.g.) are recorded for different makes of car. Altogether twelve makes of car were chosen for the experiment and they were randomly split into two groups of six. One group used the oil with the additive, and the other group used oil without the additive. Results were:

Oil with additive   26.2   35.1   43.2   36.2   29.4   37.5
Oil without additive   37.1   42.2   26.9   30.2   33.1   30.9

(a) Are the population variances of the two oils significantly different?
(b) Has the new oil increased miles per gallon?
(c) How many cars would be needed in an investigation of this type in order to be 95% certain of estimating the size of an improvement to within ±1.0 m.p.g.?
(d) What is the population referred to in (a)?

5.3 The research manager of Tertiary Oils is most dissatisfied with the experimental design given in Problem 2. He recognizes that there is large variability between cars and that this can be removed by using a two-sample paired design. He therefore authorizes an experiment in which six cars are used. The following results are obtained:

Make of car	A	B	C	D	E	F
Oil with additive	26.2	35.1	43.2	36.2	29.4	37.5
Oil without additive	24.6	34.0	41.5	35.9	29.1	35.3

(a) Has the additive significantly increased the miles per gallon?
(b) How many cars are needed to be 95% certain of estimating the size of an improvement to within ±1.0 m.p.g., when using this type of design. (You will need to use a formula from Chapter 4.)

6
Testing and estimation: qualitative data

6.1 INTRODUCTION

In the two previous chapters we have been primarily concerned with t-tests. We were looking for a significant difference between two sample means or between one sample mean and a specified value. The data used in these t-tests resulted from measuring such variables as yield, impurity and brightness.

In each of the situations we examined it was meaningful to calculate an average value. Clearly, it makes sense to speak of the mean yield, the mean impurity or the mean brightness of several batches of pigment. In many other situations we must deal with qualitative variables which cannot be averaged to produce a meaningful result. We might, for example, record whether a batch was made for export or for the home market. We might record that an employee was female or that she was located in the research department. Destination, sex and location are qualitative variables and when summarizing such variables it is useful to count and then to calculate proportions or percentages. We might speak of the proportion of batches made for export and to calculate this proportion we would need to count the total number of batches and the number made for export. We might also speak of the percentage of employees who are female or the proportion of staff located in the research department.

In this chapter we will examine a variety of significance tests which are applicable to proportions in particular and to counted data (i.e. frequencies) in general. As these tests are often used in situations which involve people, we will first examine an experiment concerned with the sensory evaluation of a product used in the home.

6.2 DOES THE ADDITIVE AFFECT THE TASTE?

Duostayer Research Laboratories have been asked by their parent company to investigate the effectiveness of a new toothpaste additive, PX

235. The inclusion of PX 235 is expected to increase the protection given to the tooth enamel when the paste is used regularly, and the extent of this increased protection is to be estimated from the results of a major investigation carried out over a long period of time.

In addition to evaluating the effectiveness of the ingredient as a protector of enamel the laboratory is required to investigate the customer acceptability of the toothpaste when PX 235 is included. This minor investigation will be carried out by the sensory evaluation department. Within this department Dr Murphy, a very experienced research chemist, has been given the responsibility for finding an answer to the question: does the inclusion of PX 235 affect the taste of the toothpaste? Dr Murphy plans his investigation as follows:

A 1 kg batch of normal toothpaste and a 1 kg batch of treated toothpaste are each subdivided into 18 samples. Twelve members of a trained tasting panel are each confronted with three samples, the samples being set out as follows:

Assessor	Left	Centre	Right
1	N	T	T
2	T	N	T
3	T	T	N
4	N	N	T
5	N	T	N
6	T	N	N
7	N	T	T
8	T	N	T
9	T	T	N
10	N	N	T
11	N	T	N
12	T	N	N

(T = treated; N = normal)

The three samples presented to any assessor contain two which are identical and one which is different. Each assessor is asked to select the sample which tastes different from the other two. The 12 assessors are assigned at random to the 12 sets of samples which are laid out in accordance with the recommendations of British Standard 54233: Sensory analysis–methodology–triangular test.

Dr Murphy carries out the experiment and he finds that 8 of the 12 assessors correctly select the odd-one-out, whilst the other 4 assessors fail to do so. What conclusion can he draw concerning the taste of the treated toothpaste?

Had he found that all 12 assessors correctly identified the odd one amongst the 3 samples, he would have been happy to conclude that the treated toothpaste did taste different. On the other hand if none of the assessors had been successful he would have readily concluded that the treated and untreated toothpastes were indistinguishable. Eight out of 12 does seem to Dr Murphy to be a borderline case. He realizes that it is quite possible for an assessor to guess correctly which of the 3 samples is the odd one.

In previous chapters we have used significance tests to help us reach decisions in borderline cases. The 12 assessors are, after all, only a sample taken from a population of assessors. Dr Murphy would really like to know what the result of his experiment would have been if the whole population of assessors had taken part.

To carry out a significance test we will need critical values. British Standard 54233 contains a table of critical values which can be used as a yardstick by anyone in Dr Murphy's position. A simplified version of this table is included with the other statistical tables (Table D). The significance test proceeds as follows:

Null hypothesis – the treated and the untreated toothpastes have the same taste.

Alternative hypothesis – the treated and the untreated toothpastes do not have the same taste.

Test statistic = 8 (the number of assessors who were correct).

Critical values – from the triangular test table (Table D), for an experiment with 12 assessors:

$7\frac{1}{2}$ at the 5% significance level
$8\frac{1}{2}$ at the 1% significance level.

Decision – because the test statistic is greater than the lower critical value we reject the null hypothesis at the 5% significance level.

Conclusion – we conclude that it is possible for assessors to distinguish between the treated and the untreated toothpaste.

Though this significance test has followed the same six-step procedure which served us well in the t-test and the F-test, it does differ in some respects from these more common tests. Perhaps the only fundamental difference is that in the triangular test we are dealing with a discrete variable. This discreteness is evident in the test statistic which can only be a whole number.

6.3 WHERE DO STATISTICAL TABLES COME FROM?

Tables of critical values must appear rather mysterious to many who use them, because they are rarely accompanied by an explanation which is comprehensible to anyone who is not a statistician. Fortunately the table of critical values for the triangular test is based on *relatively* simple statistical theory.

You will note that the word 'relatively' is emphasized in the previous sentence. Though the theory underlying the triangular test is much simpler than that underlying the *t*-test, for example, it is not trivial. To fully understand the triangular test and its table of critical values we must explore a new probability distribution. If you do not relish this prospect jump to the next section, but in doing so you may miss a valuable insight into the philosophy of significance testing.

Let us start by assuming that the null hypothesis is true. As the null hypothesis states that there is no difference in taste between the two toothpastes it follows that each assessor is seeking a difference that does not exist. Regardless of the thought processes in the minds of the assessors, a selection of the odd-one-out must be a matter of pure chance. Each assessor would therefore have a one-third chance of guessing correctly and we could calculate the probability of any number of the 12 assessors picking out the odd one in the 3 samples placed before him. The probabilities are obtained by substituting suitable values of n and r into the equation below.

Triangular tests

The probability of r assessors selecting the odd sample in a triangular test involving n assessors

$$= \frac{n!}{r!(n-r)!} \left(\frac{1}{3}\right)^r \left(\frac{2}{3}\right)^{n-r}$$

if there is actually no difference between the samples.

In Dr Murphy's experiment the probability of eight assessors making a correct selection is equal to:

$$\frac{12!}{8!4!} \left(\frac{1}{3}\right)^8 \left(\frac{2}{3}\right)^4 = 0.015$$

With n equal to 12 we can let r equal any number from 0 to 12 inclusive. Taking these values of r one at a time we can calculate the probabilities in Table 6.1.

Don't forget that the probabilities in Table 6.1 are based on the assumption that the null hypothesis is true, i.e. that the two types of toothpaste are

Table 6.1 Possible results of the experiment

Number of assessors who select the odd sample	Probability
0	0.008
1	0.046
2	0.127
3	0.212
4	0.238
5	0.191
6	0.111
7	0.048
8	0.015
9	0.003
10	0.001
11	0.000
12	0.000

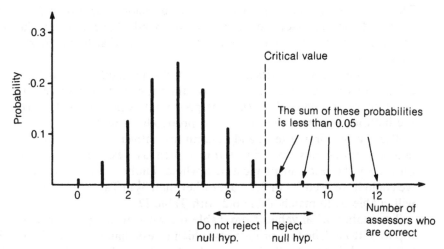

Figure 6.1 What is likely to happen if the toothpastes are indistinguishable

indistinguishable. The probability distribution that is given in Table 6.1 is also illustrated in Figure 6.1.

Marked on Figure 6.1 is the critical value (7.5) for the 5% significance level and we can see that the probability of rejecting the null hypothesis is less than 5%. When Dr Murphy found that eight assessors had identified the odd sample he rejected the null hypothesis at the 5% significance level.

Table 6.2 Cumulative probabilities from Table 6.1

X	Probability of more than X selecting the odd sample
$\frac{1}{2}$	0.992
$1\frac{1}{2}$	0.946
$2\frac{1}{2}$	0.819
$3\frac{1}{2}$	0.607
$4\frac{1}{2}$	0.369
$5\frac{1}{2}$	0.178
$6\frac{1}{2}$	0.067
$7\frac{1}{2}$	0.019
$8\frac{1}{2}$	0.004
$9\frac{1}{2}$	0.001
$10\frac{1}{2}$	0.000
$11\frac{1}{2}$	0.000

The chance of his having made a wrong decision is less than 5%. Had he found nine assessors to be correct he would have rejected the null hypothesis with even greater confidence, knowing that he incurred a risk which was less than 1%.

The whole table of critical values for the triangular test is based on calculations similar to those which gave the probabilities in Table 6.1. This point might be clearer if we sum the probabilities in Table 6.1, starting at the bottom, to obtain the cumulative probabilities in Table 6.2.

The $\frac{1}{2}$ values in the left-hand column are introduced in order to avoid complex statements involving phrases such as 'less than or equal to'. Obviously the test statistic must be a whole number in a triangular test so we could have used 3.4 or 3.7, say, rather than $3\frac{1}{2}$ in Table 6.2. The $\frac{1}{2}$ values will enable us to match Table 6.2 with Table D.

To obtain critical values from Table 6.2 we scan the probabilities from top to bottom. The first probability which is less than 0.05 (i.e. 5%) is the 0.019 which is next to $7\frac{1}{2}$. The critical value at the 5% significance level is therefore $7\frac{1}{2}$. Scanning further downwards we find the first probability which is less than 0.01 (i.e. 1%). This is the 0.004 which is next to $8\frac{1}{2}$. The critical value at the 1% significance level is therefore $8\frac{1}{2}$.

Strictly speaking we could say that $7\frac{1}{2}$ was the critical value at the 1.9% significance level and that $8\frac{1}{2}$ was the critical value at the 0.4% significance level. It might, however, cause considerable confusion to depart from the respected convention of using 5% and 1% significance levels (with occasional use of 10% and/or 0.1% to extend the repertoire).

We have examined the theoretical basis of one statistical table. The table of critical values for the triangular test (Table D) is the simplest example we could have chosen. The theory underlying the t-table, for example, is considerably more complex and the calculation of the critical values very much more difficult. Fortunately it is possible to use significance tests without any knowledge of how tables of critical values are constructed. Much more important is to be able to select the appropriate test and to be aware of the assumptions on which it is based. The assumptions underlying some common significance tests will be discussed in the next chapter.

6.4 ESTIMATING PERCENTAGE REJECTS

EPW Ltd is a subsidiary of the largest European manufacturer of polythene. Using raw materials from the parent company, EPW makes polythene bags and wrappers for customers in the food and allied industries. The Research and Development Department in EPW is strongly encouraged to develop new products which will open up new markets for polythene. A very important, ongoing project seeks to produce rigid polythene containers at a price that will challenge the supremacy of metal cans in food packaging.

Considerable progress has been made and Ron White, the project leader, is currently assessing the performance of the 'mark seven' production line. Three characteristics of the containers have given cause for concern on the earlier lines. They are wall thickness, ovality and edge blips. The first two are measured variables, whilst the third is a qualitative variable, or attribute, which is assessed by visual inspection. Each container is examined and classed as either 'good' or 'blipped'.

'What percentage of containers from the mark seven line will be blipped?' This is the question Ron White wishes to answer. He examines 200 containers and finds that 9 of the 200 are blipped, which gives a blip percentage of 4.5%. He is well aware that the 200 containers examined are only a sample from the 10 000 produced during the test run. He is also aware that the 10 000 containers produced can also be regarded as a sample from the infinite number which could have been made. Ron White requires confidence limits for the percentage he would have obtained if he had examined the whole population.

Confidence limits for a population percentage are given by

$$p \pm \{k\sqrt{[p(100 - p)/n]} + (50/n)\}$$

where p is the sample percentage, n the sample size, k is from the normal distribution table, with $k = 1.96$ for 95% confidence, or 2.58 for 99% confidence.

Note that the above formula should only be used if $np > 500$ and $n(100 - p) > 500$.

To calculate 95% confidence limits for the true percentage blipping of the mark seven production line, we would use $n = 200$, $p = 4.5$ and $k = 1.96$. The confidence limits are:

$$p \pm \{k \sqrt{[p(100 - p)/n]} + (50/n)\}$$
$$= 4.5 \pm \{1.96 \sqrt{[4.5(95.5)/200]} + (50/200)\}$$
$$= 4.5 \pm \{2.87 + 0.25\}$$
$$= 4.5 \pm 3.1$$
$$= 1.4 \text{ to } 7.6\%$$

Ron White can be 95% confident that the true percentage blipping lies between 1.4 and 7.6%. He is very surprised by the width of this confidence interval, as you may be. His training in statistics had led him to regard 30 or more as a large sample. By this standard 200 would be very large. It is true that a sample of 30 or more is often more than adequate when we are dealing with measured data, as we were in Chapters 4 and 5. However, with qualitative data, much larger samples are often required. Ron White wonders how large a sample he would need in order to estimate the population percentage to the precision he requires. For example, how many containers would he need to inspect in order to estimate the true percentage blipping to within ±1%? He suspects that several thousand would be required.

The size of sample needed to estimate a population percentage to within ±c% is given approximately by

$$n = p(100 - p)(k/c)^2$$

where p is an initial estimate of the population percentage, c is the required precision, and k is taken from the normal distribution table, with $k = 1.96$ for 95% confidence and $k = 2.58$ for 99% confidence.

For estimation of the population percentage to within ±1%, with 95% confidence, we would use $p = 4.5$, $c = 1$ and $k = 1.96$, to obtain the required sample size:

$$n = p(100 - p)(k/c)^2$$
$$= 4.5(95.5)(1.96/1)^2$$
$$= 1650.9$$

Thus, we estimate that Ron White would need to inspect at least 1651 containers to estimate the true percentage blipping to within ±1%. Why does he require this precision? He had hoped to prove, by this run of the new line, that the percentage blipping was less than the 5.9% obtained with the mark six production line. The sample percentage of 4.5% is, of course, less than 5.9%. Unfortunately, the confidence interval stretches from 1.4 to 7.6%, which contains 5.9%, so he cannot claim to have proved beyond reasonable doubt that the new line is superior. If the 4.5% had come from a much larger sample, it would have given a narrower confidence interval that did not contain 5.9%, thus proving that the mark seven production line was superior to its predecessor.

The formula we used to estimate the required sample size required an initial estimate of the percentage. If no such estimate is available we put p equal to 50 and the formula becomes

$$n = 2500(k/c)^2$$

For 95% confidence k will be equal to 1.96 which is approximately equal to 2. Substituting $k = 2$ into the formula and rearranging gives an expression which is quite easy to remember:

$$n = (100/c)^2$$

Note that this simplified formula gives only a rough estimate, especially if the true percentage is less than 10% or greater than 90%. However, the formula is safe to use as it tends to overestimate the required sample size. Substituting various values for c we get the sample sizes in Table 6.3. Keep those in mind when you next read the results of a survey of voting intentions.

6.5 HAS THE ADDITIVE BEEN EFFECTIVE?

TX Chemicals manufactures Topex. This is a high-density polymer in pellet form, used in the extrusion of heavy-duty piping. One particular customer has repeatedly expressed concern about the pellet shape. It appears that

Table 6.3 Sample size needed to estimate a percentage

Approximate confidence interval	Required sample size
$p \pm 10\%$	100
$p \pm 5\%$	400
$p \pm 3\%$	1000
$p \pm 2\%$	2500
$p \pm 1\%$	10 000

his process cannot operate at its maximum rate if a high percentage of pellets have concavities. No other customers have expressed concern about pellet shape.

It has been suggested, by an R&D chemist, that the inclusion of a particular additive may reduce the incidence of concavitation. To check this theory samples of 500 pellets are taken before and after the additive is introduced. The first sample is found to contain 68 concave pellets and the second sample only 44. Thus the percentage of pellets with concavities was 13.6% without the additive, but only 8.8% after the additive was included. Can we reasonably conclude that the use of the additive will bring a permanent reduction in the concavity rate?

To answer this question, we can calculate confidence limits for the difference between two population percentages. The two populations in question can be defined as 'all pellets which might be made without the additive' and 'all pellets which might be made with the additive'. If the confidence interval does not include zero we can reasonably conclude that the two population percentages do differ.

Confidence limits for the difference between two population percentages are given by

$$(p_1 - p_2) \pm k\sqrt{\{p(100 - p)[(1/n_1) + (1/n_2)]\}}$$

where p_1 is the larger of the two percentages, p_2 the smaller of the two percentages, p the combined percentage, n_1 and n_2 are the sample sizes, and k is taken from the normal distribution table, with $k = 1.96$ for 95% confidence and $k = 2.58$ for 99% confidence.

With the pellet data $p_1 = 13.6\%$, $p_2 = 8.8\%$, $n_2 = 500$ and $n_1 = 500$. The total number of concave pellets amongst the 1000 inspected was 112, giving a combined percentage of 11.2%. For 95% confidence $k = 1.96$ and the confidence limits are

$$(p_1 - p_2) \pm k\sqrt{\{p(100 - p)[(1/n_1) + (1/n_2)]\}}$$
$$= (13.6 - 8.8) \pm 1.96\sqrt{\{11.2(88.8)[(1/500) + (1/500)]\}}$$
$$= 4.8 \pm 3.9$$
$$= 0.9 \text{ to } 8.7\%$$

We can be 95% confident that the difference between the two population percentages will lie between 0.9 and 8.7%. This interval does not contain zero, so we can claim to have proved beyond reasonable doubt that the inclusion of the additive has resulted in a decrease in the concavity percentage. This decrease could be as little as 0.9%, but it is very unlikely to be zero. On the other hand the decrease in concavity percentage could be as great as 8.7%.

The statistical analysis has shown that a real decrease in percentage did occur. Whether or not this decrease was caused by the introduction of the additive is not a statistical question. It is a matter for scientific judgement. Scientists, quite rightly, place great faith in experiments. When we deliberately change an independent variable, such as the amount of additive, and this is immediately followed by a change in a dependent variable, such as percentage concavity, we conclude that the one caused the other. Our conviction is further strengthened if the change in the dependent variable closely matches what was predicted. We speak confidently of 'proof' when the whole experiment has been replicated. The planning of experiments and the analysis of results will be important themes in later chapters of this book.

6.6 SIGNIFICANCE TESTS WITH PERCENTAGES

In the two previous sections we have used confidence intervals to help us make decisions. In section 6.4 we calculated confidence limits for the true percentage of blipped containers. We were 95% confident that the true percentage was between 1.4 and 7.6%. However, as this confidence interval contained 5.9% we failed to prove beyond reasonable doubt that the percentage had decreased. In section 6.5 we calculated confidence limits for the reduction in cavitation percentage amongst polymer pellets. The 95% confidence interval extended from 0.9 to 8.7% and, as this interval did not include zero, we concluded that the additive had effectively reduced the cavitation percentage.

In both of these situations we could have used significance tests rather than confidence intervals. With the first example we could use a 'one-sample percentage test' and with the second sample a 'two-sample percentage test'. Both tests follow the six-step procedure we used in Chapters 4 and 5. Both tests will bring us to the same conclusion that we reached using confidence intervals. The one-sample percentage test proceeds as follows:

Null hypothesis – the percentage of containers which are blipped is equal to 5.9% (i.e. $\pi = 5.9$).

Alternative hypothesis – the percentage of containers which are blipped is not equal to 5.9% (i.e. $\pi \neq 5.9$).

$$\begin{aligned} \text{Test statistic} &= [|p - \pi| - (50/n)]/\{\sqrt{[\pi(100 - \pi)/n]}\} \\ &= [|4.5 - 5.9| - 0.25]/\{\sqrt{[5.9(94.1)/200]}\} \\ &= 1.15/1.666 \\ &= 0.69 \end{aligned}$$

Critical values – from the normal distribution table:

1.96 at the 5% significance level
2.58 at the 1% significance level.

Decision – as the test statistic is less than both critical values we cannot reject the null hypothesis.

Conclusion – we are unable to conclude that the percentage blipping has changed from 5.9%.

The one-sample percentage test has led us to the same conclusion that we reached when we used the confidence interval earlier. The two approaches are not absolutely identical but will usually give the same conclusion. Note that the one-sample percentage test should only be used if np and $n(100 - p)$ are both greater than 500. This is the same restriction that applied to the confidence interval formula.

Let us now turn to the two-sample percentage test which is used when we wish to compare the percentages found in samples taken from two populations. You will recall that the percentage concavity for the polymer pellets was 13.6% before the additive was included and 8.8% after. Are these percentages significantly different? The two-sample percentage test proceeds as follows:

Null hypothesis – the true percentage concavity is the same, before and after the inclusion of the additive.

Alternative hypothesis – the true percentage concavity differs as a result of including the additive.

$$\text{Test statistic} = |p_1 - p_2|/\sqrt{\{p(1 - p)[(1/n_1) + (1/n_2)]\}}$$
$$= |13.6 - 8.8|/\sqrt{11.2(88.8)[(1/500) + (1/500)]\}}$$
$$= 4.8/1.995$$
$$= 2.41$$

Critical values – from the normal distribution table:

1.96 at the 5% significance level
2.58 at the 1% significance level.

Decision – we reject the null hypothesis at the 5% significance level.

Conclusion – we conclude that the percentage concavity has changed for the better as a result of including the additive.

Again the conclusion from the significance test agrees with that reached when we used a confidence interval. This will not surprise you if you noticed the striking resemblance of the two formulae.

Table 6.4 Observed frequencies

	Without additive	With additive	Total
Concave	68	44	112
Not concave	432	456	888
Total	500	500	1000

Table 6.5 Expected frequencies

	Without additive	With additive	Total
Concave	56.0	56.0	112
Not concave	444.0	444.0	888
Total	500.0	500.0	1000

Before we leave the subject of significance testing with percentages, I would like to introduce a third test, known as the chi-square test. This can be used as an alternative to the two-sample percentage test, but it is more widely applicable. The chi-square test may well be the most widely used of all significance tests as it is very popular with social scientists.

The chi-square test focuses on frequencies rather than percentages. To carry out this test we must first tabulate our data as in Table 6.4. Secondly, we calculate what are known as 'expected frequencies' which are set out in Table 6.5. Note that the two tables have exactly the same row totals and column totals. There the resemblance ends, for the observed frequencies tell us what we actually found, whilst the expected frequencies show what we would expect to find, if the null hypothesis were true. Large differences between the observed and expected frequencies would suggest that the null hypothesis is false.

Null hypothesis – the additive has no effect on concavity.

Alternative hypothesis – the additive does have an effect on concavity.

$$\text{Test statistic} = \sum \frac{(\text{observed frequency} - \text{expected frequency})^2}{\text{expected frequency}}$$

$$= \frac{(68 - 56.0)^2}{56} + \frac{(44 - 56.0)^2}{56} + \frac{(432 - 444.0)^2}{444}$$

$$+ \frac{(456 - 444.0)^2}{444}$$

$$= 2.571 + 2.571 + 0.324 + 0.324$$
$$= 5.79$$

Critical values – from the chi-square table (Statistical Table E) with one degree of freedom:

3.841 at the 5% significance level
6.635 at the 1% significance level.

Decision – we reject the null hypothesis at the 5% significance level.

Conclusion – we conclude that the introduction of the additive does have an effect on concavity of the polymer pellets.

The chi-square test has led us to the same conclusion that we reached when we used the two-sample percentage test and when we used the confidence interval. The chi-square test is, however, more widely applicable as it can be used with more than two samples. Two questions remain unanswered. They concern the calculation of expected frequencies and degrees of freedom. With these data the expected frequencies were self-evident because both samples contained the same number of pellets, 500. With unequal sample sizes they might not have been so obvious. In all cases we can calculate the expected frequencies using

Expected frequency = (row total × column total)/(overall total)

Thus for the top left corner of Table 6.5 we have

Expected frequency = (112 × 500)/1000 = 56.0

Note that the chi-square test should only be used if all of the expected frequencies are greater than 5.0. There are two types of chi-square test. The chi-square goodness-of-fit test is not included in this book. With the chi-square contingency-table test that we have just carried out, the degrees of freedom are calculated from the formula

Degrees of freedom = (number of rows − 1)(number of columns − 1)

Thus with Tables 6.4 and 6.5, which have two rows and two columns, the degrees of freedom will be

Degrees of freedom = (2 − 1)(2 − 1) = 1

With the chi-square test the number of degrees of freedom does not depend upon the sample size, but on the shape of the data table. A table with 2 columns and 2 rows, known as a 2 × 2 table, will have 1 degree of freedom. This will be true regardless of the size of the frequencies within the table. A 3 × 4 table will have 6 degrees of freedom, a 2 × 7 table will also have 6 degrees of freedom, etc.

6.7 THE BINOMIAL DISTRIBUTION

The formulae we have used in this chapter are based on the 'binomial distribution'. This statement may surprise you, as the critical values of 1.96 and 2.58 were obviously taken from the normal distribution table. We used the normal table because it was very convenient to do so and this short cut was acceptable because the binomial distribution had a similar shape to the normal distribution. However, this similarity in shape will not always exist, so we need to take extra care. This is why I have asked you to note certain restrictions such as 'np must be at least 500', and 'expected frequencies should be greater than 5'.

To illustrate the nature of the binomial distribution, let us consider the manufacture of a large number of items, with the process performance being monitored by inspecting a sample of ten items. The items could be pellets, containers, insulation boards, brake pads, tyres, electronic circuits, screws, etc. Each item can be classed as good or defective. Suppose the manufactured population contains 10% of items which are defective. How many defective items would we expect to get in our sample of ten? It is reasonable to answer 'one', as 10% of 10 is equal to 1. However, we realize that the sample could contain zero defectives, or 2, or 3, etc. We might even find 10 defectives in a sample of 10, but this would appear to be unlikely. The probability of each possibility can be calculated using the formula:

With a binomial distribution the probability of finding r defective items in a sample of n items is

$$(\pi/100)^r[1 - (\pi/100)]^{n-r}n!/[r!(n - r)!]$$

where π is the percentage of defectives in the whole population.

Substituting $n = 10$, $r = 0$, and $\pi = 10\%$ into this formula will give us the probability of finding zero defectives in the sample of 10 items:

$$(10/100)^0[1 - (10/100)]^{10}10!/[0!\,10!]$$
$$= (0.1)^0[0.9]^{10}3\,628\,800/[(1)(3\,628\,800)]$$
$$= 0.348\,68$$

By substituting $r = 1$, then $r = 2$, etc., we could calculate all 11 probabilities given in Table 6.6.

Table 6.6 Binomial probabilities, $n = 10$, $\pi = 10\%$

No. of defectives	0	1	2	3	4	5	6	7	8	9	10
Probability	0.35	0.39	0.19	0.06	0.01	0.00	0.00	0.00	0.00	0.00	0.00

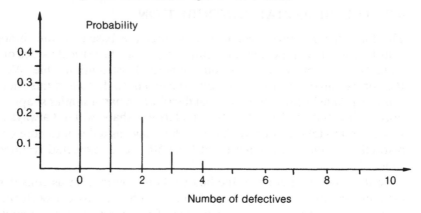

Figure 6.2   Binomial distribution, $n = 10$, $\pi = 10\%$

As we expected, the most likely outcome of this sampling exercise is to find one defective item. However, we see in Table 6.6 that zero defectives are almost as likely with a probability of 0.35 compared with 0.39. The chance of finding five or more defectives is very small and appears as 0.00 in Table 6.6 where the probabilities have been rounded to two decimal places. Perhaps the most striking feature of the binomial distribution is its skewness. This is very clear in Figure 6.2.

Compare the skewness of Figure 6.2 with the symmetry of Figure 6.3, which represents a binomial distribution with $n = 10$ and $\pi = 50\%$. This is the distribution that would apply if we were taking a sample of ten items from a population in which 50% of the items were defective. Obviously the shape of the binomial distribution depends very much on the value of π. It also depends on the value of n.

Clearly Figure 6.3 resembles a normal distribution but Figure 6.2 does not. We could draw a normal curve on to Figure 6.3 and it would fit well.

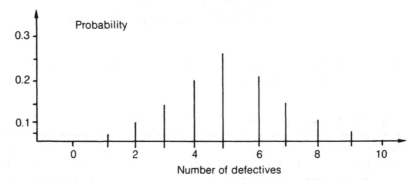

Figure 6.3   Binomial distribution, $n = 10$, $\pi = 50\%$

We could not do so with Figure 6.2. In practice we find that the binomial distribution is 'very normal' when $n\pi$ is greater than 500 and when $n(100 - \pi)$ is also greater than 500. In Figure 6.3, $n\pi = 500$, and $n(100 - \pi) = 500$, but in Figure 6.2, $n\pi = 100$ and $n(100 - \pi) = 900$. Having compared Figures 6.2 and 6.3 you will, I am sure, heed the warnings that have appeared throughout this chapter about the need to check np and $n(100 - p)$ before calculating confidence intervals or carrying out significance tests on qualitative data.

Another point which may worry you concerning the normal and binomial distributions is the obvious fact that only *continuous* variables can have normal distributions and only *discrete* variables can have binomial distributions. In this chapter, for example, we have examined only discrete variables, such as 'number of concave pellets' and 'number of defective items'. In earlier chapters we dealt with continuous variables such as yield, impurity, etc.

Because of this need to match a continuous distribution with a discrete distribution we must think of the area under the normal curve as being divided into vertical strips. Thus the bar in Figure 6.3 representing five defective items would become a strip, from 4.5 to 5.5, under the normal curve. This introduces into some formulae what is known as a 'continuity correction'. The $(50/n)$ in the confidence interval formula is a continuity correction.

6.8 SUMMARY

In this chapter we have used significance tests and confidence intervals with qualitative data. This type of data arises in the assessment of quality because there are many characteristics which are important to the customer, but which cannot be measured. Such features as discoloration, blemishes, bad shape, surface finish, etc., are qualitative variables.

We have seen that very large samples are required when analysing qualitative data. Whereas 20 measured values would often be more than adequate, for a t-test, say, a sample of 20 items might be grossly inadequate for assessing a defect rate. The behaviour of qualitative variables can often be described by either the Poisson distribution or the binomial distribution. In this chapter we have focused on the latter. In particular, we have used binomial distributions which were similar in shape to the normal curve. Thus, we have to observe certain restrictions when using the formulae in this chapter.

PROBLEMS

6.1   In order to test a colour-matcher's ability to discriminate, a triangular test was arranged with two fabrics whose difference in colour was only a shade. The colour matcher picked the odd-one-out correctly seven times out of fifteen.

(a) Can the colour-matcher discriminate between the fabrics?
(b) Is this a suitable design?

6.2 With electrical insulators intended for outdoor use, the surface finish of the insulator is extremely important. In the manufacture of a particular high-quality insulator the surface coating is formed by heating as the insulators pass through a continuous belt furnace. Every insulator is visually inspected as it leaves the furnace to detect 'hazing' of the surface. We know from experience that even when the furnace is working at its best we can expect 10% of insulators to be hazed. If the operating conditions of the furnace change from the optimum, the percentage of insulators which are hazed will be higher than 10%. Hazing appears to occur at random, and no 'serial patterns' of hazing have ever been substantiated.

(a) Examination of the inspection results from a single day's production revealed that 31 insulators were hazed out of a total of 200. Carry out a significance test to confirm that the furnace has deteriorated below its optimum.
(b) Calculate a 95% confidence interval for the true percentage of hazed insulators.
(c) What is the population for the significance test used in part (a)?
(d) A closer analysis of the results of the day's production revealed that two inspectors were used and that inspector A found only eight hazed insulators out of the 80 which he examined. The remainder of the 200 insulators were examined by inspector B. Complete the following table:

	Inspector		
Insulator	A	B	Total
Hazed			
Not-hazed			
Total			200

(e) Carry out a test using the table in part (d) to check whether the inspectors have obtained a significantly different proportion of hazed insulators.
(f) Use the binomial distribution to complete the following table

which refers to the probability of obtaining hazed insulators for a sample of size ten with the furnace operating at its optimum.

Number of hazed insulators	0	1	2	3	4	5 and above
Probability	0.349		0.194	0.057	0.011	

What is the probability of finding two or more hazed insulators in a sample of ten under this condition?

(g) Use the binomial distribution to complete the following table which refers to the probability of obtaining hazed insulators for a sample of size ten when the furnace is producing 20% hazed insulators.

Number of hazed insulators	0	1	2	3	4	5 and above
Probability		0.268	0.302	0.201		0.034

What is the probability of finding less than two hazed insulators in the sample when the furnace is producing 20% hazed insulators?

(h) It is decided to control the quality of the day's production using the first ten insulators. If there are less than two hazed insulators in this sample it is concluded that the furnace is working at its optimum, otherwise it is concluded that the furnace has deteriorated.

(i) What is the probability of concluding that the furnace has deteriorated when it is working at its optimum?
(ii) What is the probability of concluding that the furnace is working at its optimum when the true proportion of hazed insulators is 20%?

What conclusions can be drawn about the above decision criteria?

7

Testing and estimation: assumptions

7.1 INTRODUCTION

After reading the last three chapters you may well be convinced that it is easy to carry out a significance test. True, there is the initial difficulty of deciding which test to do and then there is the chore of having to think about the population when it is so much easier to confine one's attention to the sample, but these problems shrink into insignificance with practice.

Unfortunately there is another problem which has been carefully avoided in earlier chapters but will now be brought out into the open. This concerns the assumptions on which significance tests are based. It is essential that the user of a significance test should be aware of the assumptions which were used in the production of the table of critical values. Any statistical table is intended to be used in certain circumstances. The use of a table of critical values in a situation for which it was not intended can lead to the drawing of invalid conclusions.

Before we probe into assumptions, however, we will examine more sophisticated ways of estimating the variability which must be taken into account when doing a t-test.

7.2 ESTIMATING VARIABILITY

In Chapter 5 we used the F-test to compare the precision of two analytical test methods. Both methods had been used to obtain repeat measures of impurity on the same batch and the standard deviation of the five results by method 1 was 0.3341 whilst the standard deviation of the six results by method 2 was 0.2190. The ratio of the two variances was less than the 5% critical value and we were therefore unable to conclude that the precision of the second method was superior to that of the first. We then went on to explore the possible bias of the two methods using a t-test.

In passing so quickly from the F-test to the t-test we missed the opportunity to estimate the variability of each test method. We will now do so, by using the two sample standard deviations to calculate 95% confidence intervals for the population standard deviations. To do this we make use of the following formula:

A confidence interval for a population standard deviation is given by:

Lower limit = $L_1 s$
Upper limit = $L_2 s$

where L_1 and L_2 are obtained from Statistical Table F.

Five results were obtained by method 1 and therefore the sample standard deviation (0.3341) has 4 degrees of freedom. Table F gives us $L_1 = 0.60$ and $L_2 = 2.87$ for a 95% confidence interval with 4 degrees of freedom.

$$\text{Lower limit} = 0.60 \times 0.3341 = 0.200$$

$$\text{Upper limit} = 2.87 \times 0.3341 = 0.959$$

We can therefore be 95% confident that the population standard deviation lies between 0.200 and 0.959. If we were to continue repeatedly measuring the impurity of this particular batch using this test method we could be confident that the standard deviation of the impurity measurements would eventually settle down to a figure between 0.200 and 0.959.

With the second test method six measurements were made, giving a standard deviation of 0.2190 with 5 degrees of freedom. Table F gives us $L_1 = 0.62$ and $L_2 = 2.45$ for a 95% confidence interval with 5 degrees of freedom.

$$\text{Lower limit} = 0.62 \times 0.2190 = 0.136$$

$$\text{Upper limit} = 2.45 \times 0.2190 = 0.537$$

We can, therefore, be 95% confident that the population standard deviation for the second method of test lies between 0.136 and 0.537. The two confidence intervals are depicted in Figure 7.1 which illustrates two points very clearly:

(a) There is considerable overlap between the two intervals.
(b) Neither of the confidence intervals is symmetrical (i.e. the sample standard deviation is not in the centre of the interval).

It should not surprise us that there is overlap between the two confidence intervals since we have already failed to demonstrate, by means of the F-test, that the two population variances are unequal. The absence of symmetry in the confidence intervals can come as a surprise to many who are familiar with the more common confidence interval for the population mean. The asymmetry arises because the sampling distribution of sample

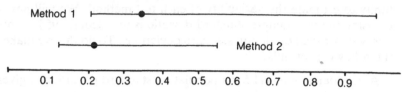

Figure 7.1   Confidence intervals for population standard deviations

standard deviations is skewed. (For a discussion of sampling distributions, see Appendix C).

A third point which may have struck you forcefully when looking at Figure 7.1 is the enormous width of the confidence intervals. Had the samples been larger the intervals would have been narrower, of course. A browse down the columns of Table F will soon convince you, however, that a very large sample would be needed to get a much narrower interval. For example a sample standard deviation of 0.3 would give a 95% confidence interval from 0.255 to 0.366 if based on 60 degrees of freedom. At the other extreme the same standard deviation (0.3), if based on only 1 degree of freedom, would give rise to a 95% confidence interval which extended from 0.135 to 9.57.

7.3   ESTIMATING VARIABILITY FOR A t-TEST

When carrying out a one-sample t-test we calculated the test statistic using:

$$\frac{\bar{x} - \mu}{s/\sqrt{n}}$$

The value of the population mean (μ) was not known, but a value was given to μ in the null hypothesis. The other three ingredients ($\bar{x}$, s and n) were obtained from the sample. We had taken one sample which yielded n observations and the values of $\bar{x}$ and s were calculated from these observations.

It is not necessary, however, to calculate the standard deviation (s) and the mean ($\bar{x}$) from the same sample. In fact there are many situations in which it is desirable to obtain $\bar{x}$ and s from different sets of data. Perhaps the most common example occurs when a sample of data has been obtained in order to explore a possible change in mean (μ). Clearly it is possible to calculate $\bar{x}$ and s from these data and then carry out a one-sample t-test. Whilst this would be quite valid it may not be the *best* course of action, especially if the sample is small. If we are able to assume that a change in mean (μ) will not have been accompanied by a change in standard deviation (σ) and we have an estimate of σ from past data then it may be better to use this estimate rather than the sample standard deviations (*s*). It will certainly be tempting to use this estimate if it is based on a much larger number of degrees of freedom than the $(n - 1)$ on which the sample standard deviation is based. We will be in a better position to detect a change in mean (μ) if we have an estimate of σ based on a large number of degrees of freedom than we would be if we were using the standard deviation of a small sample.

Estimates of population standard deviations do not occur by magic! They can only come from data gathered in the past. We are not advocating that you should make indiscriminate use of all available data in a mad scramble to boost your degrees of freedom. We are simply advising that you should make the best use of any available data which are known to be suitable.

In some situations it is possible to group together the data from several small samples in order to obtain a good estimate of the population standard deviation. The same practice can be followed if we have one sample from each of several populations which are known to have the same standard deviation. The point will be illustrated by the following example.

Example 7.1

The research centre of a well-known oil refining company has been asked to investigate the biodegradation of hydrocarbons in water. The aim of the investigation is to compare the performance of several micro-organisms and to examine ways of accelerating the degradation process. Each experiment involves the introduction of a specified level of a hydrocarbon into a vessel of water and the measurement of the level at regular intervals over a three-month period. The determination of the hydrocarbon level in a sample of water is a time-consuming and expensive operation, with each determination requiring the undivided attention of a laboratory assistant for approximately three hours. It is only practicable therefore to perform two, or at the most three, repeat determinations on each sample.

In an experiment designed to compare six micro-organisms, a concentra-

Table 7.1 Hydrocarbon level after 14 days

Vessel	A	B	C	D	E	F
Determination of	12.2	8.6	17.4	13.0	14.6	12.7
hydrocarbon level	14.3	10.4	16.1	16.2	11.3	12.9
(p.p.m.)		9.9			13.1	
Mean	13.25	9.63	16.75	14.60	13.00	12.80
Standard deviation	1.485	0.929	0.919	2.263	1.652	0.141

tion of 25 parts per million of hydrocarbon is introduced into each of six vessels. Fourteen days later the hydrocarbon levels are found to be as in Table 7.1. As a first step in the analysis of the data in Table 7.1 it is required to calculate a 95% confidence interval for the mean hydrocarbon level in each vessel.

The calculation of a confidence interval for a population mean is not difficult. We first attempted the task in Chapter 4, when we made use of the formula:

$$\text{Confidence interval for } \mu \text{ is} \quad \bar{x} \pm ts/\sqrt{n}$$

Using this formula with the data in Table 7.1 we will need $t = 12.71$ for vessels A, C, D and F, on which only two determinations were made, and $t = 4.30$ for vessels B and E, on which three determinations were carried out. Each vessel is treated quite separately from the other five vessels and the 95% confidence intervals are as in Table 7.2.

Several of the confidence intervals in Table 7.2 are staggeringly wide. This is a consequence of basing a confidence interval on such little information. With vessels A, C, D and F we have used the absolute minimum number of observations ($n = 2$) from which a confidence interval can be calculated and the resulting t value of 12.71 for 1 degree of freedom gives rise to the very wide intervals. (The interval for vessel F is surprisingly narrow, however, because of the very small sample standard deviation.) Figure 7.2 illustrates the six confidence intervals and the determinations on which they are based.

Visual inspection of Figure 7.2 might give you the impression that the confidence intervals are unduly pessimistic. Common sense may suggest to you that the calculated intervals are not a fair reflection of the uncertainty which prevails. Surely the cause of our uncertainty results from the fact that repeat determinations on any vessel do not give identical results. Whether this is caused by sampling variation or by testing error or by some other factor it is the same cause which is affecting all six vessels. It is

Table 7.2 95% confidence intervals for true hydrocarbon level

Vessel	Mean	SD	t	Confidence interval	
A	13.25	1.485	12.71	13.25 ± 13.35	−0.10 to 26.60
B	9.63	0.929	4.30	9.63 ± 2.31	7.32 to 11.94
C	16.75	0.919	12.71	16.75 ± 8.26	8.49 to 25.01
D	14.60	2.263	12.71	14.60 ± 20.34	−5.74 to 34.94
E	13.00	1.652	4.30	13.00 ± 4.10	8.90 to 17.10
F	12.80	0.141	12.71	12.80 ± 1.27	11.53 to 14.07

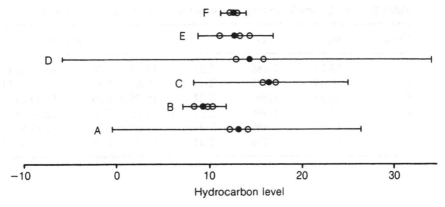

Figure 7.2   Confidence intervals for true hydrocarbon level

reasonable therefore to calculate one estimate of variability and to use it when calculating each of the six confidence intervals. The standard deviations of the six samples can be combined using the following formula:

Combined estimate of the population standard deviation

$$s_c = \sqrt{\frac{\Sigma(\text{d.f.} \times s^2)}{\Sigma \text{d.f.}}}$$

where d.f. is degrees of freedom and s is sample standard deviation.

The combined estimate of σ is given by

$$\sqrt{\{[(1 \times 1.485^2) + (2 \times 0.929^2) + (1 \times 0.919^2)}$$
$$+ (1 \times 2.263^2) + (2 \times 1.652^2) + (1 \times 0.141^2)]/$$
$$[1 + 2 + 1 + 1 + 2 + 1]\}$$

$$= \sqrt{\frac{15.36}{8}}$$

$$= 1.386$$

This combined estimate of σ is simply the square root of a weighted average of the sample variances. Not surprisingly, then, its value lies between the smallest sample standard deviation (0.141) and the largest (2.263). The important feature of this combined estimate of σ is the fact that it has 8 degrees of freedom. Combining the sample standard deviations in this way has given us an estimate which is as good as one would get from nine repeat determinations on any one vessel. When using this estimate to calculate a confidence interval we would therefore use $t = 2.31$ which is the appropriate value for 8 degrees of freedom.

Recalculating the 95% confidence intervals we obtain Table 7.3. The

Table 7.3 95% confidence intervals using the combined estimate of σ

Vessel	Mean	SD	t	Confidence interval	
A	13.25	1.386	2.31	13.25 ± 2.26	10.99 to 15.51
B	9.63	1.386	2.31	9.63 ± 1.85	7.78 to 11.48
C	16.75	1.386	2.31	16.75 ± 2.26	14.49 to 19.01
D	14.60	1.386	2.31	14.60 ± 2.26	12.34 to 16.86
E	13.00	1.386	2.31	13.00 ± 1.85	11.15 to 14.85
F	12.80	1.386	2.31	12.80 ± 2.26	10.54 to 15.06

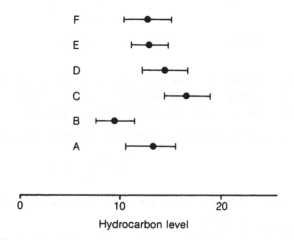

Figure 7.3 Revised confidence intervals for true hydrocarbon level

confidence intervals in this table are illustrated in Figure 7.3. Comparing these intervals with those depicted in Figure 7.2 we see considerable differences. Using the combined estimate of the population standard deviation has given us much narrower intervals which are surely a better indication of the uncertainty in our estimates of the six population means.

This example has taken us one step beyond the straightforward application of the one-sample t-test, towards a very powerful technique known as analysis of variance. This is discussed in more detail in the later chapters of this book.

7.4 ASSUMPTIONS UNDERLYING SIGNIFICANCE TESTS – RANDOMNESS

In all of the significance tests described in this book, it is assumed that the sample from which the test statistic is calculated is drawn at random from the population referred to in the null hypothesis. To obtain a random sample we must engage in random sampling which was earlier described as any process which ensures that every member of the population has the same chance of being included in the sample. To complete the definition we must add that every possible sample of size n must have the same chance of being selected.

One way to ensure these equal chances is to number each member of the population and then to select by means of random numbers taken from tables which are readily available.

If the population is infinitely large it will not be possible to number each member and it is difficult to imagine how one might take a random sample from such a population. In other situations it will not be possible to obtain a random sample because some members of the population do not exist at this point in time.

It may be possible to circumvent some of these difficulties by redefining the population in such a way that it becomes possible to take a random sample from the redefined population. Unfortunately this can be self-defeating as the conclusions may become so particular as to be of no interest whatsoever. Perhaps it is more useful to have a non-random sample from an interesting population than to have a random sample from a population which has been so narrowly defined that it is sterile.

It seems intuitively obvious that a significance test will lead us to a valid conclusion if the sample is representative of the population. Perhaps significance testing would be easier to understand if statisticians spoke of representative samples rather than random samples. Unfortunately there are two major obstacles:

(a) It does not seem possible to define satisfactorily what we mean by the word representative, especially when speaking of a small sample.
(b) It is hard to imagine how one should set about taking a sample in order to get representativeness, without resorting to randomness.

Despite the difficulty, or even impossibility, of obtaining a random sample in many situations we are stuck with the assumption underlying all significance tests that a random sample has been taken from the population in question. If you cannot take a random sample what are you to do? Perhaps the best course of action is to carry out the most appropriate significance test and then to qualify your conclusions with comments on the possible representativeness of the sample.

Before we pass on to other assumptions it must be stated that in many areas of research a random sample is the exception rather than the rule. Charles Darwin and Isaac Newton managed to draw conclusions of great generality without the benefit of random sampling. Darwin's theory of evolution relates to all living creatures but he never attempted to take a random sample from even one species. Newton didn't need to be bombarded by a random sample of apples before he saw the light.

7.5 ASSUMPTIONS UNDERLYING SIGNIFICANCE TESTS – THE NORMAL DISTRIBUTION

The F-test and the various types of t-test are all based on the assumption that the observations have been taken from a normal distribution. In other words, it is assumed that the population referred to in the null hypothesis has a normal distribution.

When we discussed probability distributions in Chapter 3 it was pointed out that the normal distribution is a mathematical model which, strictly speaking, is only applicable to certain, very abstract situations. On the other hand it is also true that:

(a) many sets of data give histograms which resemble normal distribution curves;
(b) the normal distribution has proved its usefulness for describing variability in an enormous variety of situations.

In practice, then, the F-test and the t-test are used in circumstances where, strictly speaking, they are not applicable. The user of significance tests should always question the assumptions underlying a test but it is surely better to ask: How badly are the assumptions violated? rather than: Are all the assumptions satisfied?

A further question which the user might ask of a significance test is: To what extent is this test dependent on the assumptions being satisfied? In doing so he would be questioning what is known as the robustness of the significance test. All statistical techniques are based on assumptions. A robust technique is one which is not very dependent upon its assumptions being satisfied. When using a technique which is robust we are likely to draw a valid conclusion even if the assumptions are quite badly violated. On the other hand, when using a technique which is not robust we are likely to draw invalid conclusions if the assumptions are not well satisfied.

It is difficult to quantify robustness. The t-test, however, is generally regarded as being quite robust as far as the normality assumption is concerned, especially if the sample is large. The F-test, unfortunately, is known to be very dependent on the normality assumption. Despite this well-known fact the F-test is widely used especially in the computer analysis of large sets of data, as we will see in the later chapters of this book. A technique which is less than perfect may nonetheless be very useful especially if used with caution.

The t-test, then, is not very dependent on the assumption of an underlying normal distribution. You can feel quite safe in using the t-test, provided the population distribution is not very skewed. It is always possible to check, by examination of the sample distribution, whether or not the population distribution is reasonably close to the shape of a normal curve. This can be done by carrying out a chi-square goodness-to-fit test. Unfortunately this is a much bigger task than the t-test itself and in practice it is much more convenient to examine only the tails of the sample distribution. For this purpose there exists a variety of tests known as outlier tests or tests for outliers. These will be considered later.

7.6   ASSUMPTIONS UNDERLYING SIGNIFICANCE TESTS – SAMPLE SIZE

In the previous chapter we used the triangular test and the one-sample proportion test. Though the two tests were used in slightly different situations you may have spotted a similarity and it may have occurred to you that the triangular test is only a special case of the one-sample percentage test in which the null hypothesis was $\pi = 33.3\%$. This is not quite true, however, because of the assumptions concerning the sample size in the two tests. Whilst the one-sample percentage test must only be used with a large sample (i.e. greater than 30) the triangular test can be used with any size of sample. It would only be valid to replace the triangular test with a proportion test, therefore, if the sample were large.

Many statistics books refer to the triangular test as an exact test and to the proportion test as an approximate test. Other books make a slightly different distinction, between small sample tests on the one hand and large sample tests on the other.

Regardless of how the material is presented there are several significance tests which depend upon the fact that certain probability distributions can be approximated by the normal distribution. Unfortunately the approximation will only be a good one if certain conditions are satisfied. The one-sample percentage test is an approximate test based upon the assumption that a binomial distribution can be approximated by a normal distribution. It can be shown that the approximation is better:

(a) with larger samples;
(b) with values of π close to 50%.

It is generally agreed, therefore, that the one-sample percentage test should only be used if the sample size (n) is at least 30 and furthermore both $n\pi$ and $n(100 - \pi)$ are greater than 500. Thus it would be reasonable to test the null hypothesis $\pi = 20\%$ with a sample size of 30 but to test $\pi = 10\%$ we would need a sample of at least 50.

Another approximate test dealt with in the previous chapter is the chi-square test. This again is dependent upon a normal approximation and for this reason a chi-square test should never be carried out if any of the expected frequencies are less than 5. Clearly if we are to satisfy this condition our sample size will need to exceed a minimum figure. For example, if we have a two-way frequency table with four cells (like Table 6.5) then we cannot have all four expected frequencies greater than 5 unless our sample size exceeds 20.

7.7 OUTLIERS

The idea of an outlier appeals to the practical scientist/technologist. It seems intuitively obvious to him or her that a set of data may contain one or more observations which need to be singled out as 'not belonging' and may need to be discarded before analysis can begin. We will define an outlier, then, as an observation which does not appear to belong with the other observations in a set of data.

Picking out observations which 'do not appear to belong' is a very subjective process and, when practised by the unscrupulous (or the weak), could lead to undesirable results. Fortunately statisticians have devised significance tests which can be used to determine whether or not an 'apparent outlier' really is beyond the regular pattern exhibited by the other observations. Tests for outliers, like all other significance tests, are carried out using the purely objective six-step procedure. They can, nonetheless, be abused by persons who:

(a) either make use of outlier tests only when it suits their purpose to do so;
(b) or are unaware of the assumptions underlying outlier tests;
(c) or reject outliers indiscriminately, without first seeking non-statistical explanations for the cause of the outlier(s).

Underlying every test for outliers is the usual assumption that the sample was selected at random. In addition there is an assumption concerning the distribution of the population from which the sample was taken. This second assumption is not appreciated by everyone who makes use of outlier tests and ignorance of this assumption can lead to the drawing of ridiculous conclusions.

One of the best known of all outlier tests is Dixon's test, the use of which will be illustrated with reference to the data introduced in Chapter 2. To clarify the presentation we will consider only the first six batches of pigment and the yield and impurity of these batches is given in Table 7.4.

You may recall that in Chapter 4 we used the yield values of the first six batches to test the null hypothesis that the mean yield was still equal to 70.3. Many statisticians would advocate that before carrying out this t-test we should have used Dixon's test to check for outliers.

Table 7.4   Yield and impurity of six batches of pigment

Batch	Yield	Impurity
1	69.0	1.63
2	71.2	5.64
3	74.2	1.03
4	68.1	0.56
5	72.1	1.66
6	64.3	1.90

To calculate the test statistic we first rearrange the observations into ascending order, with the smallest value (x_1) at the left and the largest value (x_n) at the right. For the yields of the six batches this would give:

$$
\begin{array}{cccccc}
x_1 & x_2 & x_3 & x_4 & x_5 & x_6(x_n) \\
64.3 & 68.1 & 69.0 & 71.2 & 72.1 & 74.2
\end{array}
$$

The formula used to calculate the test statistic depends upon the sample size, as in Table 7.5.

Table 7.5   Test statistic for Dixon's test

The value of the test statistic is given by A or B, whichever is the greater.

Sample size	A	B
3 to 7	$\dfrac{x_2 - x_1}{x_n - x_1}$	$\dfrac{x_n - x_{n-1}}{x_n - x_1}$
8 to 12	$\dfrac{x_2 - x_1}{x_{n-1} - x_1}$	$\dfrac{x_n - x_{n-1}}{x_n - x_2}$
13 or more	$\dfrac{x_3 - x_1}{x_{n-2} - x_1}$	$\dfrac{x_n - x_{n-2}}{x_n - x_3}$

For the yields of the six batches we calculate:

$$A = \frac{x_2 - x_1}{x_6 - x_1} = \frac{68.1 - 64.3}{74.2 - 64.3} = 0.384$$

$$B = \frac{x_6 - x_5}{x_6 - x_1} = \frac{74.2 - 72.1}{74.2 - 64.3} = 0.212$$

Test statistic = greater of A or B = 0.384

Critical values – are taken from Statistical Table G and for a sample size of six are:

0.628 at the 5% level of significance
0.740 at the 1% level of significance.

Decision – we cannot reject the null hypothesis.

Conclusion – we are unable to conclude that the yields of the six batches contain an outlier.

The reader will have noticed that the Dixon's test was carried out without a null hypothesis or an alternative hypothesis. Obviously this is unwise and we will return to this point later.

We can, of course, apply the same procedure to the impurity determinations of the six batches. If we intended to use a one-sample t-test to make a decision about the mean impurity of all batches, it would be reasonable to precede the t-test with a Dixon's test. Putting the impurity determinations into ascending order gives:

x_1	x_2	x_3	x_4	x_5	x_6
0.56	1.03	1.63	1.66	1.90	5.64

$$A = \frac{x_2 - x_1}{x_6 - x_1} = \frac{1.03 - 0.56}{5.64 - 0.56} = 0.093$$

$$B = \frac{x_6 - x_5}{x_6 - x_1} = \frac{5.64 - 1.90}{5.64 - 0.56} = 0.736$$

Test statistic = greater of A or B = 0.736

Critical values – are taken from Table G and for a sample size of six are:

0.628 at the 5% significance level
0.740 at the 1% significance level.

Decision – we reject the null hypothesis at the 5% level of significance.

Conclusion – we conclude that the impurity determinations of the six batches do contain an outlier.

A graphical representation of the data in Figure 7.4 clearly indicates which of the six determinations does not appear to belong with the other five.

When we say that the 5.64 does not appear to belong with the other five determinations we are basing this judgement on our preconception of how we expect data to cluster together. It may be difficult, or even impossible, to state explicitly what we do expect but the extreme value (5.64) only

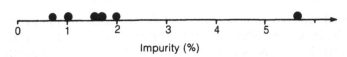

Figure 7.4

appears to be an outlier because it conflicts with what we expect to find when we make six observations. Dixon's test is also based on a preconception, or assumption. With a significance test it is possible, indeed essential, to state explicitly the assumptions on which it is based. Dixon's test is based on the assumption that the observations have come from a normal distribution. (There are actually other versions of Dixon's test based on other distributions.)

Perhaps at this point it would be wise to correct the earlier omission, by stating a null hypothesis and an alternative hypothesis for the Dixon's test. These hypotheses can be presented in such a way that the assumption of normality is clear:

Null hypothesis – all six yield values came from the same population, which has a normal distribution.

Alternative hypothesis – five yield values came from the same population, which has a normal distribution, whereas the most extreme yield value does not belong to this population.

Changing the word 'yield' to 'impurity' would give us the hypothesis for the second Dixon's test.

Having concluded that the impurity value of 5.64 is an outlier it would not be valid to carry out a *t*-test on these data, and we now have two difficulties to overcome:

(a) How are we to reach a decision about the mean impurity of all batches, now that we know the *t*-test is invalid?
(b) What do we do with the extreme impurity determination (5.64)?

One solution to both problems would be to ignore the outlier and to press on with the *t*-test. This is at best unwise and at worst dishonest.

An alternative solution to both problems would be to reject the outlier and to carry out a *t*-test using the other five determinations. Perhaps this is the most natural reaction to an outlier – reject it! In this case, however, it would be a ridiculous course of action.

This assertion is made in the light of further information. We have already examined the impurity determinations of the first 50 batches in Chapter 2. The distribution of impurity is illustrated in Figure 2.2, which depicts a very skewed distribution. The use of Dixon's test has in effect proved what we already know, that the variation of impurity from batch to batch does not follow a normal distribution. You may recall that in Chapter 2 we transformed the impurity data by taking logarithms and that the result of the transformation was to give a symmetrical distribution.

We have therefore a third course of action that we might take. Transformation of the impurity data would appear to be more sensible than either ignoring or rejecting the outlier. If we were to carry out a Dixon's test on the six log impurity values we would conclude that there were no outliers in the data and it would then be valid to carry out a t-test using the transformed data.

In conclusion we note that all significance tests mentioned in this book are based on assumptions concerning the distribution of the population referred to in the null hypothesis. In many cases these assumptions can be checked by means of a test for outliers. We can, for example, precede a one-sample t-test with a Dixon's test. It is important to be aware, however, that the outlier test itself is based on an assumption about the population distribution. In fact there exists an enormous variety of outlier tests which are discussed at length in Barnett and Lewis (1979).

7.8 SUMMARY

In this chapter we have discussed the assumptions which underlie significance tests and confidence intervals. These assumptions fall into two sets, which relate to:

(a) how the sample is selected from the population;
(b) the distribution of the population from which the sample is taken.

All of the tests and confidence interval formulae covered in this book are based on the assumption that a random sample has been taken and most of the tests share the further assumption that the population has a normal distribution. These assumptions are very restricting but they cannot be ignored. Many researchers, especially in the social sciences, find that these assumptions cannot be satisfied and turn to a set of significance tests known as non-parametric tests. These are dealt with in Caulcutt (1989).

PROBLEMS

7.1 A random sample of size eight was chosen from a week's production of a particular make of car. Road tests on the cars gave the following m.p.g.:

      30.7   32.4   32.3   34.1   27.3   32.5   31.7   32.7

(a) Carry out Dixon's test to determine whether the data contain an outlier.

(b) Should the outlier be rejected?

7.2 Source: *The Analyst*, Jan. 1980, p. 71.

In a collaborative study, each of five laboratories analysed the same grass sample (2 g) for antimony, using the nitric-acid/sulphuric-acid digestion method. Each laboratory carried out three replicate determinations to obtain the results in the table below.

(a) Calculate a combined estimate of the standard deviation.

(b) Use this estimate to obtain a 95% confidence interval for the true mean of laboratory B.

(c) For each laboratory use Dixon's test to determine whether there is a possible outlier.

(d) By examining the pattern of results given in Figure 7.5, consider whether the analyses in parts (a), (b) and (c) fairly represent the data.

Laboratory	Determination (μg antimony)			Mean	SD
A	2.34	2.10	1.96	2.13	0.192
B	1.97	1.92	1.91	1.93	0.032
C	2.01	2.01	2.07	2.03	0.035
D	2.06	2.02	2.12	2.06	0.050
E	2.38	1.77	2.47	2.20	0.380
F	1.94	2.06	2.06	2.02	0.069

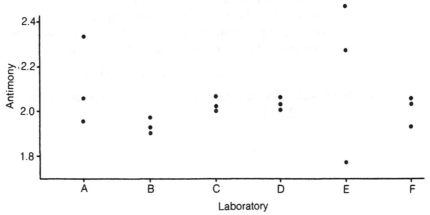

Figure 7.5   Three determinations from each of five laboratories

(e) Which of the following assumptions may have been violated in parts (a), (b) or (c):

    (i) The results follow a normal distribution.
    (ii) The population standard deviations are equal.
    (iii) The results are independent of each other.
    (iv) A random sample has been taken.
    (v) A continuous variable has been measured.

7.3 On 11 March 1980, *Newsnight*, a BBC television programme, carried out an investigation into the death rate due to cancer at the Atomic Weapons Research Establishment at Harwell. One worker who had recently died of cancer had been subjected to whole-body monitoring for plutonium radiation level at four different time periods. The following radiation levels were obtained:

Time period	1	2	3	4
Radiation level	58	16	0	52

A level of above 50 is considered dangerous and a level of 16 is considered acceptable. After the first and fourth time periods the worker was removed from contact with radiation.

(a) Calculate a 95% confidence interval for the mean radiation level.
(b) Calculate a 95% confidence interval for the true proportion of results above 50.
(c) Which of the following assumptions (or conditions) have been violated in parts (a) and (b):

    (i) results are a random (or representative) sample of the population;
    (ii) results are independent of each other;
    (iii) the results follow a normal distribution;
    (iv) the number of successes is greater than five. This is a necessary condition for the use of the formula:

$$p \pm k \sqrt{[p(100 - p)/n]}$$

(d) In the light of your conclusions from (c) comment on the analysis in parts (a) and (b).

8
Statistical process control

8.1 INTRODUCTION

In 1931 an American industrial scientist, Walter Shewhart, wrote a book entitled *Economic Control of Quality of Manufactured Product*. In this text he introduced the idea that a statistical process control (SPC) chart could be used to decide whether or not a production process had changed. This would help us to detect, as early as possible, any deterioration in the product being manufactured.

I would be happy to report that the SPC chart, or Shewhart chart, was readily adopted throughout the USA and UK, thus helping these countries to dominate world trade. However, that would not be true. Shewhart charts were used in the USA and the UK during the 1930s and during the Second World War, but were then largely abandoned. Later it emerged that Japanese industry picked up what we had discarded and used control charts to such good effect that we are now reintroducing Shewhart's techniques in the hope of regaining lost ground.

In this chapter we will examine the SPC chart and how it is used in practice. We will also look at several adaptations which can be incorporated to improve its performance. Naturally, you will also wish to consider the assumptions underlying this technique, so that you can use it with safety.

8.2 A CONTROL CHART FOR THE MEAN YIELD

In previous chapters we have analysed several sets of data. Perhaps you remember the first data that we examined in Chapter 2. It contained the yield and impurity of 50 batches of digozo blue pigment. We used histograms to display the distributions of these two variables, and then in Chapter 4 we re-examined the data as we attempted to estimate the mean yield of future batches. Perhaps, at that time, we should have explored the possibility that yield or impurity might have increased or decreased during the period when these 50 batches were produced. If, for example, there had been a gradual increase in yield throughout the period, then the average yield of the 50 batches would not have provided the best predictor

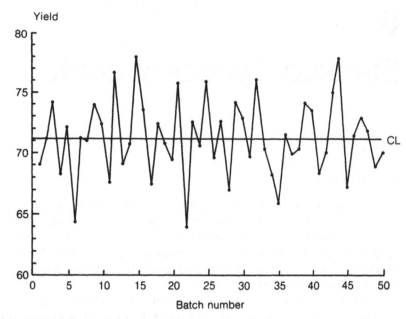

Figure 8.1   Fifty batches of digozo blue pigment

of future yield. Similarly, if there had been a sudden increase in yield half-way through the period, then it might have been better to base our prediction on the mean yield of the later batches. Perhaps it would have been wise to check out these possibilities before we attempted to estimate future yield. The best way to carry out such a check would be to examine a run chart, or time series graph, such as we find in Figure 8.1.

The points in Figure 8.1 are based on the yield data in Table 2.1. Note that 'time', in this case 'batch number', is on the horizontal axis and that consecutive points are joined by straight lines. These two conventions are always observed when plotting run charts. The horizontal line in Figure 8.1 is known as the centre line (CL). It is drawn through a yield value of 71.23, which is the mean yield for the 50 batches. In some cases it might be more useful to draw the centre line through some target value rather than the mean of the data. This target value could be midway between two specification limits. If we now include two additional horizontal lines, our run chart will become a control chart; a 'Shewhart control chart for the mean yield', would be its complete title (Figure 8.2). The additional lines will be placed three standard deviations above and below the centre line. They are known as action lines or control lines. The standard deviation of the 50 yield values is 3.251. Thus the upper action line (UAL) is drawn through a

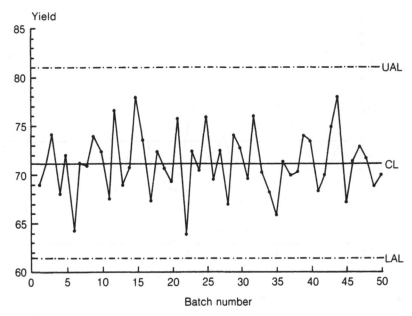

Figure 8.2   A control chart for mean yield

yield value of 81.0 and the lower action line (LAL) is drawn through a yield of 61.5.

What can be concluded from Figure 8.2? We see that 26 of the 50 points lie below the centre line whilst 24 lie above. This should not surprise us. We would expect approximately 50% of our data points to lie below the mean if our data were coming from a normal distribution. We also note that all 50 points lie between the two action lines. This again should not cause us any surprise. With a normal distribution we expect most of the data to lie within three standard deviations of the mean. Finally, we note that the points fluctuate randomly, or unpredictably, with no discernible patterns in the data. Shewhart used the phrase 'in control' to describe any process that gives us data similar to that in Figure 8.2.

If we removed the data points from Figure 8.2, or if we transferred the centre line and the action lines to a clean sheet of graph paper, we would have a control chart on which we could monitor the yield of future batches. This monitoring would be achieved by plotting on the chart the yield of each batch, as soon as this became available. Immediately after plotting a point we would make a decision as to whether or not a change in the level of yield had occurred. If the new point lay between the two action lines we would take no action, on the assumption that fluctuations within three standard deviations of the mean could be regarded as normal process

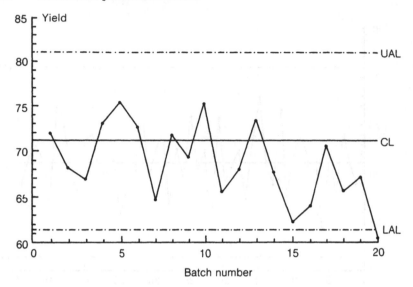

Figure 8.3 The control chart in use

variation. On the other hand, if the new point lay outside the action lines, we would conclude that the mean yield had changed. Figure 8.3 illustrates how the control chart might evolve over a period of time.

We can see in Figure 8.3 that the first 19 points all lie between the action lines. Thus no action is called for until we plot the point for batch number 20, at which time we conclude that the mean yield has decreased. The control chart has, therefore, alerted us to the need for a process adjustment. Obviously the chart does not tell us what to adjust or by how much. The action we take to increase the mean yield will be based upon our understanding of the production process.

It is important to realize that the control chart is operating in 'real time'. As each batch is produced we obtain a yield determination which is plotted as an additional point on the chart and a decision is made about the process, depending upon the position of this new point. If you have studied Chapters 4 to 7 of this book you will realize that the person plotting and interpreting the chart is carrying out a succession of significance tests. In fact he or she is doing a one-sample t-test as each point is plotted. It is a rather unusual t-test, in which the sample mean is based on only one observation, hence $n = 1$, and the sample standard deviation is based on the data in Figure 8.1 which has 49 degrees of freedom. Obviously, the assumptions underlying the t-test, which were discussed in Chapter 7, will apply equally to the Shewhart control chart for the mean. We will return to these assumptions later. Let us first examine how well the control chart is likely to perform.

8.3   HOW QUICKLY WILL CHANGE BE DETECTED?

You will no doubt recall from your knowledge of earlier chapters that there is a risk associated with every significance test. To be more precise, there is a 5% risk of concluding that a change has occurred when in fact it has not, if we use a 5% significance level. With the control chart in Figure 8.3 there is obviously a risk of getting a point outside the action lines even though the long-term mean yield has not changed.

Clearly the size of the risk will depend on the position of the action lines. As these were placed three standard deviations from the mean the risk will be equal to 0.27%. This figure was calculated by doubling the probability obtained from Table A, corresponding to a standardized value of 3.00. The logic of this procedure is illustrated in Figure 8.4.

Thus there is a small risk of error every time we plot a point on the control chart. In practice we plot point after point after point, so it is inevitable that we will, sooner or later, make a wrong decision. 'How soon?' you may wonder. We can estimate how often we are likely to find a point outside the action lines, by using the formula

$$\text{Average run length} = 1/(\text{probability of action})$$

With a probability of 0.0027 from Table A we get an average run length of 370. Thus we can expect a false alarm to occur every 370 batches on average. By 'false alarm', I mean a decision to take action when, in fact, no action is needed because the mean yield has not changed.

Perhaps you are not very interested in false alarms, provided they do not occur too often of course. You may prefer to talk about real changes and

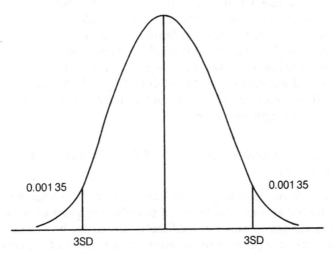

Figure 8.4   Probability of a false alarm

Table 8.1   Average run length and size of change

Size of change as a multiple of the SD	Probability of an action indication	Average run length
0.0	0.002 70	370
0.5	0.006 53	153
1.0	0.028 23	44
1.5	0.066 8	15
2.0	0.158 7	6.3
2.5	0.308 5	3.2
3.0	0.500 0	2.0

how long it is likely to take the control chart to indicate a real change, after it has occurred. Common sense would suggest that a control chart, or any other system, is likely to detect a large change quite quickly but may take much longer to detect a small change. If we use our formula to calculate average run lengths for different sizes of change we find that common sense has served us well on this occasion.

We can see in Table 8.1 that a change of two standard deviations has an average run length of 6.3. This implies that a sudden change in mean of two standard deviations is likely to be detected after 6 or 7 points have been plotted. It might be detected earlier of course, or it might be detected later. Let us return to our digozo blue example and see what this implies in practical terms. The standard deviation of the data in Figure 8.1 is 3.251. Thus two standard deviations would be equal to 6.5. If the mean yield suddenly decreased by 6.5 units, from the target value of 71.23 to 64.73, we might well manufacture a further 6 or 7 batches before we were made aware of the change by a point falling below the lower action line on our control chart. You may be surprised and somewhat disappointed to learn that the Shewhart chart takes so long to detect a change which is so large. However, if we were interested in detecting a change in yield of only one standard deviation, 3.251, we would suffer an even longer delay, as the average run length would be 44.

8.4   INCREASING THE SENSITIVITY OF THE CONTROL CHART

The average run lengths given in Table 8.1 might provide a sobering dash of reality for anyone who believed that the Shewhart control chart would detect a change in mean immediately, no matter how small the change. Fortunately, it is possible to increase the sensitivity of the control chart so that it will detect small changes more quickly. By introducing additional rules for interpretation we can increase the usefulness of the chart without greatly increasing its complexity.

Let us re-examine Figure 8.3. I think you will agree that there is some evidence of change in Figure 8.3 even before the twentieth point is plotted. We observe, for example, that only one of the last ten points is above the centre line, though we would expect to find approximately half of the points on each side of this line if the long-term mean were equal to the central value of 71.23. We also observe that several of the later points are quite close to the lower action line, whereas none are close to the upper action line. Further detailed examination of Figure 8.3 would no doubt strengthen our increasing belief that the mean yield was less than 71.23 long before the twentieth point was plotted. In fact the mean yield of batches 11 to 19 is 67.14, which is found to be significantly less than 71.23 if we carry out a one-sample t-test.

The foregoing discussion of Figure 8.3 is based on one person's subjective impressions. Perhaps it would be unwise to put such freedom of interpretation into the hands of process operators. In practice we require a more formal and unambiguous statement of the rules we intend to use, so that our control charts will be interpreted consistently by all of our staff.

Let us examine some of the many supplementary rules that have been recommended by various writers since Shewhart charts were first introduced in the 1920s. The purpose of these supplementary rules is to increase the sensitivity of the control chart. I offer the following selection for your consideration:

(a) Take action if one point lies outside the action lines. (The action lines are three standard deviations from the centre line.)
(b) Take action if two consecutive points lie between a warning line and the adjacent action line. (Warning lines are drawn two standard deviations from the centre line.)
(c) Take action if any two of three consecutive points lie between a warning line and the adjacent action line.
(d) Take action if five consecutive points lie between one and three standard deviations from the centre line.
(e) Take action if any four of five consecutive points lie between one and three standard deviations from the centre line.
(f) Take action if seven consecutive points lie on the same side of the centre line.
(g) Take action if eight consecutive points lie on the same side of the centre line.

The above list is not exhaustive. Other supplementary rules can be found in other texts on SPC. Clearly, you could devise your own rules for inter-

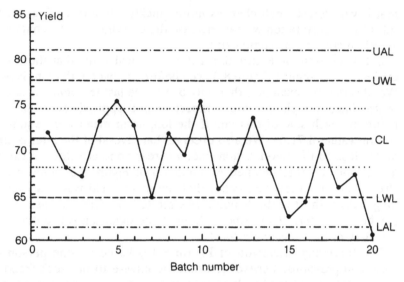

Figure 8.5   The control chart with warning lines and one-sigma lines

preting a Shewhart control chart. However, you would be wise to consider the likely consequences of adopting an additional rule before doing so. Later we will explore the reduction in average run length that can be achieved by using different combinations of the supplementary rules, but first I would like to see what effect the use of these rules would have on our interpretation of Figure 8.3.

Figure 8.5 is identical with Figure 8.3 except for the additional lines, which have been drawn at one standard deviation and two standard deviations from the centre line. The two-sigma lines are known as 'warning lines' and often appear on control charts in the UK but are not popular in the USA. The one-sigma lines are rarely shown on charts. They are included in Figure 8.5 to help you reinterpret the data, using the seven supplementary rules that I have listed.

Rule number 1 gives us an action indication at batch number 20, as we have already noted. Rule number 2 gives an action indication at batch number 16, since the fifteenth and sixteenth points are both outside the lower warning line. Rule number 3 also gives an action indication at batch number 16. Clearly rules 2 and 3 are very similar and no one would use both simultaneously. Neither rule number 4 nor rule number 5 give any action indications with this set of data. Again we note that these two rules are very similar and would not be used together. Rule number 6 gives an action indication at batch 20, as did rule number 1. However, rule number

Table 8.2 Average run length is reduced by the use of supplementary rules

Size of change as a multiple of the SD	Average run length		
	Rule 1 only	Rules 1 and 2	Rules 1, 2, 4 and 7
0.0	370	278	133
0.2	308	223	96
0.4	200	134	52
0.6	120	75	29
0.8	72	43	18
1.0	44	26	12
1.2	28	16	8.9
1.4	18	11	6.8
1.6	12	7.4	5.4
1.8	8.7	5.4	4.4
2.0	6.3	4.1	3.6
2.2	4.7	3.2	3.0
2.4	3.7	2.6	2.5
2.6	2.9	2.2	2.2
2.8	2.4	1.9	1.9
3.0	2.0	1.7	1.7

7, which is clearly less sensitive to change than rule 6, does not give any action indication.

Obviously by using the supplementary rules we have obtained earlier indication of change than we got when we used only the action lines. 'Would this be the case with every set of data?' you may wonder. I am sure you would agree that the use of extra rules can only reduce the average run length. Furthermore, the reduction we achieve will depend upon which of the rules we include. Two examples are illustrated in Table 8.2 and Figure 8.6.

We can clearly see in Table 8.2 and Figure 8.6 that the use of supplementary rules reduces average run lengths. By using warning lines, in addition to the action lines (i.e. rules 1 and 2), we are likely to detect change more quickly. This is particularly true if the size of the change is about one standard deviation. The inclusion of rules 4 and 7 gives further reduction in average run length, thus further increasing the speed with which we are likely to detect a change. Clearly it is beneficial to use supplementary rules when you attempt to interpret a control chart for the mean and in practice most people do so. Let me repeat, however, that users of Shewhart charts do not all use the same rules, and the list I have presented

Average run length

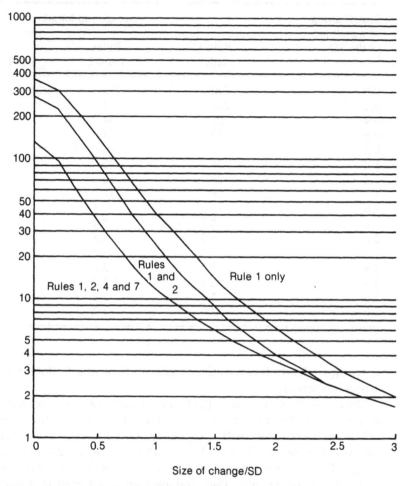

Figure 8.6   Average run lengths from Table 8.2

is not exhaustive. For a more detailed discussion of the effect of using supplementary rules see Champ and Woodall (1987).

A further point to note, whilst studying Figure 8.6, is that the increased sensitivity to small change has to be paid for by increased frequency of false alarms. Thus the introduction of rules 2, 4 and 7 almost trebles the false alarm rate. Perhaps this is an acceptable price to pay for reducing the average run length from 44 to 12, when a change of one standard deviation occurs.

Obviously you would be interested to learn of any modification to the

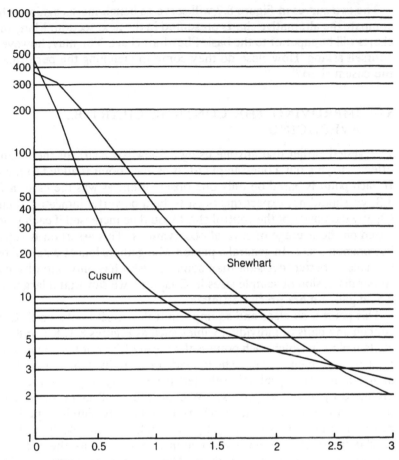

Figure 8.7   The cusum chart is more powerful

Shewhart control chart which would give quicker detection of real change without also giving increased frequency of false alarms. We will consider several such alternatives in the next two sections. Before we do so, however, I would like to show you the magnitude of improvement we might expect to strive for. Figure 8.7 allows us to compare the average run length curve for the basic Shewhart chart with that for a cusum chart. The cusum chart is very important and will be fully described in Chapter 9. Clearly it is a more powerful detector of change than the Shewhart chart, for any size of change up to three standard deviations. Unfortunately the cusum chart has not found favour as a real-time control chart for reasons we will discuss later.

The two curves in Figure 8.7 will serve as bench-marks against which to compare the alternative charts presented in the next section. Many of these charts will be superior to the basic Shewhart chart. You may, however, find yourself asking 'How close do they come to matching the performance of the cusum chart?'

8.5 IMPROVING THE CONTROL CHART BY AVERAGING

I suggested earlier that anyone using a Shewhart chart for the mean is, in effect, carrying out a one-sample t-test as each point is added to the chart. If each point is based on only one observation, as was the case in Figure 8.3, we should not expect this t-test to be a powerful indicator of change. Clearly the power of the control chart would be increased if each point was based on the average of several observations. This assertion is supported by common sense. In general a person who gathers lots of data will be able to make a 'better' decision than someone who has only one observation. In our discussion of sample sizes in Chapter 4 we saw that a larger sample would give a more precise estimate.

This point is so clear and so important that most books on SPC do not contain any plots of individual values such as Figures 8.1, 8.3 or 8.5. They contain plots in which each point is the mean of four or five measurements. Each measurement will have been made on an individual item and the four or five items were probably selected from a very large number of items produced by a 'widget process'. Widgets are small metal or plastic components produced in large numbers by relatively simple repetitive processes. The manager of a widget process takes a small sample of items at regular intervals, measures each item and then plots the mean of the measurements on a control chart. Very wisely the manager wishes to obtain the benefit of averaging. This benefit is highlighted in Figure 8.8.

Unfortunately we have very few widget processes in the chemical industry. If you attempt to apply to chemical production processes, the conventional wisdom presented in many SPC texts, you may experience great difficulty. For example, how are you to get the benefit of averaging if you are managing a chemical batch process? The quality of the product cannot be measured until the reaction is completed, at which time one determination of quality becomes available. To increase the number of determinations you could ask the analytical laboratory to make several determinations on the one sample of product. You would then be able to plot the mean of these results on a control chart. Unfortunately the variability within this group of results would be relatively small, reflecting only analytical error and not batch-to-batch variation in quality. Alternatively you could take several samples of product from the batch and

Average run length

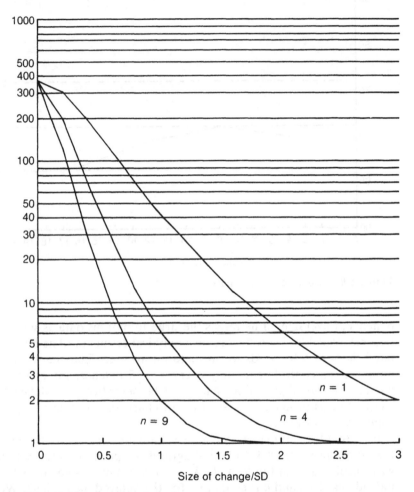

Figure 8.8   Average run length is dependent on sample size

request one determination on each. These results would be more variable, reflecting product heterogeneity within batch, in addition to the analytical error, but again they would not encompass the batch-to-batch variation in quality. The effect of using a control chart based on this reduced variation would be increased action as you responded to random fluctuations in quality in addition to the real long-term changes that you wished to detect.

The manager of the chemical batch process can have only one meaningful determination of quality on each batch. Nonetheless, he can enjoy the

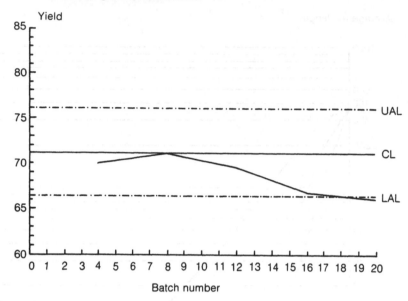

Figure 8.9   A delayed mean chart

benefits of averaging if he resists the urge to plot his data as soon as they become available. He can store his analytical results until he has accumulated the 4 or 5 he needs to calculate an average. Thus he would plot a point on the control chart only every 4 or 5 batches, rather than every batch. This would give him what could be described as a 'delayed mean chart'. Let us produce a delayed mean chart for the data in Figure 8.3, using groups of four.

Note that the action lines in Figure 8.9 are much closer to the centre line than those in Figure 8.3. In both diagrams the position of the action lines was calculated from the formula $\bar{X} \pm 3\sigma/\sqrt{n}$, but n is equal to 1 for the individuals chart and n is equal to 4 for the delayed mean chart. We have seen in earlier chapters that sample means are not so variable as individual values. This point is discussed more formally in Appendix C. The first point plotted in Figure 8.9 is based on the mean yield of batches 1 to 4. Thus the results for batches 1, 2 and 3 were stored whilst we awaited the fourth yield value, without which we could not proceed. The second point is based on the mean yield of batches 5 to 8, the third point on batches 9 to 12 and the fourth point on batches 13 to 16. When the result for batch 20 becomes available we can plot the fifth point and we find that this lies below the lower action line. Thus the delayed mean chart has indicated a decrease in mean at batch number 20 as did the individuals chart, Figure 8.3.

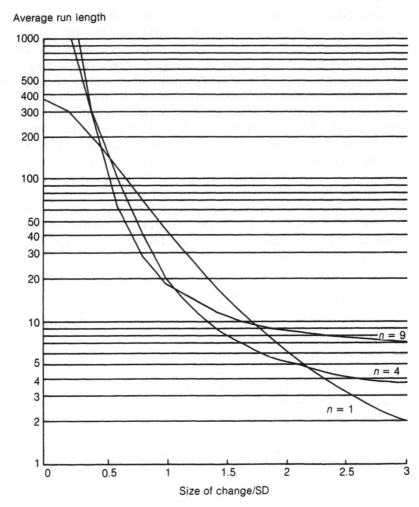

Figure 8.10   The performance of delayed mean charts

At this point we could conclude that the delayed mean chart is no better or worse than the individuals chart. That would be rash. We cannot reasonably claim to have compared two alternative control charts after simply observing their performance with one particular set of data. We need to compare their average run length curves.

We turned to the delayed mean chart in order to gain the advantage of averaging. Does the increased power derived from averaging outweigh the possible delay in detection introduced by failing to plot data as early as possible? We can see in Figure 8.10 that the answer is 'yes and no'. Yes, the delayed mean chart is likely to detect a small change, say one standard deviation, more quickly than an individuals chart. No, the delayed mean chart is not likely to detect a large change as quickly as an individuals chart. Thus we benefit from the power of averaging when a small change occurs, but we pay the penalty for delayed plotting when a large change occurs.

The delayed mean chart has been used in the chemical industry. Perhaps it was seen as a way of adapting the conventional SPC approach to the special needs of chemical processes. Had the users of the delayed mean chart studied Figure 8.10, it is unlikely they would have been happy to continue storing data rather than acting upon them. Common sense suggests that the best control chart will be one which makes use of data as soon as they become available. But common sense also suggests that decisions should be based on averages rather than individual values. Fortunately there is an alternative control chart which is based on averages and which allows us to plot our data without delay. It is known as a moving average chart or moving mean chart.

8.6 MOVING MEAN CHARTS

To set up a moving mean chart we must first decide how many results will be incorporated in each average. Let us say we choose four, in which case we will be using a four-point moving average. Each new result will be added to the previous three and the total divided by four to give an average which is plotted. Thus each result is used on four occasions and will influence the position of four consecutive points on the control chart. Let us apply this moving average technique to the data in Figure 8.3. The four-point moving average is plotted in Figure 8.11.

The action lines in Figure 8.11 are identical to those in Figure 8.9, the delayed mean chart. There is a further similarity. The five points plotted on the delayed mean chart also appear on the moving average chart, supplemented by 12 other points. As the moving mean chart contains all that is on the delayed mean chart and more, we can reasonably expect it to be more sensitive to change. With this particular set of data that is certainly the case, for the moving mean chart gives an action indication at batch 17 compared with batch 20 on the delayed mean chart. Can we reasonably expect that this greater sensitivity would manifest itself with any set of data? The average run length curves in Figure 8.12 offer a better basis for comparison than the consideration of particular data sets.

We can see in Figure 8.12 that the moving mean chart gives lower average run lengths than the delayed mean chart, for any size of change. Furthermore, the moving mean chart gives lower average run lengths than the individuals chart for changes up to 2.8 standard deviations in magnitude. Thus for smaller changes the moving mean chart is likely to give an

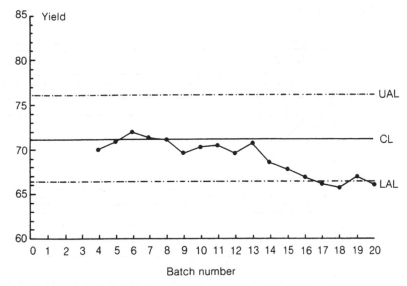

Figure 8.11   A moving average chart

earlier indication than would the other two. Clearly the moving mean chart should be given serious consideration by the manager of any batch process in the chemical industry.

When deciding how many observations to include in the moving average we must compromise. We wish to put as many data as possible into each average, but we also wish to base our decisions on information which is up to date. In practice the four-point moving mean is considered by many users to offer a balance between these conflicting objectives. Obviously a five-point moving mean would be more sensitive to small changes, whilst a three-point moving mean would respond more quickly to large changes.

No matter how many observations we include in the moving average it is usual to regard all the observations as being equally important. Indeed, the formula is such that it gives each observation the same weight. However, it could be argued that we should give more weight to more recent data, if we wish our average to reflect the current situation. This could be achieved in many ways. For example, with a four-point moving average, we could multiply by 2 the most recent result, before adding it to the previous three and then divide by 5. Alternatively, we could multiply the four results by 0.4, 0.3, 0.2 and 0.1 before adding them. Either of these methods would give us a four-point weighted moving average, with greater importance being attached to the most recent data. Many other formulae exist for calculating weighted averages. One of the most useful gives us what is known as an 'exponentially weighted moving average', or EWMA. Despite

Average run length

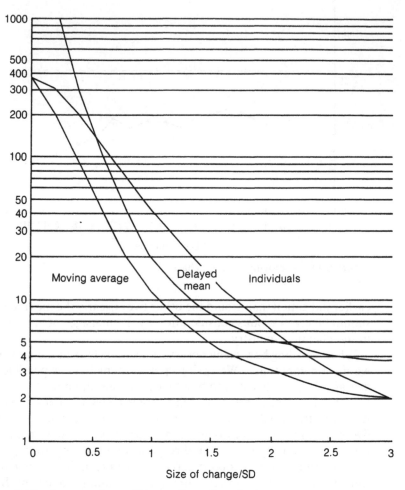

Figure 8.12   A comparison of three alternative charts

its rather forbidding title this is very simple to use. As each new observation becomes available we calculate

New EWMA = A(new value) + $(1 - A)$(old EWMA)

In the formula A is known as the 'smoothing constant'. It can have any
value between 0 and 1, but a good value for the smoothing constant is 0.2,
when plotting a control chart. Let us return to the data in Figure 8.3 and
calculate an EWMA.

Table 8.3   Calculation of EWMA

Batch no.	Yield	EWMA
0	—	71.23
1	71.9	71.36
2	68.1	70.71
3	66.9	69.95
4	73.0	70.56
5	75.4	71.53
6	72.6	71.74
7	64.7	70.33
8	71.8	70.62
9	69.3	70.36
10	75.5	71.39
11	65.5	70.21
12	68.0	69.77
13	73.6	70.54
14	67.0	69.83
15	62.4	68.34
16	64.2	67.51
17	70.6	68.13
18	65.7	67.64
19	67.3	67.57
20	60.6	66.18

For the calculation of the EWMA in Table 8.3 I have used a smoothing constant of 0.2. Thus the formula becomes

New EWMA = (0.2)(new yield) + (0.8)(old EWMA)

Each new value for the EWMA is calculated from the previous value. In order to get this process started we put the first EWMA equal to the target value of 71.23, and then the calculation proceeds as follows:

$$EWMA\ 1 = (0.2)(yield\ 1) + (0.8)(EWMA\ 0)$$
$$= (0.2)(71.9) + (0.8)(71.23)$$
$$= 71.36$$

$$EWMA\ 2 = (0.2)(yield\ 2) + (0.8)(EWMA\ 1)$$
$$= (0.2)(68.1) + (0.8)(71.36)$$
$$= 70.71$$

I am sure you will appreciate that the EWMA values in Table 8.3 do not fluctuate so wildly as the yield values on which they are based. Whilst

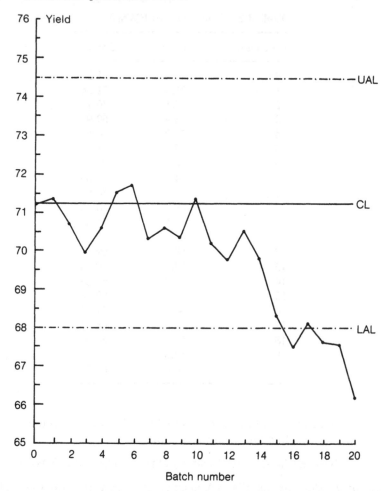

Figure 8.13   An exponentially weighted moving average chart

calculating the EWMA we are smoothing the run chart. This will be clear if you compare Figure 8.3 with the EWMA control chart in Figure 8.13.

The action lines in Figure 8.13 are much closer to the centre line than those in the individuals chart, Figure 8.3. This should not surprise us, for the EWMA is an average, and we are well aware that averages are not so variable as individual values. The position of the action lines on an EWMA chart are dependent upon the variability of the process, of course, but they are also dependent upon the smoothing constant A. The appropriate values for the action lines are calculated from the formula

$$\bar{X} \pm k\sigma\sqrt{[A/(2 - A)]}$$

With the smoothing constant of 0.2, $\sqrt{[A/(2 - A)]}$ is equal to 0.3333. If we let $k = 3$, the formula becomes

$$\bar{X} \pm \sigma$$

Substituting $\bar{X} = 71.23$ and $\sigma = 3.251$ we get the action lines in Figure 8.13.

The EWMA chart is just as easy to set up as a moving average chart, and is just as easy to use in practice. How quickly will it detect change? Will its performance prove to be superior to the alternative charts we have examined earlier? We can see in Figure 8.13 that, with this particular set of data, an action indication is given when the point for batch 16 is plotted.

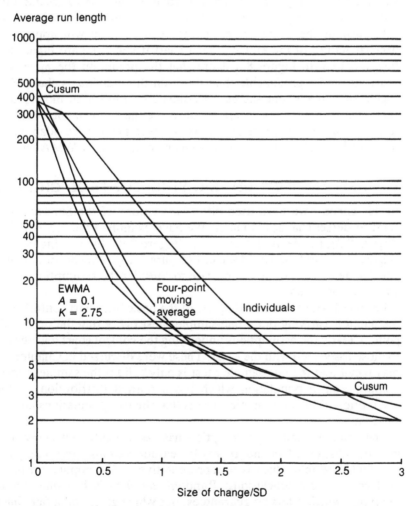

Figure 8.14 Comparison of EWMA chart with alternatives

You may recall that the four-point moving mean chart in Figure 8.11 did not give an action indication until batch 17. Thus we have some evidence that the EWMA chart may be superior. Much stronger evidence is provided by the average run length curve in Figure 8.14.

Before we close this discussion of the benefits of averaging in SPC, I should point out that Figures 8.12 and 8.14 are based on the use of action lines only. This is also true of Figures 8.8 and 8.10. Thus it is fair to compare the four diagrams with each other, and it is also fair to compare the curves within Figures 8.12 and 8.14. However, it should be realized that the benefits of using supplementary rules are not so easily obtained with moving average charts. Thus it would be unwise to assume that the reductions in average run length illustrated in Figure 8.6 could be achieved with each of the three curves in Figure 8.12. You will appreciate the peculiar nature of the moving mean chart if you compare Figure 8.11 or Figure 8.13 with Figure 8.3. Note that the points on the moving mean charts are more predictable than those on the individuals chart. Because each result affects four successive points on a four-point moving average chart, the points are not independent. Since the supplementary rules refer to two or more successive points, these rules need to be adapted before they can be applied to the moving mean chart or the EWMA chart.

8.7   ASSUMPTIONS UNDERLYING CONTROL CHARTS

In this chapter we have used different versions of the Shewhart control chart to detect changes in mean. We have examined the individuals chart (Figure 8.3), the delayed mean chart (Figure 8.9), the moving mean chart (Figure 8.11) and the EWMA chart (Figure 8.13). Whichever one of these you decide to use, you would be wise to consider the assumptions on which your chart is based.

I pointed out earlier that the user of the basic Shewhart control chart for the mean is carrying out a one-sample t-test every time he or she plots a point. Clearly the assumptions underlying the control chart must be similar to those associated with the t-test. These were discussed in Chapter 7. You may recall that the one-sample t-test is only valid if the sample is drawn at random from a population which has a normal distribution. A further assumption of the two-sample t-test is that the two populations are equally variable.

The random sampling assumption has been mentioned many times in previous chapters. I will not repeat the earlier warnings, but simply remind you that a sample which is unrepresentative of the population may well lead you to false conclusions. Random sampling is just one method of sampling, which offers no guarantees, but which gives you a fair chance of getting a representative sample.

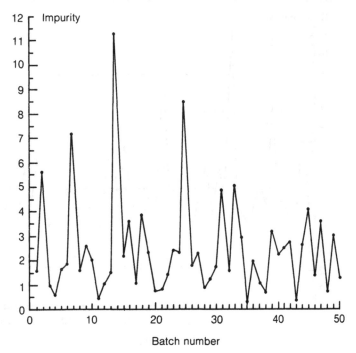

Figure 8.15   A run chart with skewed data.

I suggest that the normal distribution assumption is not so important if you have a large sample. Unfortunately, when plotting a control chart we are very likely to have a small sample. We may even resort to the ultimately small sample if we decide to base our control chart on individual values. Clearly, then, we need to give some consideration to the shape of the distribution as we set up the control chart. Hopefully at that time we will have gathered together a large set of data from which to estimate the process variability. The distribution of these data should be assessed very carefully by drawing a histogram and/or a normal probability plot, as we did in Chapters 2 and 3.

Let us see what happens if we try to set up a control chart using data which are obviously not normal. You may recall that we compared two histograms in Chapter 2 when we examined the batch-to-batch variation in yield and impurity. Figure 2.1 had a very symmetrical shape, indicating that the variations in yield might have a normal distribution. On the other hand, the histogram in Figure 2.2 was so very skewed that it would have been foolish to assume a normal distribution for the variation in impurity. The impurity determinations for the 50 batches are displayed as a run chart in Figure 8.15.

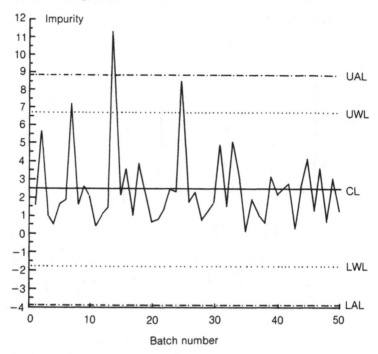

Figure 8.16 A misleading control chart

Examine Figure 8.15 carefully. Compare it with Figure 8.1 and note the distinctive differences between the two run charts. Note particularly that two-thirds of the 50 points lie below the centre line in Figure 8.15, whereas they are split almost equally 26 : 24, in Figure 8.1. The centre line is drawn at the mean, of course. As we have seen in Chapter 2 it is often better to use the median when dealing with very skewed data. Clearly we need to take extra care when interpreting Figure 8.15, but it is, none the less, a very useful diagram. However, if we add action and warning lines to the run chart, we produce a control chart which can be very misleading.

The warning and action lines in Figure 8.16 are placed two and three standard deviations from the centre line as was the case in Figure 8.2. However, it is quite clear that these lines would not be very useful in any attempt to monitor the impurity of future batches. The lower warning and action lines would never be crossed, as we cannot have negative percentage impurity. The upper lines might prove more useful, but one suspects that any supplementary rules we adopted would give spurious indication of change because of the unusually high proportion of points below the centre line.

The longer you study Figure 8.16 the more convinced you will become

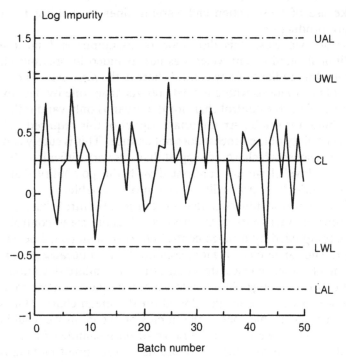

Figure 8.17 A more useful control chart

that the conventional control chart should only be used with data that are coming from a normal distribution. So, what are we to do when we have data like those in Figure 8.15? You may recall that, in Chapter 2, we took logarithms of the 50 impurity determinations in an attempt to normalize the distribution. A comparison of Figures 2.2 and 2.3 confirmed that this transformation had been successful. Perhaps a control chart based on log impurity would prove more useful than Figure 8.16, which we will now consign to the wastepaper bin.

To plot the points on Figure 8.17 we calculate the log of the impurity values. Alternatively, we could use a logarithmic scale on the vertical axis and plot impurity directly. The centre line is drawn at the mean of the log impurities which is 0.266. The warning and action lines are placed two and three standard deviations from the centre line, using the standard deviation of the log impurities which is 0.3488.

I am sure you will agree that this control chart, based on the transformed data, is more satisfactory than Figure 8.16. After examining this new chart we could reasonably claim that the process was in control during the period when these 50 batches were manufactured. Thus we could reasonably make use of these action and warning lines to monitor the impurity of future production.

Before we close this discussion of assumptions I must mention an additional requirement which was not included in previous chapters. An essential ingredient of any control chart is the *time axis*. The control chart is used in situations where the data points arise, one by one, over a period of time. Thus the control chart for the mean is only valid if the population distribution is stable. An important aspect of this required stability is that the variability of the process does not change. This is surely consistent with common sense which would suggest that it is very difficult to detect changes in mean level if we are confronted with the possibility that the process might suddenly or gradually become more variable.

This point is so important that an additional control chart is often used to monitor the variability of the process. Thus the mean control chart would be accompanied by a range control chart or a standard deviation control chart. The range chart is more commonly used because, until recently, it was much easier to calculate a range than a standard deviation. The ranges plotted on the range chart would usually be based on the same groups of data which gave the means plotted on the mean chart. Thus with a conventional mean chart based on samples of five items we would use a range chart on which we plotted the range of each sample of five. With a four-point moving mean chart we would use a four-point moving range chart. Unfortunately the person who monitors the mean level by plotting individual values cannot adhere to this principle, and it is usual to accompany the individuals chart with a two-point moving range chart. Figure 8.18 is such a chart, which is based on the successive differences between the points in Figure 8.3.

We see that none of the 19 points in Figure 8.18 lie outside the action lines and only one falls outside the warning lines. Clearly there is no strong evidence in the range chart that the process became more variable, or less variable, during the period when these 20 batches were produced. Thus we can have confidence in our interpretation of the individuals chart, Figure 8.3, to which this range chart was an accompaniment. In many situations we might regard the range chart as being just as important as the mean chart. Changes in process variability are a matter of concern in their own right, regardless of the assumptions underlying the mean chart.

The first point plotted in Figure 8.18 is based on the difference between the first and second yield values plotted in Figure 8.3. The second point in the range chart is based on the difference between the second and the third yield values, etc. The position of the warning and action lines in a range chart are, of course, related to the variability of the process. You may recall that we estimated the standard deviation of yield to be 3.251 from the data in Figure 8.1. This standard deviation is multiplied by the appropriate constants from Table J to give us the position of the control lines.

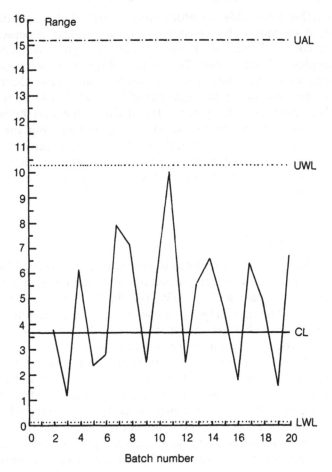

Figure 8.18 A two-point moving range chart

The lower action line is at zero. You will have noted that the control lines are not symmetrical about the centre line as they were in all the mean charts we examined earlier.

One of the five columns in Table J is of particular interest. The numbers used to calculate the position of the centre line (CL) are often referred to as Hartley's constants. For a sample size of 2, Hartley's constant is equal to 1.13. Thus we multiply the standard deviation by 1.13 to obtain the mean range, which we use as the centre line for our mean chart. In the setting up of control charts this procedure can be used in reverse. We can estimate the process standard deviations from the mean of the ranges of several samples using

Standard deviation = (mean range)/(Hartley's constant)

In fact, this is probably the most common method of estimating the process standard deviation, for those who control widget processes. Typically samples of 4 or 5 items would be inspected at regular intervals until at least 20 samples had been taken. The mean and range of each sample would be calculated and then these would be plotted on a mean chart and a range chart. The mean of the 20 ranges would be divided by Hartley's constant to obtain an estimate of the process standard deviation and this would then be used to calculate the position of the action and warning lines.

I am sure you appreciate that the usefulness of a control chart is crucially dependent upon our obtaining a good estimate of what I have described as 'the process standard deviation'. I will have more to say about process variability in Chapter 15 when we discuss a technique known as analysis of variance.

8.8 SUMMARY

In this chapter I have attempted to give you an overview of SPC. We have considered the conventional Shewhart chart, the delayed mean chart, the moving mean chart and the EWMA chart. These are four alternative charts for detecting changes in mean if you are attempting to control a process from which every item is inspected. This would be true for a batch process in the chemical industry, for example.

To help you choose one of the four alternatives, I have presented average run length curves. These tell us how quickly a chart is likely to indicate that a change has occurred and they confirm, what common sense would suggest, that a large change is likely to be detected more quickly than a small change. Indeed, the conventional Shewhart chart using individual values detects large changes very quickly, but its insensitivity to small change forces us to consider the use of averages or the use of supplementary rules.

We have also considered the assumptions underlying SPC charts. These focused our attention on random sampling, the normal distribution and the stability of the process variation. We saw the benefits of data transformation with a very skewed distribution. We explored the use of the range control chart to check the stability of the variation in the process. In many situations, of course, process variability is worthy of study in its own right.

PROBLEMS

8.1 Regardless of which type of control chart you intend to use, before you can set up the chart, you must assess the variability of the process. This should be done with great care. In this chapter we plotted the yield data from Table 2.1 as a run chart and drew action lines at $\pm 3\text{SD}$. The standard deviation, 3.251, was obtained by typing the data into a calculator which used a routine based on the usual formula $\sqrt{\{[\Sigma x^2 - nx^2]/(n-1)\}}$. This could reasonably be described as the 'long-term standard deviation'. We will now calculate the 'short-term standard deviation'.

(a) Put the yield results for the first 48 batches in Table 2.1 into 12 groups of 4. Put the results for batches 1 to 4 in the first group, batches 5 to 8 in the second group, etc. Calculate the mean, the range and the standard deviation for each group of four.

(b) Calculate the mean of the 12 ranges and then calculate the short-term standard deviation using

Short-term SD = (mean range)/(Hartley's constant)

Hartley's constant is obtained from the CL column of Table J.

(c) Use the 12 standard deviations from part (a) to calculate the combined standard deviation:

Combined SD = $\sqrt{\{\Sigma[(\text{df})(\text{SD}^2)]/[\Sigma \text{df}]\}}$

(d) We now have *three* standard deviations calculated from the same data. It is reasonable to suggest that they are roughly equal. What would you have concluded if the short-term SD and the combined SD were much less than the long-term SD?

8.2 Having obtained a standard deviation which quantifies the variability of the process, whilst it is in control, we can now set up control charts to monitor the future performance of the process. We will use SD = 3.5 and a target yield of 71.

(a) Draw a centre line, warning lines and action lines for a four-point moving average chart.

(b) Draw a centre line, warning lines and action lines for a four-point moving range chart.

8.3 (a) On your control charts plot moving means and moving ranges using the following data:

Batch no.	51	52	53	54	55	56	57	58	59	60
Yield	69.1	73.2	73.1	71.4	76.1	72.0	68.5	70.5	66.3	74.1

Batch no.	61	62	63	64	65	66	67	68	69	70
Yield	62.0	67.0	71.1	66.4	68.7	63.2	64.1	63.7	66.5	69.8

(b) What conclusions do you draw from the points you plotted in part (a)?

9
Detecting process changes

9.1 INTRODUCTION

In later chapters of this book we will be concerned with experiments. We will discuss the advantages to be gained by carrying out a well-designed experiment and we will struggle with the difficulties that arise when we try to analyse the results of a bad experiment. For the researcher who wishes to draw conclusions about cause and effect relationships in a production system the most powerful approach is that based on the designed experiment but the researcher is unwise to completely ignore the vast amount of data which is gathered during routine production. By means of a simple statistical technique it is possible to look back at the data recorded for recent batches of product and to decide:

(a) if changes have taken place; and
(b) roughly when the changes occurred.

The technique will be illustrated by an example in which complaints have been received concerning the quality of recent batches of a pigment.

9.2 EXAMINING THE PREVIOUS 50 BATCHES

A particular pigment is manufactured by Textile Chemicals Ltd, using a batch process. This pigment has been sold in varying quantities to a variety of customers throughout Europe and the Middle East, for many years. Recently several complaints have been received from local customers concerning what they considered to be an unacceptably high level of a certain impurity in the pigment. The research and development manager of Textile Chemicals has been asked to investigate the problem and, with the co-operation of the plant manager, he has extracted from the plant records the analytical results on the last 50 batches. These include determinations of yield and impurity which are given in Table 9.1. The impurities are presented in Figure 9.1.

The determinations of yield and impurity were both recorded to three significant figures but they have been rounded to whole numbers in Table

Table 9.1   Yield and impurity of 50 batches of pigment

Batch number	Yield	Impurity	Batch number	Yield	Impurity
1	88	5	26	88	2
2	92	8	27	87	5
3	88	5	28	89	1
4	92	4	29	89	4
5	91	7	30	90	2
6	87	6	31	92	3
7	88	4	32	88	1
8	91	7	33	90	5
9	93	6	34	93	3
10	89	5	35	91	6
11	92	7	36	91	3
12	90	5	37	86	2
13	91	3	38	87	4
14	94	2	39	89	4
15	95	3	40	88	5
16	89	4	41	91	2
17	95	6	42	90	5
18	93	3	43	86	7
19	94	2	44	88	6
20	89	1	45	89	3
21	90	4	46	91	5
22	88	3	47	88	4
23	87	1	48	90	2
24	92	3	49	89	3
25	90	4	50	92	5

9.1. This has been done to simplify the calculations and other data used in this book will be similarly rounded. In a real-life situation one would certainly not round off the data in this way.

Clearly the percentage impurity and the yield both vary from batch to batch. This variation may be quite random but it might be possible to detect patterns of change amongst the randomness. For example we might find, in either or both of the variables:

(a) An upward or a downward trend;
(b) A sudden increase or decrease (i.e. a step change).

Trends will be dealt with in later chapters but we will now concern ourselves with a statistical technique that is ideally suited to this type of situation in which we are looking back at recent data in order to detect

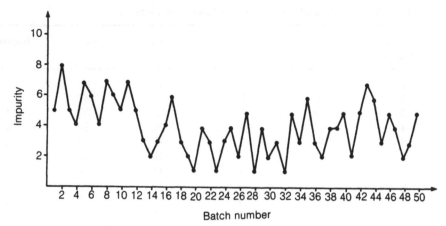

Figure 9.1   Impurity of 50 batches of pigment

whether or not a step-like change has occurred. By means of this technique, known as cusum analysis, we can produce the graphical representation of our impurity data in Figure 9.2. (Cusum is an abbreviation of cumulative sum.)

When we examine Figure 9.2 we are struck most forcibly by a very clear pattern. This same pattern is actually embedded in Figure 9.1, but it is not so easily discernible in the simple graph. Though Figure 9.1 and Figure 9.2 transmit the same message, it is more easily received from the latter.

Two questions need to be answered at this point:

(a) What must we do to the data to convert Figure 9.1 into Figure 9.2?
(b) How are we to interpret Figure 9.2?

The calculations carried out in order to transform the impurity data of Figure 9.1 into the cusum of Figure 9.2 are set out in Table 9.2. Two steps are involved; first we subtract a target value from each impurity determination to obtain the deviations in column 3, then we add up these deviations from target in order to obtain the cusum in column 4. The cusum column is simply a running total of the deviation column. Figure 9.2 is a graph of column 4, plotted against the batch number in column 1.

Looking back over a whole set of data in this way is often referred to as a *post-mortem investigation* but cusums are also used in quality control work when we have a set of points which grows larger as each sample is inspected. In the quality control application the target value might well be obtained from a production specification and this is discussed in section 9.6 and more fully in Oakland and Followell (1990). In the post-mortem investigation we use the sample mean as a target value. As the mean impurity of the 50 batches is 4.0 we subtract 4.0 from each entry in column 2 to obtain

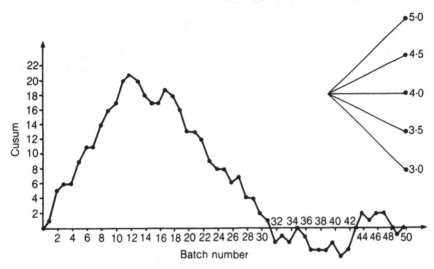

Figure 9.2   Cusum plot of impurity, with slope indicator

the deviations in column 3. One consequence of using the sample mean is that the final entry in the cusum column is zero and a second consequence is that the cusum graph (Figure 9.2) starts and finishes on the horizontal axis. It is possible that a cusum plot will wander up and down in a random manner, never deviating far from the horizontal axis. It may, on the other hand, move well away from the axis only to return later. If this is the case there will be changes in the slope of the graph. Visual inspection of Figure 9.2 might lead us to conclude that there are two points at which the slope changes; at batch number 12 and at batch number 32. These two change points split the cusum graph (Figure 9.2) into three sections and hence split the 50 batches into three groups:

(a) from batch 1 to batch 12 the graph has a positive slope;
(b) from batch 13 to batch 32 the graph has a negative slope;
(c) from batch 33 to batch 50 the graph is approximately horizontal (i.e. zero slope).

The slope of a section of the cusum graph is related to the mean impurity for the batches in that section. Thus the positive slope for batch 1 to 12 tells us that the mean impurity of these batches is above the target value (4.0) and the negative slope for batches 12 to 32 tells us that these batches have a mean impurity below the target value.

The radiating lines in the top right-hand corner of Figure 9.2 help us to relate the slope of a section of the cusum plot to the mean impurity of the batches in that section. Clearly a sequence of batches which has a mean

Table 9.2 Calculation of cusum for impurity data

Batch number	Impurity	Deviation from target value	Cusum	Deviation from previous batch	Squared deviation
1	5	1	1	—	—
2	8	4	5	−3	9
3	5	1	6	3	9
4	4	0	6	1	1
5	7	3	9	−3	9
6	6	2	11	1	1
7	4	0	11	2	4
8	7	3	14	−3	9
9	6	2	16	1	1
10	5	1	17	1	1
11	7	3	20	−2	4
12	5	1	21	2	4
13	3	−1	20	2	4
14	2	−2	18	1	1
15	3	−1	17	−1	1
16	4	0	17	−1	1
17	6	2	19	−2	4
18	3	−1	18	3	9
19	2	−2	16	1	1
20	1	−3	13	1	1
21	4	0	13	−3	9
22	3	−1	12	1	1
23	1	−3	9	2	4
24	3	−1	8	−2	4
25	4	0	8	−1	1
26	2	−2	6	2	4
27	5	1	7	−3	9
28	1	−3	4	4	16
29	4	0	4	−3	9
30	2	−2	2	2	4
31	3	−1	1	−1	1
32	1	−3	−2	2	4
33	5	1	−1	−4	16
34	3	−1	−2	2	4
35	6	2	0	−3	9
36	3	−1	−1	3	9
37	2	−2	−3	1	1
38	4	0	−3	−2	4
39	4	0	−3	0	0
40	5	1	−2	−1	1
41	2	−2	−4	3	9
42	5	1	−3	−3	9

Table 9.2 Cont.

Batch number	Impurity	Deviation from target value	Cusum	Deviation from previous batch	Squared deviation
43	7	3	0	−2	4
44	6	2	2	1	1
45	3	−1	1	3	9
46	5	1	2	−2	4
47	4	0	2	1	1
48	2	−2	0	2	4
49	3	−1	−1	−1	1
50	5	1	0	−2	4
Total	200.0	0.0	—	—	230
Mean	4.0	0.0	—	—	—
SD	1.784	1.784	—	—	—

impurity of 4.0 can be expected to give a section of graph which is horizontal. A change in mean impurity will give a change in slope and the steepness of the slope will depend upon the deviation of the mean from the target value of 4.0.

The changes in mean impurity which are indicated by the changes in slope in Figure 9.2 are summarized in the stark simplicity of Figure 9.3.

9.3 SIGNIFICANCE TESTING

Before we translate into practical terms what we appear to have found we will examine its statistical significance. If the pattern that we have detected in the cusum graph (Figure 9.2) is telling us anything useful then the continuous line in Figure 9.3 should fit the data much better than does the dotted line. In other words the batch impurity values should be

Table 9.3 Changes in mean impurity

Group of batches	Mean impurity	Standard deviation
1 to 12	5.75	1.288
13 to 32	2.85	1.387
33 to 50	4.11	1.491
1 to 50	4.00	1.784

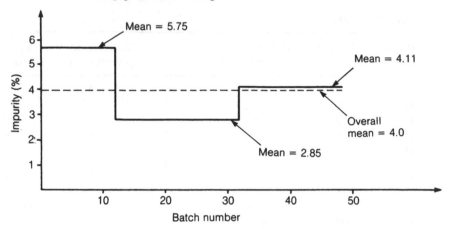

Figure 9.3 Changes in mean impurity

much closer to their local means (5.75, 2.85 and 4.11) than to the overall mean (4.0). Whether this is true or not can be judged by the summary in Table 9.3.

We see in Table 9.3 that the within group standard deviations (1.288, 1.387 and 1.491) are noticeably smaller than the overall standard deviation (1.784). We also notice that the three 'within group standard deviations' are approximately equal to each other indicating that the batch to batch variation is roughly the same in each group of batches. So, even if the process mean impurity has changed twice during this period, the process variability appears to have remained constant and we can combine the three standard deviations to obtain one estimate of the batch to batch variability in impurity:

Combined standard deviation

$$= \sqrt{\{\Sigma[(\text{degrees of freedom})(\text{within group SD})^2]/\Sigma(\text{degrees of freedom})\}}$$
$$= \sqrt{\{[(11 \times 1.288^2) + (19 \times 1.387^2) + (17 \times 1.491^2)]/[11 + 19 + 17]\}}$$
$$= 1.404$$

This combined standard deviation (1.404) is a good estimate of the batch to batch variation in impurity that we would find if there were no changes in mean impurity during the period in which the 50 batches were produced. An alternative method of estimating this variability is illustrated in the last two columns of Table 9.2. The entry of 230 at the bottom of the last column is the sum of the squared deviations in the penultimate column. These deviations are obtained by subtracting the impurity of each batch from that of the previous batch. This method is an alternative to using deviations from the overall mean (4.0) which would be increased by any local changes in mean which might occur. Using the 230 from Table 9.2 we can substitute into the formula:

$$\text{Localized standard deviation} = \sqrt{\left[\frac{\Sigma(x_c - x_{c+1})^2}{2(n - 1)}\right]}$$

$$= \sqrt{\left[\frac{230}{2(49)}\right]}$$

$$= 1.532$$

The value of the localized standard deviation (1.532) is a little higher than the combined standard deviation (1.404) but it is much lower than the overall standard deviation (1.784), as we would expect. The localized standard deviation (1.532) is more difficult to calculate than the other two since its very nature does not allow us to make use of the standard deviation facility on a calculator. Nonetheless it is very important because we use it in testing the statistical significance of the pattern we have observed in our cusum graph. This is achieved as follows:

Null hypothesis – there was no change in process mean impurity during the period in which the 50 batches were manufactured.

Alternative hypothesis – there was a change in process mean impurity during this period.

$$\text{Test statistic} = \frac{\text{maximum cusum}}{\text{localized standard deviation}}$$

$$= \frac{21.0}{1.532}$$

$$= 13.71$$

Critical values – from Statistical Table I, for a span of 50 observations:

9.1 at the 5% level of significance
10.4 at the 1% level of significance.

Decision – as the test statistic is greater than the 1% critical value we reject the null hypothesis.

Conclusion – we conclude that a change of process mean impurity did occur during the period in question and it is reasonable to suggest that the change occurred between batches 12 and 13.

We have established beyond reasonable doubt that one of the changes in mean impurity indicated by our cusum graph (Figure 9.2) did actually occur. We now split the whole series of 50 batches into two sets:

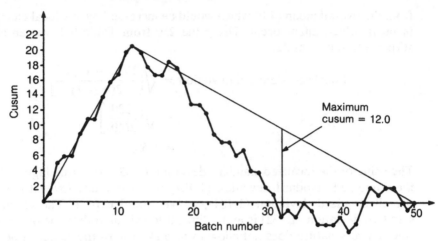

Figure 9.4 Modified cusum plot – to detect further changes

(a) batch numbers 1 to 12;
(b) batch numbers 13 to 50.

We could now make a fresh start and produce a cusum plot for each of the two sets of batches. This is not necessary, however, as we can obtain the same results by drawing two straight lines on the old cusum plot (Figure 9.2) and measuring deviations from these lines. This is illustrated in Figure 9.4.

Examination of Figure 9.4 reveals that batch numbers 13–50 have a maximum cusum of 12 and this value can be used in a second significance test as follows:

Null hypothesis – there was no change in process mean impurity during the period in which batches 13 to 50 were manufactured.

Alternative hypothesis – there was a change in process mean during this period.

$$\text{Test statistic} = \frac{\text{maximum cusum}}{\text{localized standard deviation}}$$

$$= \frac{12.0}{1.532}$$

$$= 7.83$$

Critical values – from Table I for a span of 38 observations:

7.74 at the 5% significance level
9.04 at the 1% significance level.

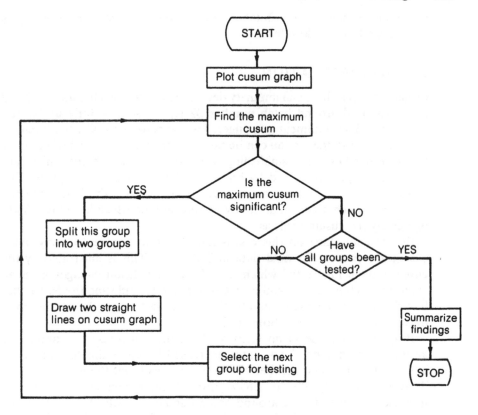

Decision – as the test statistic is greater than the 5% critical value we reject the null hypothesis.

Conclusion – we conclude that a change in process mean impurity did occur during the period in question and it is reasonable to suggest that the change occurred about the time that batch number 32 was produced.

We have now established that two changes in mean occurred during the period in which batches 1 to 50 were manufactured. Further significance testing would proceed as follows:

(a) Examine the maximum cusum for batches 1 to 12.
(b) Split batches 13 to 50 into two groups (i.e. 13 to 32 and 33 to 50).
(c) Draw two more straight lines on the cusum graph.
(d) Examine the maximum cusum for batches 13 to 32.
(e) Examine the maximum cusum for batches 33 to 50.

Following through these five steps would, however, fail to reveal any more significant changes and the analysis would cease at this point. The whole procedure of splitting the batches into groups and testing each group is summarized by the flow chart.

9.4   INTERPRETATION

Concluding that the mean impurity decreased around batch number 12 and later increased around batch number 32 is, in itself, of little use to the research and development mananger. He wants to know why the impurity level changed so that action can be taken to ensure a low level of impurity in future batches. The establishing of cause and effect relationships can, however, be greatly helped by knowing when a change took place. The research and development manager can now discuss with plant personnel the events which occurred at the two points in time which have been highlighted by the cusum analysis.

Before embarking on this discussion it would be wise to carry out a cusum analysis on the yield data in Table 9.1. This is illustrated by the cusum plot in Figure 9.5 which reveals two significant changes in mean yield; at batch number 7 and at batch number 19. Splitting the 50 batches into the three groups suggested by the cusum plot we get the within group means and standard deviations in Table 9.4.

It appears that the mean yield of the process increased by about 3% around batch number 7 and later decreased by approximately the same amount around batch number 19. You will recall that the mean impurity decreased from 5.75% to 2.85% around batch number 12 and then increased again to 4.11% around batch number 32. All of these detected changes are illustrated in Figure 9.6. This useful figure is often referred to as a Manhattan diagram because of its resemblance to the skyline of that part of New York City.

The research and development manager is now in a position to discuss these findings with the plant manager and other interested parties with the objective of deciding why these changes took place. Such discussions are not always fruitful, of course, especially if information concerning unscheduled occurrences is not revealed. After a prolonged interchange of views the following conclusions were reached:

Table 9.4

Group of batches	Mean yield	Standard deviation
1 to 7	89.4	2.149
8 to 19	92.2	2.167
20 to 50	89.3	1.811
1 to 50	90.0	2.268

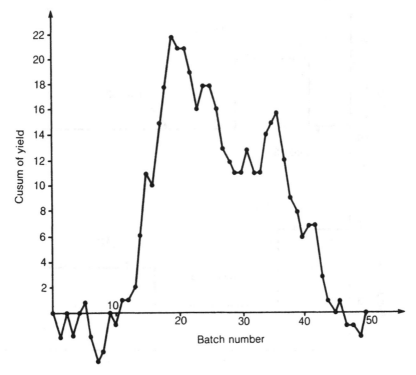

Figure 9.5   Cusum plot of yield

(a) The decrease in impurity (around batch 12) was attributed to the intro-
    duction of a special ingredient which had not been used previously.
    This ingredient was first used in batch number 10 and was introduced
    with the specific aim of reducing impurity. It is not at all surprising,
    then, that we found a decrease in mean impurity around batch 12.
(b) No chemical explanation could be found for the increase in yield
    around batch number 7. It was felt by the research and development
    manager that the special interest shown in the plant by his chemists, at
    about the time that the special ingredient was introduced, may have
    been indirectly responsible for the increase in mean yield. The later
    withdrawal of research and development support could also account
    for the decrease in mean yield around batch number 19. It appeared
    that no one was aware that the mean yield was higher during this
    period even though an increase of 3% is regarded as quite important in
    profit terms.
(c) The increase in impurity around batch number 32 was the main focus
    of attention. This finding confirmed what had been suggested by the

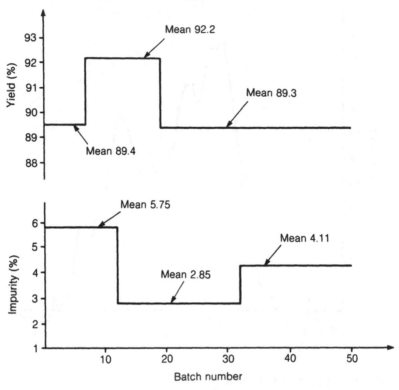

Figure 9.6 Changes in mean impurity and mean yield

customer complaints, that a recent increase in impurity had occurred. Plant records indicated that the same weight of special ingredient had been used in every batch since the successful introduction of this additive in batch number 10. Indeed, the plant records showed that production of this pigment had proceeded very smoothly for some time.

The knowledge that mean impurity increased at around batch number 32 had focused attention on the operation of the plant during quite a short time period but no concrete explanation for the change had arisen. Thus the field was wide open for each member of the meeting to ride his or her personal hobby-horse and to suggest what might have occurred. The research and development manager had great faith in the special ingredient, the use of which had successfully reduced the impurity. He felt sure that the right quantity was not being used, though he had to admit that this very expensive substance needed to be used sparingly. The plant manager suggested that impurity could be reduced and yield increased if only he had the improved control systems that he needed to control the feedrate and the inlet temperature of the main ingredient. The assistant plant manager recalled that ageing of the catalyst had at one time been held responsible for excess impurity. He felt that it had been a mistake to introduce the new working arrangements which allowed the plant operators more discretion to choose when catalyst changes would take place. The night shift foreman yawned, and muttered that few batches of pigment would ever reach the warehouse if the operators did not use their discretion to bend the rules laid down by management. He added, in a louder voice, that there was some doubt amongst plant personnel concerning the best setting for the agitator speed.

The discussion might have continued much longer but it was brought to a conclusion when the research and development manager suggested that the most urgent need was for a better understanding of why the yield and impurity varied from batch to batch. It was his opinion that this understanding would only be obtained by carrying out a controlled experiment on the plant. The plant manager, who had anticipated this suggestion, agreed that an experiment should take place provided that it was entirely under his control.

9.5   HAS THE PROCESS VARIABILITY INCREASED?

We have used the cusum technique to detect changes in yield and impurity. To be more precise, the changes we detected were changes in the mean level of yield and the mean level of impurity. Of course, it was necessary to take account of the process variability as we checked the statistical significance of the changes in mean.

It could be argued that this batch-to-batch variation is important in its own right, and changes in variability should also be sought. If the process becomes more variable we wish to know of this and find the reason why. Fortunately, we can easily adapt our cusum procedure so that it will detect changes in variability as well as changes in mean.

To seek out changes in variability we first calculate differences between successive values, and then apply the cusum procedure to these differences. An increasing cusum will result from large differences and will thus be indicative of higher variability. Conversely a decreasing cusum will signify a period of lower variability.

If we wish to search for changes in the variability of our impurity data in Table 9.1 we can make use of the differences in Table 9.2. We would ignore the minus signs, for it is absolute differences that we need. The mean of the absolute differences would be subtracted from each difference and then a running total of the deviations would give us the cusum. These calculations are displayed in Table 9.5.

Table 9.5 Calculation of a cusum for changes in variability

Batch number	Impurity	Absolute difference	Deviation from target	Cusum
1	5	—	—	—
2	8	3	1.04	1.04
3	5	3	1.04	2.08
4	4	1	−0.96	1.12
5	7	3	1.04	2.16
6	6	1	−0.96	1.20
7	4	2	0.04	1.24
8	7	3	1.04	2.28
9	6	1	−0.96	1.32
10	5	1	−0.96	0.36
11	7	2	0.04	0.40
12	5	2	0.04	0.44
13	3	2	0.04	0.48
14	2	1	−0.96	−0.48
15	3	1	−0.96	−1.44
16	4	1	−0.96	−2.40
17	6	2	0.04	−2.36
18	3	3	1.04	−1.32
19	2	1	−0.96	−2.28
20	1	1	−0.96	−3.24
21	4	3	1.04	−2.20
22	3	1	−0.96	−3.16
23	1	2	0.04	−3.12
24	3	2	0.04	−3.08
25	4	1	−0.96	−4.04
26	2	2	0.04	−4.00
27	5	3	1.04	−2.96
28	1	4	2.04	−0.92
29	4	3	1.04	0.12
30	2	2	0.04	0.16
31	3	1	−0.96	−0.80
32	1	2	0.04	−0.76
33	5	4	2.04	1.28
34	3	2	0.04	1.32
35	6	3	1.04	2.36
36	3	3	1.04	3.40
37	2	1	−0.96	2.44
38	4	2	0.04	2.48
39	4	0	−1.96	0.52
40	5	1	−0.96	−0.44
41	2	3	1.04	0.60
42	5	3	1.04	1.64

Table 9.5 Cont.

Batch number	Impurity	Absolute difference	Deviation from target	Cusum
43	7	2	0.04	1.68
44	6	1	−0.96	0.72
45	3	3	1.04	1.76
46	5	2	0.04	1.80
47	4	1	−0.96	0.84
48	2	2	0.04	0.88
49	3	1	−0.96	−0.08
50	5	2	0.04	−0.04
Total	200	96	—	—
Mean	4.0	1.96	—	—

You can see that the final value in the cusum column is not exactly equal to zero. This is due to using a target value of 1.96 rather than the mean of the absolute differences which is 1.959 183 673. Rounding to two decimal places has greatly simplified the calculation but has had little effect on the cusum.

The cusum of the absolute differences is plotted in Figure 9.7. The negative slope from batch 8 to batch 25 indicates a period of lower variability. The subsequent positive slope indicates that impurity was more variable between batches 26 and 36. Combining these two findings, we conclude that the batch-to-batch variation in impurity increased at about the time that batch 25 was produced. We can now seek out the causes of this increase in variability, but before doing so we should check its statistical significance. If you carried out a significance test on the maximum cusum in Table 9.5 you would find that it was not significant at the 5% level. With the cusum post-mortem technique, however, we often use the 10% significance level, or we investigate possible changes which are not statistically significant. Of course, this relaxation of constraints increases the likelihood of false alarms, but this may be an acceptable price to pay for increasing the chance of detecting real change.

With this thought in mind let us re-examine the impurity data in Figure 9.1. Does it appear that the batch-to-batch variation in impurity was reduced from batches 8 to 25 and increased from 26 to 36? You may answer 'no'. In fact you may feel that the variability was greater in the earlier of the two periods. Calculation of the standard deviations reveals that this is indeed so. The standard deviation of impurity for batches 8 to 25 is 1.855 whilst the standard deviation for batches 26 to 36 is only 1.662. What is the explanation for this apparent contradiction?

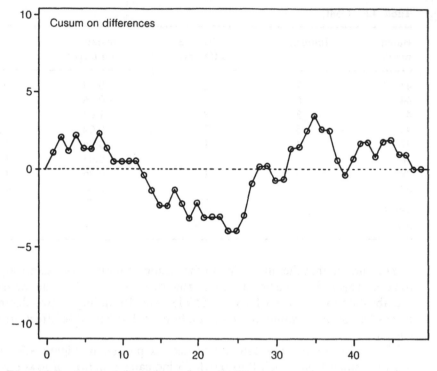

Figure 9.7 A cusum for changes in variability

The standard deviations were calculated by the usual formula and therefore reflect the variability about the mean in each set of data. The absolute differences are based on successive observations as were the localized standard deviations calculated earlier. We have already noted that the conventional standard deviation and the localized standard deviation can differ substantially in certain circumstances. This would be true, for example, if there was a large step change in mean level within the data. It would also be true if the data exhibited what is known as serial correlation. There is evidence of serial correlation in the impurity of batches 8 to 26. If you re-examine Figure 9.1 you will see the upward and downward 'runs' in impurity. Thus each batch tends to be similar to its neighbours and the absolute differences underestimate the total variation.

The cusum has revealed something which could be important, though it is not what we might have expected. We could attempt to find the reasons why the runs appear in the impurity graph. We might learn something of importance, but we should bear in mind that the serial correlation might simply be due to chance.

We will now transfer our attention to the other variable, yield. By applying the cusum post-mortem technique to absolute differences in the yield data we could answer the question 'Was there any change in batch-to-batch variation in yield, during the period when these 50 batches were being produced?' If you carry out the analysis you are unlikely to answer 'yes', for the maximum cusum is not statistically significant, nor does the cusum graph reveal any interesting patterns.

I hope you are not disappointed with the failure of the adapted cusum technique to find dramatic changes in the variability of yield and impurity. This lack of success could be regarded as welcome news. Any significant change in variability would invalidate the earlier cusum analyses we did on mean yield and mean impurity, for they are based on the assumption that the process variability is constant. You may recall that we encountered a similar assumption with the two-sample t-test, when it was required that the two populations should be equally variable. We will meet a similar assumption when we use regression analysis in Chapter 10. In general, any techniques which focus on differences in mean will also take account of variability and will require that magnitude of this variability be constant.

9.6 THE USE OF CUSUMS IN STATISTICAL PROCESS CONTROL

In Chapter 8 we used average run length curves to compare statistical process control (SPC) charts. We examined in some detail the conventional Shewhart chart, the moving average chart and the exponentially weighed moving average (EWMA) chart. We used the average run length curves to show how quickly each chart would detect changes of different sizes. Amongst the average run length curves I included one for the cusum control chart. You may recall that it showed how powerful the cusum chart is, when faced with relatively small changes. I promised that the cusum technique would be introduced in Chapter 9, and so it has been. However, we have so far discussed only the post-mortem application of cusums. Let us now turn to the use of cusums in SPC.

SPC operates in real time, unlike the cusum post-mortem analysis. When using the latter we have all our data to hand before we start the analysis. With SPC, however, the data arrive in small quantities at regular intervals and we wish to make a decision about the process as each point is added to the chart. So it is with the cusum graph, when it is used in SPC. As each new point is added to the cusum chart we place upon it a so-called 'V-mask' which helps us to decide whether or not the mean has changed.

To understand Figure 9.8 you need to imagine that we are using the cusum chart to control the digozo blue process and that the yield for batch number 14 has just been determined. The target value of 89 has been

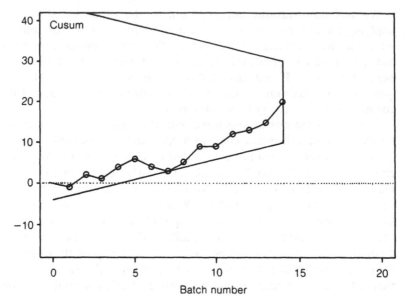

Figure 9.8   A cusum chart with a V-mask

subtracted from the yield to give a deviation of 5.0 which has been added to the previous cusum value to give a new cusum of 20.0. This allows us to plot the fourteenth point on the cusum chart and the V-mask is placed on the new point.

You can see that the whole of the cusum plot is within the arms of the V-mask. Only when part of the plot lies outside the mask do we conclude that a change in mean has occurred. Clearly, at this time, we are unable to conclude that the mean yield has changed. If you imagine how the cusum chart would have looked as it grew, you will realize that the same conclusion was reached after batch 1, batch 2, etc. However, it is also clear that this decision will be reversed if the upward slope of the cusum continues. This is indeed the case, for the yield of batch 15 is 95, which gives a deviation of 6, and the cusum increases to 26.0. When the V-mask is moved to point number 15 we find that several points lie outside the arms of the mask, as you can see in Figure 9.9.

The point which lies furthest outside the V-mask is point number 7. Thus we conclude that an increase in mean yield occurred after batch number 7. To assess the size of this increase we calculate the mean yield for batches 8 to 15. This is equal to 91.88 which is an increase of 2.88 when compared with the target value of 89. The outcome of all this activity is that immediately after producing batch 15 we conclude that the mean yield

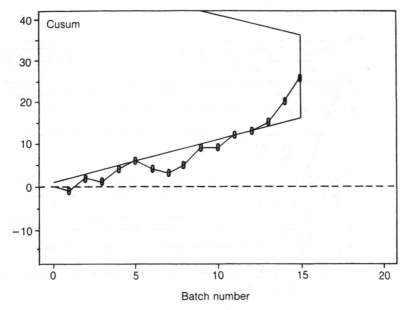

Figure 9.9 The V-mask indicates a change

increased after batch 7 and the size of the increase was approximately 3 units. As the standard deviation of yield is approximately 2 units this is a change of 1.5 SD.

I feel sure you want to know how a V-mask is produced. Clearly the sensitivity of the cusum chart will depend on the shape of the mask. If, for example, we make the mask narrower it will detect change more quickly, but it will give more frequent false alarms. On the other hand, we could reduce the frequency of false alarms by widening the V-mask, but the average run length for real changes would be increased. This point is illustrated by Figure 9.10.

The average run length curves in Figure 9.10 are for cusum control charts using different V-masks. The 'standard' V-mask, which is widely used, has a decision interval of 5 standard deviations and a decision slope (or reference value) of 0.5 standard deviations. The meaning of these two terms is explained in Figure 9.11.

9.7 SCIENTIFIC METHOD

Much research and development work is concerned with investigating production processes and other complex systems. Frequently such investigations involve sampling and often take place in an environment in which

Average run length

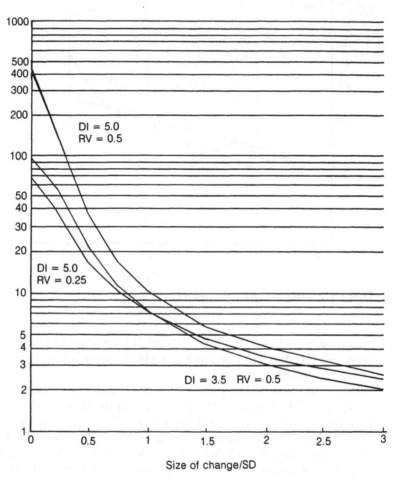

Figure 9.10   Average run length depends on the shape of the V-mask

there is random variation. It is not surprising then, that the scientist or technologist finds use for statistics when he is attempting to reach decisions about cause–effect relationships within the system under investigation.

Though the techniques of statistical analysis may not be used until the later stages of an investigation it is highly desirable that many of the ideas put forward in this book should be taken into account during the planning stage, before any data are obtained. Perhaps this point will be made clear if we consider the methodological framework in which an investigation takes place. Figure 9.12 illustrates what is often referred to as the scientific method or the hypothetico-deductive method.

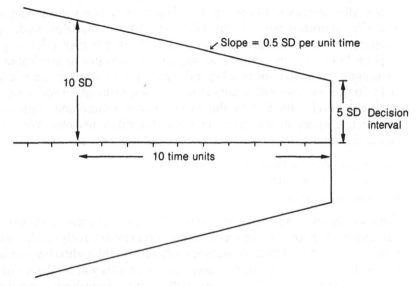

Figure 9.11   Construction of a V-mask

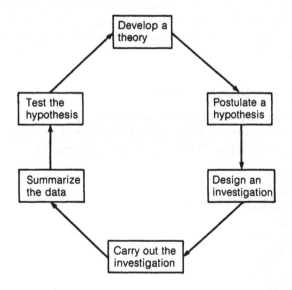

Figure 9.12   The scientific method

Not all scientists would accept that Figure 9.12 is a valid model of the scientific approach but it is probably true to say that this model could command as much acceptance as any other. Certainly Figure 9.12 is just as applicable to a survey as it is to an experiment and just as applicable to an investigation carried out in a laboratory as it is to one carried out in the field. In this book we will be concerned with experiments in which variables are deliberately changed by the experimenter rather than with surveys in which changes in the variables occur for other reasons. We will pay particular attention to three of the boxes in Figure 9.12:

(a) design an investigation;
(b) summarize the data;
(c) test the hypothesis.

For data summary and hypothesis testing we will make use of a very powerful statistical technique known as multiple regression analysis. An attempt will be made to illustrate the dangers inherent in the indiscriminate use of this technique and ways of avoiding certain pitfalls will be demonstrated. On the other side of the coin we will discuss some basic principles of experimental design. We will attempt to analyse the results from several well-designed experiments and the results from two badly-designed experiments. Our inability to draw clear-cut conclusions from the latter will serve to emphasize the need for statistical wisdom at the planning stage of an investigation.

9.8 SOME DEFINITIONS

The performance of a production process or the outcome of a laboratory experiment can often be quantified in terms of one or more responses or dependent variables. These responses might include:

(a) yield
(b) taste
(c) throughput
(d) impurity
(e) brightness
(f) shelf life
(g) cycle time
(h) quality.

In order to control a process or a reaction, it may be useful to understand the effect on a response of one or more independent variables (or factors or treatments). These independent variables might include:

(i) temperature of reaction
(j) speed of agitation

(k) time of addition of raw material
(l) type of catalyst
(m) concentration of an ingredient
(n) duration of reaction
(o) stirred or not stirred
(p) particular operators.

Of the variables listed above, (i), (j), (k), (m) and (n) can be described as *quantitative* variables since they can be set at different values. Variables (l), (o) and (p) on the other hand are described as *qualitative* variables since they cannot be given different values or even quantified at all.

An experiment consists of two or more trials and for each trial we might use a different treatment combination. The simplest experiment would consist of two trials with one independent variable (or treatment) being set at different values in each trial. For example, we might produce a batch of pigment using 3.6 litres of acid and then produce a second batch using 3.8 litres of acid, recording the brightness of the pigment in each case. The single independent variable (volume of acid) has been deliberately set at different levels (3.6 and 3.8) in the two trials, whilst the response (brightness) has been observed at each trial. We also speak of levels of a qualitative factor. If, for example, three different catalysts are used in an experiment which consists of nine trials, we say that the particular independent variable, catalyst type, has three levels.

The effect of an independent variable is the change in response which occurs when the level of the independent variable is changed. In analysing the results of an experiment we wish to estimate the effects of the independent variables which have been deliberately (or perhaps accidentally) changed in value from trial to trial. In designing an experiment we must decide:

(a) How many trials we will carry out.
(b) Which treatment combination we will use at each trial.

Many scientists and technologists would use the words *experiment* and *trial* rather differently to the way we have defined them. A biologist, for example, might speak of a large-scale trial involving many experiments whereas we will speak of a large-scale experiment involving many trials.

There may be no limit to the many different experiments one might design in any given situation. On the other hand there will be strict limits on the conclusions one might draw from the results of any particular experiment after it has been carried out. It is essential, therefore, to regard design and analysis as being interdependent activities. This point will be illustrated in subsequent chapters, by considering the difficulties in analysis which result from imperfect design.

9.9 SUMMARY

This chapter has been focused on one technique, cusum analysis. We have seen that this can be used in two ways. The cusum post-mortem analysis is an excellent way of detecting whether or not changes occurred over a period of time. To know *when* a change occurred may well be the first step towards discovering *why* it occurred. The cusum control chart is an alternative to the Shewhart chart we examined in Chapter 8. In terms of its ability to detect small changes the cusum is superior to its older brother, but it has not enjoyed the popularity that this superiority would suggest it deserves. Perhaps this is because the vertical axis on a cusum chart is not so meaningful, making it very difficult for some people to progress from the run chart to the cusum chart. Clearly the progression from run chart to Shewhart chart is much simpler, as we saw in Chapter 8.

In the next chapter we will return to the situation discussed in section 9.4. You may recall that the plant manager agreed to an experiment being carried out on the digozo blue process. We will analyse the results of that experiment and attempt to draw conclusions which can be translated into a practical strategy for future production. We will use a very powerful statistical technique to help us decide which of the variables included in the experiment can be used to control the level of impurity.

PROBLEM

9.1   Mr Jones always buys exactly 4 gallons of petrol at the nearest convenient garage after his petrol gauge registers half full. He always records the mileage since his last visit to a petrol station. Given below is a sequential record of the mileage recorded at the time of each purchase.

Purchase	Mileage	Purchase	Mileage	Purchase	Mileage	Purchase	Mileage
1	135	11	129	21	104	31	119
2	123	12	121	22	123	32	126
3	134	13	148	23	115	33	134
4	141	14	137	24	103	34	126
5	127	15	132	25	103	35	128
6	126	16	136	26	122	36	110
7	132	17	120	27	115	37	114
8	120	18	103	28	131	38	122
9	122	19	116	29	120	39	123
10	141	20	131	30	111	40	137

(a) Calculate the mean mileage.
(b) Calculate the cusum and plot it on a graph.
(c) Carry out significance tests to determine whether there are any different sections within the data, using a localized standard deviation of 8.848.
(d) Using the three different sections found in part (c), calculate a within group standard deviation for each section and hence compute a combined standard deviation. Compare this value with the localized standard deviation used in part (c).

10

Investigating the process – an experiment

10.1 INTRODUCTION

In Chapter 9 we used a cusum post-mortem analysis in order to discover when and why the impurity and the yield of a process had changed. By examining the plant data concerning past batches of pigment, we hoped to acquire some information which would help us to exercise better control over future batches. The exercise was not entirely successful. We detected changes in mean yield and in mean impurity but we were unable to account for all of these by reference to incidents in the plant.

The cost of the cusum analysis was minimal as it made use of existing data; but herein lies both the strength and the weakness of the technique. The 50 batches examined were not produced in order to facilitate learning about the process. They were simply produced under normal operating conditions. We may be much more successful in gaining a better understanding of the production process if we examine the results of a planned experiment.

It was with these thoughts in mind that the research and development manager suggested, in Chapter 9, that an experiment be carried out. The plant manager agreed to organize such an investigation and Chapter 10 will be devoted to analysing the results. Though the data resulting from the experiment will be more expensive to obtain than those used in the cusum analysis we can anticipate that they will be more useful to us in our search for better operating conditions.

10.2 THE PLANT MANAGER'S EXPERIMENT

You may recall that several variables were referred to in the meeting which was arranged for discussion of the cusum analysis. Each participant had a favourite variable which he or she felt was related to the percentage impurity and which would need to be controlled if impurity were to be reduced. The variables were:

(a) weight of special ingredient;
(b) catalyst age;
(c) feedrate of the main ingredient;
(d) inlet temperature of the main ingredient;
(e) agitation speed.

The plant manager decides to include all five as independent variables in his experiment. The two dependent variables will be impurity and yield. He also decides that the experiment will be limited to ten batches of pigment and the results of the ten trials are given in Table 10.1. (Yield values have been omitted from Table 10.1 as we will concentrate on only one of the dependent variables. In practice we would consider each of the dependent variables separately and then attempt to draw 'compromise' conclusions.) The distinction between independent variables on the one hand, and the response or dependent variable on the other hand, is vitally important. When recommending a production strategy to the plant manager we must specify values for the independent variables and predict the level of response (i.e. impurity) which can be expected if the strategy is adopted.

The research and development manager asks one of his young chemists to work with the plant manager in the analysis of these data. The purpose of this analysis is to:

Table 10.1 The plant manager's experiment

Impurity in pigment y	Weight of special ingredient x	Catalyst age w	Main ingredient		Agitation speed s
			Feedrate z	Inlet temp. t	
4	3	1	3	1	3
3	4	2	5	2	3
4	6	3	7	3	3
6	3	4	5	3	2
7	1	5	2	2	2
2	5	6	6	1	2
6	1	7	2	2	2
10	0	8	1	3	4
5	2	9	3	1	4
3	5	10	6	2	4
Mean 5.0	3.0	5.5	4.0	2.0	2.9
SD 2.357	2.000	3.028	2.055	0.816	0.876

(a) identify those variables which could be changed in order to reduce the impurity in the pigment;
(b) ascertain the nature of the relationship between these chosen variable(s) and the percentage impurity;
(c) recommend a strategy for the production of future batches.

10.3 SELECTING THE FIRST INDEPENDENT VARIABLE

From our five independent variables we will first select the one which appears to be most closely related to the dependent variable. Before we perform any calculations let us examine the scatter diagrams in Figure 10.1, which illustrate these relationships. In each of the five diagrams the dependent variable is on the vertical axis and one of the independent variables is on the horizontal axis. Each diagram illustrates the association between the percentage impurity in the pigment and one of the variables which might be having some influence upon it.

A visual inspection of the five diagrams reveals that:

(a) Those batches which contain larger quantities of special ingredient tend to have a low level of impurity in the pigment.
(b) Those batches which were manufactured with a high feed rate tend to have a low level of pigment impurity.
(c) Use of a higher agitation speed may result in a higher level of impurity, but the association between the two variables is not so strong and the visual impression is perhaps over influenced by the batch which had 10% impurity.
(d) A high level of impurity in the pigment may be associated with the use of a high inlet temperature for the main ingredient.
(e) It is difficult to decide whether or not the dependent variable is influenced by the age of the catalyst.

The drawing of scatter diagrams, and the careful inspection of these diagrams, can often bring to light features of a set of data which might otherwise go unnoticed, and the utility of a simple scatter diagram should not be underestimated. Despite this, there is also a need to express the degree of association between two variables in a numerical form. This is achieved by calculating the correlation coefficient.

If you have access to a suitable pocket calculator or computer you can obtain a correlation coefficient very easily. The calculation is rather tedious if you have to substitute into formulae, so you are strongly advised to make an investment. For the reader who is stranded on a desert island I will demonstrate the calculation of the correlation between impurity (y) and weight of special ingredient (x).

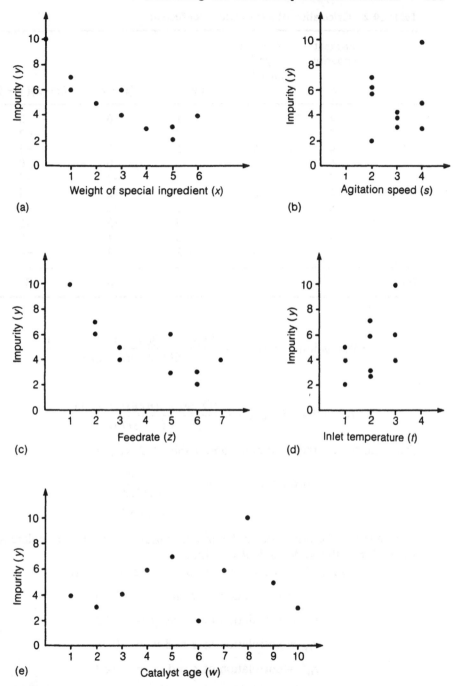

Figure 10.1   Is impurity related to the independent variables?

Table 10.2   Calculation of a correlation coefficient

Pigment impurity y	Weight of special ingredient x	$(y - \bar{y})$	$(x - \bar{x})$	$(x - \bar{x})(y - \bar{y})$
4	3	−1	0	0
3	4	−2	1	−2
4	6	−1	3	−3
6	3	1	0	0
7	1	2	−2	−4
2	5	−3	2	−6
6	1	1	−2	−2
10	0	5	−3	−15
5	2	0	−1	0
3	5	−2	2	−4
Total 50	30	0	0	−36

$$\text{Correlation of } x \text{ and } y = \frac{\{\Sigma[(x - \bar{x})(y - \bar{y})]\}/(n - 1)}{(\text{SD of } x)(\text{SD of } y)}$$

or

$$\text{Correlation of } x \text{ and } y = \frac{[(\Sigma xy) - (n\bar{x}\bar{y})]/(n - 1)}{(\text{SD of } x)(\text{SD of } y)}$$

Using the first of these formulae and Table 10.2 we get

$$\text{Correlation of } x \text{ and } y = \frac{[-36]/9}{(2.00)(2.357)}$$

$$= -0.8485$$

For each of the other four independent variables we could calculate its correlation with the dependent variable.

The calculations have been carried out and the results are:

$$r_{xy} = \text{correlation of } x \text{ and } y = -0.85$$

$$r_{wy} = \text{correlation of } w \text{ and } y = 0.23$$

$$r_{zy} = \text{correlation of } z \text{ and } y = -0.78$$

$$r_{ty} = \text{correlation of } t \text{ and } y = 0.52$$

$$r_{sy} = \text{correlation of } s \text{ and } y = 0.11$$

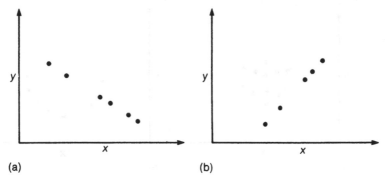

Figure 10.2 Extreme values of correlation coefficient

These correlation coefficients tell us how strongly the independent variables are associated with the dependent variable. Since the largest of the correlations is that between x and y (ignoring the minus signs) we can say that the independent variable most strongly associated with the percentage impurity in the pigment (y) is the weight of special ingredient (x). This statement refers only to the sample of ten batches and not to the population of all batches about which the research and development manager would like to draw conclusions or make predictions. Before we turn our attention to the correlation coefficients that one might find if one examined all possible batches we will take a closer look at the meaning one can extract from a correlation coefficient.

It can be shown that a correlation coefficient will have a value between -1 and $+1$. A value of -1 corresponds to a perfect negative association between the two variables [Figure 10.2(a)] whilst a value of $+1$ corresponds to a perfect positive association [Figure 10.2(b)].

Between the two extreme cases of Figure 10.2 we have the five scatter diagrams in Figure 10.1. Note that Figure 10.1(a) and Figure 10.1(c) have negative values of correlation coefficient, whilst the weakness of the association between w and y is reflected in the correlation of 0.11 which is close to zero. A correlation of 0 is an indication of the absence of a linear association between the two variables. It would be unwise, however, to assume that a correlation of 0 was evidence of the independence of the two variables, as Figure 10.3(b) demonstrates.

Both parts of Figure 10.3 illustrate sets of data which give a correlation of 0. Whilst Figure 10.3(a) strongly suggests that the variables x and y are unrelated, Figure 10.3(b) suggests that x and y are very closely related but the relationship is not linear. In short, the correlation coefficient is a measure of linear association between two variables.

Further examples of the dangers of interpreting a correlation coefficient

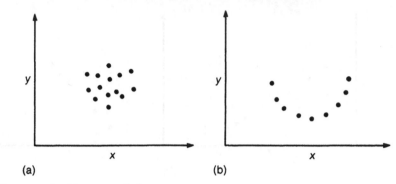

Figure 10.3 Zero correlation

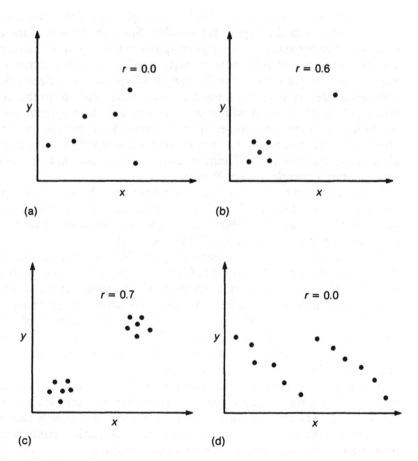

Figure 10.4 Grouping and/or outliers can affect correlation

without drawing a scatter diagram are given in Figure 10.4. Data which fall into two or more groups and data which contain outliers can both give rise to values of correlation coefficient which are spuriously high or spuriously low.

A further danger in the interpretation of correlation lies in the assumption that a correlation coefficient is necessarily an indication of the existence of a cause–effect relationship between the two variables. If we were to list the annual sales in the UK of Scotch whisky alongside mean annual salaries of Methodist ministers for the past 20 years, we would find a high positive correlation between the two variables, but no one would suggest that there is a direct cause–effect relationship between the two. A voluntary reduction in ministers' salaries will not result in a reduction in the sales of whisky any more than a drop in the sales of whisky will cause a decrease in the ministers' salaries.

10.4 IS THE CORRELATION DUE TO CHANCE?

We have selected weight of special ingredient (x) as the independent variable which is most closely associated with the percentage impurity in the pigment. This decision was based on the values of the five correlation coefficients. These were calculated from data gathered from the ten batches in the sample. They are sample correlation coefficients. Having selected this independent variable we can now turn to the population and ask: is it possible that the population correlation coefficient is equal to zero? If we had examined all possible batches would we have found that there was no association between weight of special ingredient and the percentage impurity in the pigment?

To answer this question we carry out a simple hypothesis test. In doing so we will represent the sample correlation coefficient by r and the population correlation coefficient by ρ.

Null hypothesis – within the population of batches the correlation between weight of special ingredient and pigment impurity is equal to zero (i.e. $\rho = 0$).

Alternative hypothesis : $\rho \neq 0$

Critical values – from Statistical Table H (with a sample size of ten, for a two-sided test):

0.632 at the 5% significance level
0.765 at the 1% significance level.

Test statistic = 0.845 (the modulus of the sample correlation coefficient, $|r|$).

Decision – reject the null hypothesis at the 1% significance level.

Conclusion – within the population of batches the percentage impurity in the pigment is related to the weight of special ingredient used in the batch.

Note: Strictly, the use of Table H is only valid if the sample is a random sample from the population referred to in the null hypothesis, but it is clearly not possible to satisfy this condition. A further assumption underlying Table H is that both x and y have normal distributions, though the table is often used in situations where this assumption is not satisfied.

10.5   FITTING THE BEST STRAIGHT LINE

We are now confident that the variation from batch to batch of the weight of special ingredient (x) is associated with the variation in pigment impurity (y). It is possible that there is a cause–effect relationship between the two variables and that a deliberate attempt to change the weight of special ingredient in future batches will result in a reduction in pigment impurity. We know from the negative sign of the correlation coefficient that an increase in x will give a reduction in y but we do not know the precise nature of the relationship. It would be useful to know what particular weight of special ingredient should be used to give a particular level of impurity. In order to quantify this relationship we will fit a straight line to the scatter diagram of Figure 10.1(a).

If each reader of this book drew on the diagram what he or she considered to be the best straight line we might well get a wide difference of opinion within a general area of agreement.

One way in which we can reach complete agreement is for everyone to fit what is known as the least squares regression line. This is, in one sense, the best straight line, but we will return to this point later after we have done the actual fitting.

If we let the equation of the line be $y = a + bx$ then fitting the line to the data simply involves calculating a value for the slope b and a value for the intercept a.

If you have obtained a calculator which has an automatic routine for the correlation coefficient, it will also give you the slope and intercept of the regression line. For the benefit of any reader without a suitable calculator, they can be calculated as follows:

$$\text{Slope, } b = \frac{[(\Sigma xy) - (n\bar{x}\bar{y})]/(n - 1)}{(\text{SD of } x)^2}$$

$$= \frac{-4}{(2.000)^2}$$

$$= -1.0$$

Intercept, $a = \bar{y} - b\bar{x}$
$$= 5.0 - (-1.0)\,3.0$$
$$= 8.0$$

Note: In these formulae x represents the independent variable and y represents the dependent variable.

Thus the equation of the least squares regression line is $y = 8.0 - x$, and this equation can be used to predict the y value which corresponds to any particular x value. We can predict the percentage impurity in the pigment of future batches if we know the weight of special ingredient that is to be used in their manufacture. Substituting $x = 6$ in the equation gives $y = 2$, so we would expect batches which contain 6 units of special ingredient to have 2% impurity in the pigment on average. Our prediction would be aimed at the population of batches and not at the sample of ten batches on which the equation is based. Having studied estimation and hypothesis testing in the earlier chapters of this book you would expect that any such prediction would be in error. Later we will calculate confidence intervals for such predictions, but at this point it might help you to appreciate the need for caution in the use of a regression equation if you attempt to predict the percentage impurity that we would get if we used 9 units of

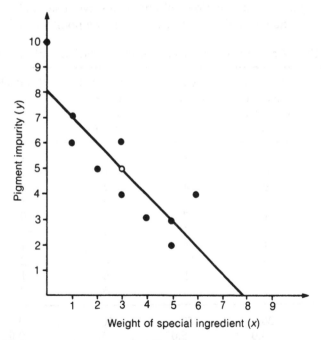

Figure 10.5   Least squares regression line

special ingredient in the manufacture of a batch. Substituting $x = 9$ into the equation gives $y = -1$, but an impurity of -1% is clearly not possible.

The regression line has been drawn on the scatter diagram in Figure 10.5. We will examine this picture in some detail in an attempt to quantify how well the line fits the data points and to explain how this particular line fits better than all other lines.

10.6 GOODNESS OF FIT

You will notice that the regression line passes through the point ($x = \bar{x}$, $y = \bar{y}$). This point is known as the *centroid* and it has been marked with a circle to distinguish it from the data points. You will also notice that two of the data points lie on the line, whilst three of the points are above the line and five are below. The vertical distances of the points from the line are known as residuals and they are listed in Table 10.3. Also listed are the predicted impurity values for each of the ten batches. These predicted values are obtained by substituting each x value in the regression equation.

The residual for any data point can be found from the graph or it can be calculated from:

$$\text{Residual} = \text{actual } y \text{ value} - \text{predicted } y \text{ value}$$

Carrying out this calculation will give a positive residual for a point which lies above the line and a negative residual for a point which lies below. You

Table 10.3 Calculation of residuals and the residual sum of squares

Weight of special ingredient	Actual impurity	Predicted impurity	Residual	Squared residual
x	y	$a + bx$		
3	4	5	−1	1
4	3	4	−1	1
6	4	2	2	4
3	6	5	1	1
1	7	7	0	0
5	2	3	−1	1
1	6	7	−1	1
0	10	8	2	4
2	5	6	−1	1
5	3	3	0	0
Total 30	50	50	0	14
Mean 3.0	5.0	5.0	0.0	1.4

will notice that the sum of the residuals is equal to 0 in Table 10.3. For the least squares regression line, or for any other line which passes through the centroid, the residuals will sum to zero. The sum of the squared residuals is equal to 14 in Table 10.3. This total is called the *residual sum of squares* and the line we have fitted is the best in that it gives a smaller residual sum of squares than any other line. We have calculated values for a and b which minimize the residual sum of squares and the procedure we have used to fit the line is known as the method of least squares.

The residual sum of squares will prove extremely useful when we come to calculate a confidence interval for the impurity level we can expect to get with different operating conditions. Fortunately the residual sum of squares can be calculated by other methods which avoid the tedium of tabulating the residual for each batch, as we did in Table 10.3.

Residual sum of squares $= (n - 1)(\text{SD of } y)^2(1 - r^2)$

where y is the dependent variable and r is the correlation between the dependent and independent variables.

Using SD of $y = 2.357$, $r = -0.8485$ and $n = 10$ we get:

$$\text{Residual sum of squares} = 9(2.357)^2[1 - (-0.8485)^2]$$
$$= 50.00(0.2800)$$
$$= 14.00$$

This result is in agreement with that obtained earlier but the reader should note the need to carry many significant figures throughout the calculation. It would be unwise, for example, to substitute into the formula a correlation coefficient which had been rounded to two decimal places.

The residual sum of squares is a measure of the variability in the dependent variable (y) which has not been explained, or accounted for, by fitting the regression equation. Had the regression line passed through all the data points the residual sum of squares would have been equal to zero. This would be an extreme case in which the line fitted perfectly and the correlation coefficient would be equal to -1 or $+1$. In such a case we would say that the variability in the independent variable (x) accounted for 100% of the variability in the dependent variable (y). In our example the variation in the weight of special ingredient (x) from batch to batch has accounted for only part of the variation in pigment impurity. The ability of the fitted regression equation to explain, or account for, the variation in the dependent variable can be quantified by calculating a very useful figure known as the percentage fit:

Percentage fit $= 100\,r^2$

where r is the correlation coefficient between the dependent variable and the independent variable.

You will recall that the correlation between y and x was -0.8485. Using this value we get a percentage fit of $100(-0.8485)^2$ which is 72.0%. (Note again the need to carry many significant figures in statistical calculations. Using a rounded correlation coefficient of 0.85 would give an erroneous result of 72.3% fit.)

A percentage fit of 72% tells us that the batch to batch variation in weight of special ingredient (x) has accounted for 72% of the variation in impurity. The remaining 28% of the variation in impurity amongst the ten batches remains unaccounted for at this point. Perhaps we will be able to demonstrate in the next chapter that this additional variation is caused, at least in part, by the other independent variables.

Another calculation which can be introduced at this point is one which converts the residual sum of squares into the very useful residual standard deviation.

$$\text{Residual standard deviation (RSD)} = \sqrt{\left(\frac{\text{residual sum of squares}}{\text{residual degrees of freedom}}\right)}$$

where residual degrees of freedom is equal to $(n - 2)$ for a simple regression equation with one independent variable.

In the earlier chapters of this book we divided a sum of squares by a degrees of freedom to obtain a variance and we then took the square root to get a standard deviation. At that time we used $(n - 1)$ degrees of freedom but we must now use $(n - 2)$ degrees of freedom whilst calculating the residual standard deviation (RSD) of a simple regression equation. By using the calculated slope b and the calculated intercept a to obtain the residuals we lose 2 degrees of freedom compared with the 1 degree of freedom we lost earlier when we took deviations from the sample mean. Considering the list of residuals in Table 10.3, if we were given any eight of the residuals it would be possible to calculate the other two, so we have only 8 degrees of freedom amongst the ten residuals. Using a residual sum of squares of 14.0 and 8 degrees of freedom we get a residual standard deviation (RSD) equal to $\sqrt{(14.0/8)}$ which is 1.3229.

In many situations the residual standard deviation is much more meaningful than the residual sum of squares. In our pigment impurity problem the residual standard deviation is an estimate of the variation in impurity that would have been found amongst the ten experimental batches if weight of special ingredient (x) had been kept constant throughout. Note that the calculated RSD (1.323) is much less than the standard deviation of the ten batch impurity values (2.357). Obviously we can expect less batch to batch variation in impurity if weight of special ingredient (x) is held constant than if it is deliberately changed as it was in the ten-batch experiment.

10.7 THE 'TRUE' REGRESSION EQUATION

The meaning of the residual standard deviation may be clarified further if we consider the true equation or the population equation as it might be called:

$$y = \alpha + \beta x \qquad (10.1)$$

This equation expresses the true relationship between pigment impurity (y) and weight of special ingredient (x). The Greek letters alpha (α) and beta (β) remind us that the equation describes a population of batches and that the values of α and β could only be obtained by investigating the whole population. Furthermore, we would need to keep all other variables constant during this investigation as only one independent variable x is included in the true equation. A much better model of our practical situation is the equation:

$$y = \alpha + \beta x + \text{error} \qquad (10.2)$$

The 'error' in this equation represents two causes of impurity variation which were not accounted for in equation (10.1). These two sources of variation are:

(a) errors of measurement in the recorded values of pigment impurity (y);
(b) batch to batch variation in impurity due to variables other than the weight of special ingredient (x).

Equation (10.2) tells us that the impurity of any particular batch, made with a known value of special ingredient, may well differ from the theoretical value predicted by equation (10.1). Furthermore, equation (10.2) can account for the fact that a succession of batches made with the same weight of special ingredient will be found to have differing values of impurity. It can be shown, in fact, that the residual standard deviation (1.323) is an estimate of the impurity variation we would get if the weight of special ingredient did not change from batch to batch. We have, therefore, estimated all three unknowns in equation (10.2):

(a) the calculated intercept a is an estimate of the true intercept α;
(b) the calculated slope b is an estimate of the true slope β;
(c) the residual standard deviation (RSD) is an estimate of the standard deviation (σ) of the 'errors'.

Rather than single value estimates we could calculate confidence intervals for α, β and σ. Before we do so, however, let us pose a very important question: could β be equal to zero? In practical terms this is equivalent to asking: if we had examined a very large number of batches would we have found that the pigment impurity (y) was not related to the weight of special ingredient (x)? To answer this question we will carry out a t-test in which the null hypothesis is $\beta = 0$.

Null hypothesis – there is no relationship between weight of special ingredient (x) and pigment impurity (y) (i.e. $\beta = 0$).

Alternative hypothesis – there is a relationship between weight of special ingredient (x) and pigment impurity (y) (i.e. $\beta \neq 0$).

$$\text{Test statistic} = \frac{|b - \beta|(\text{SD of } x)}{\text{RSD}/\sqrt{(n - 1)}} \quad \text{or} \quad \sqrt{\left[\frac{\% \text{ fit}(n - 2)}{100 - \% \text{ fit}}\right]}$$

$$= \frac{|-1.000 - 0.000|(2.000)}{1.3229/\sqrt{9}} \qquad = \sqrt{\left[\frac{72(8)}{100 - 72}\right]}$$

$$= 4.5355 \qquad\qquad\qquad = 4.5356$$

Critical values – from the two-sided t-table with 8 degrees of freedom:

2.31 at the 5% significance level
3.36 at the 1% significance level.

Decision – we reject the null hypothesis at the 1% level of significance.

Conclusion – we conclude that, within the population of batches, pigment impurity (y) is related to the weight of special ingredient (x).

We did, of course, reach this very same conclusion earlier when we tested the significance of the correlation between x and y. Why, you might ask, should we carry out the t-test when we already know the answer to the question? The t-test, as you will see in the next chapter, is easily extended to cover equations which contain two or more independent variables. The correlation test can also be extended but not so easily.

For similar reasons we have offered two formulae with which to calculate the test statistic in the t-test. Whilst the first formula may appear more meaningful the second formula, based on the percentage fit, can be easily modified for use in situations where we have several independent variables in the equation.

10.8 ACCURACY OF PREDICTION

Having decided that the true slope β is not equal to zero it is reasonable to ask for a more positive statement about what value the true slope is likely to have. We have calculated that b is equal to -1.0 and this is our best estimate of β. We can, however, calculate a confidence interval as follows:

A confidence interval for the true slope (β) is given by:

$$b \pm t(\text{RSD}) \sqrt{\left[\frac{1}{(n-1)(\text{SD of } x)^2}\right]}$$

where t has $(n-2)$ degrees of freedom.

Using $b = -1.0$, $t = 2.31$, RSD $= 1.323$, SD of $x = 2.00$ and $n = 10$ we get a 95% confidence interval for the true slope to be:

$$-1.0 \pm 2.31(1.323) \sqrt{\left[\frac{1}{9(2.0)^2}\right]} = -1.0 \pm 0.51$$

$$= -0.49 \text{ to } -1.51$$

We can therefore, be 95% confident that the true slope lies between -0.49 and -1.51. In practical terms, this means that we can expect that an increase of 1 unit in the weight of special ingredient will result in an impurity decrease of between 0.49% and 1.51%. This is a very wide interval and the plant manager is left in some doubt. To achieve a 3% reduction in impurity does he need to increase the weight of special ingredient by approximately 2 units (i.e. 3/1.51) or by approximately 6 units (i.e. 3/0.49)? It would be easier for the plant manager to estimate the cost of a change in operating conditions if he knew more precisely the effect of such a change.

We can also calculate a confidence interval for the true intercept α by means of the formula:

A confidence interval for the true intercept (α) is given by:

$$a \pm t(\text{RSD}) \sqrt{\left[\frac{1}{n} + \frac{\bar{x}^2}{(n-1)(\text{SD of } x)^2}\right]}$$

where t has $(n-2)$ degrees of freedom.

Using $a = 8.0$, $t = 2.31$, RSD $= 1.323$, $\bar{x} = 3.0$, SD of $x = 2.00$ and $n = 10$ we get a 95% confidence interval for the true intercept to be:

$$8.0 \pm 2.31(1.323) \sqrt{\left[\frac{1}{10} + \frac{(3.0)^2}{9(2.0)^2}\right]} = 8.0 \pm 1.80$$

We can be 95% confident that the true intercept α lies between 6.20 and 9.80. Thus we can advise the plant manager that he can expect an average impurity between 6.2% and 9.8% if he does not include any special ingredient (i.e. $x = 0$) in future batches. Once again the confidence interval is disappointingly wide, but this particular estimate is not very important as the plant manager fully intends to use the special ingredient (x) in future batches. What he would appreciate, much more than a confidence interval for α, is a confidence interval for the mean impurity in future batches for a particular level of special ingredient.

A confidence interval for the true value of y for a particular value of x (say $x = X$) is given by:

$$a + bX \pm t(\text{RSD}) \sqrt{\left[\frac{1}{n} + \frac{(X - \bar{x})^2}{(n - 1)(\text{SD of } x)^2}\right]}$$

where t has $(n - 2)$ degrees of freedom.

Suppose that the plant manager intends to use 6 units of special ingredient in future batches. By substituting $X = 6$ into the formula we get a confidence interval for the true impurity (or mean impurity) as follows:

$$8.0 - 1.0(6.0) \pm 2.31(1.323) \sqrt{\left[\frac{1}{10} + \frac{(6.0 - 3.0)^2}{9(2.0)^2}\right]}$$

$$= 2.00 \pm 1.80$$

Thus the use of 6 units of special ingredient can be expected to result in a mean impurity which lies between 0.20% and 3.80%. Whilst the manager would be very happy with the 0.2% impurity at the lower end of this range he would not be at all pleased to find 3.8% impurity if he had incurred the high cost of using 6 units of special ingredient. Using other values of x we could calculate several confidence intervals and then plot the confidence bands shown in Figure 10.6.

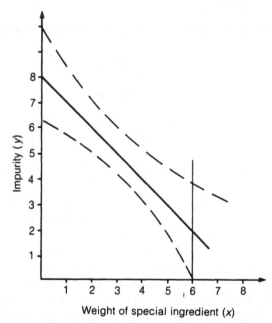

Figure 10.6   Confidence bands for predicted impurity of future batches

The plant manager's position is actually even weaker than Figure 10.6 would suggest. He must bear in mind that the confidence intervals in Figure 10.6 are for the true mean impurity of future batches in which 6 units of special ingredient are used, and that individual batches will be scattered above and below the true mean. If we want a confidence interval for the pigment impurity in a particular batch (the next batch to be produced, say) we must use the following formula:

$$a + bX \pm t(\text{RSD}) \sqrt{\left[1 + \frac{1}{n} + \frac{(X - \bar{x})^2}{(n - 1)(\text{SD of } x)^2}\right]}$$

Comparing this formula with the previous one we note the extra 1 under the square root sign. This gives a wider interval and if we again substitute $X = 6$ we get a 95% confidence interval of 2 ± 3.54. The plant manager can therefore expect that the use of 6 units of special ingredient in the next batch will give an impurity level between -1.54% and 5.54%. The lower confidence limit is clearly impossible whilst the upper limit is alarming.

The plant manager would appear to have got little benefit from the use of regression analysis in this chapter. We have fitted a regression equation ($y = 8.00 - x$) which he himself could have fitted by eye using Figure 10.1(a). It is true that he might have ended up with an equation which differed slightly from our best equation, but this error would have been negligible in the light of the confidence intervals we have just calculated. Fortunately, regression analysis has much more to offer than has so far been revealed. In the next chapter we will take into account the other four independent variables. The effect will be to increase the percentage fit of the equation and consequently reduce the residual standard deviation. This in turn will give a much narrower confidence interval for the predicted impurity of any production strategy. Before we consider the introduction of additional independent variables we will briefly examine an alternative to the simple equation ($y = 8.0 - x$) that we have already fitted.

10.9   AN ALTERNATIVE EQUATION

You may recall that we selected weight of special ingredient (x) as our first independent variable because of its high correlation (-0.85) with the dependent variable, pigment impurity (y). We later found this sample correlation coefficient to be significant at the 1% level. Amongst the other independent variables the only serious rival for inclusion in the equation was the feedrate (z). The correlation between z and y is -0.78 which, when tested, is found to be significant at the 5% level.

Had we decided to select feedrate (z) as our first independent variable the fitted equation would have been:

$$y = 8.6 - 0.90z$$

and the percentage fit would have been 61%. Clearly the use of z as an independent variable does not give as good a percentage fit as we got when we used x (72%). The difference is not very large, however, and this alternative equation may be far more attractive in terms of plant operation. The original equation offered a reduction in impurity of 1% for a unit increase in the weight of special ingredient (x). The alternative equation offers a reduction in impurity of 0.9% for a unit increase in the feedrate (z). When the plant manager considers the relative costs of increasing the weight of special ingredient and of increasing the feedrate he may conclude that the latter is more desirable. He may contemplate making an increase in x and also making an increase in z. In the next chapter we will explore this possibility.

A warning

It would be very unwise of the plant manager to change either the weight of special ingredient (x) or the feedrate or both, until he has studied multiple regression and the design of experiments which are dealt with in later chapters.

10.10 SUMMARY

In this chapter we have examined the least squares method of fitting simple regression equations. The equations were used to predict the level of impurity that might result from using specified weights of special ingredient (x) and specified levels of feedrate (z). Confidence intervals were calculated for the predicted impurity using the residual standard deviation. These confidence intervals were rather wide but we will see in the next chapter that the use of two independent variables will give a higher percentage fit and consequently a lower residual standard deviation with correspondingly narrower confidence intervals.

PROBLEMS

10.1 Edlington Chemicals produces monocylate using a batch process. The plant manager is concerned about high reaction times of certain batches and a detailed examination of past records convinces him that this is due to high levels of hexanol impurity in the feedstock. For statistical analysis he chooses a sequence of 12 previous batches in which the results look promising.

   (a) Decide which is the dependent variable (y) and which is the independent variable (x).

Concentration of hexanol	Reaction time
10	300
13	380
10	350
11	320
7	280
14	400
9	330
13	370
10	330
9	350
12	310
14	360

(b) Calculate SD of x and SD of y.

(c) Calculate the correlation coefficient between reaction time and concentration.

(d) Test the significance of the correlation coefficient (consider carefully whether it is a one-sided or two-sided test).

(e) Calculate the slope and intercept of the regression line, $y = a + bx$.

(f) Plot a scatter diagram, with the independent variable on the horizontal axis, and draw your regression line on the diagram.

(g) Calculate the percentage fit for the regression equation.

(h) Calculate the residual standard deviation.

(i) Test the statistical significance of the relationship between the two variables using a t-test.

(j) Calculate 95% confidence intervals for the true intercept and true slope.

(k) Calculate 95% confidence intervals for the true mean reaction time of batches with concentrations of:

(i) 7;
(ii) 11.

(l) Calculate 95% confidence intervals for the reaction time of a single batch with concentrations of:

(i) 7;
(ii) 11.

(m) Has the plant manager carried out an investigation which is in accord with scientific method?

10.2   Yorkshire Spinners Ltd buys triacetate polymer and converts it into different types of fabric. This process involves many stages, all of which affect the quality of the final product, but it has been shown that the critical stage is the first one in which polymer is heated, forced through minute orifices and wound onto bobbins as a fibre. The quality of the fibre may be represented by many parameters including birefringence, which is a measure of orientation of polymer molecules in the fibre and can be controlled by changing the speed of winding onto the bobbin – referred to as wind-up speed. The wind-up speed can be adjusted quickly but other associated adjustments – flow rate and temperature – cannot.

These adjustments, necessary to keep parameters other than birefringence constant, lead to a loss of production. It is necessary that the correct adjustment be made to birefringence since over- or under-correction will lead to a further loss of production.

Making the correct adjustment depends upon knowing accurately the relationship between birefringence and wind-up speed. This relationship is, however, dependent upon the type of machine.

A new machine is being developed by Yorkshire Spinners and the research manager has, absent-mindedly, independently asked four scientists – Addy, Bolam, Cooper and Dawson – to evaluate the relationship between birefringence and wind-up speed. He particularly wants two results:

(a) An estimate of the slope which will be used in conjunction with hourly quality control checks on birefringence. If a change in birefringence is necessary the slope will be used to calculate the required change in wind-up speed.

(b) An estimate of the true relationship between the two variables to be used in the calculation of start-up conditions for a new product.

The four scientists work in isolation from each other and produce different experimental designs to tackle the same problem. The designs and the experimental results are given below:

Addy

Wind-up speed	150	160	170	180	190	200
Birefringence	70.1	75.3	77.0	80.6	87.2	89.8

Bolam

Wind-up speed	150	175	175	175	175	200
Birefringence	70.0	81.3	78.1	79.9	80.5	90.2

Cooper

Wind-up speed	150	150	150	200	200	200
Birefringence	70.4	70.8	69.0	91.5	88.9	89.4

Dawson

Wind-up speed	165	169	173	177	181	185
Birefringence	76.2	78.9	78.1	79.5	83.5	83.8

Summary statistics of the four experiments are as follows, where x is wind-up speed and y is birefringence:

	$\bar{x}$	$\bar{y}$	SD (x)	SD (y)
Addy	175	80	18.708	7.448
Bolam	175	80	15.811	6.475
Cooper	175	80	27.386	10.933
Dawson	175	80	7.483	3.040

Complete the following table by calculating:
(a) the correlation coefficients for Addy and Bolam;
(b) the least squares regression line for Addy and Bolam;
(c) the residual standard deviation and hence the 95% confidence limits for the slope for Addy and Bolam;
(d) 95% confidence limits for the true value of birefringence at wind-up speeds of 150, 175 and 200 for Addy. Plot the intervals given in the table for all five conditions and four experiments on the scatter diagrams (Figure 10.7). Join the points to give 95% confidence bands.

	Addy	Bolam	Cooper	Dawson
Correlation coefficient			0.995	0.935
Intercept			10.5	13.5
Slope			0.397	0.380
Residual standard deviation			1.24	1.20
95% confidence interval for slope			±0.056	±0.199
95% CI for true line at:				
150		69.9 ± 2.7	70.2 ± 2.0	70.5 ± 5.2
160	74.1 ± 1.9	73.9 ± 1.9	74.1 ± 1.6	74.3 ± 3.3
175		80.0 ± 1.3	80.0 ± 1.4	80.0 ± 1.4
190	85.9 ± 1.9	86.1 ± 1.9	85.9 ± 1.6	85.7 ± 3.3
200		90.1 ± 2.7	89.8 ± 2.0	89.5 ± 5.2

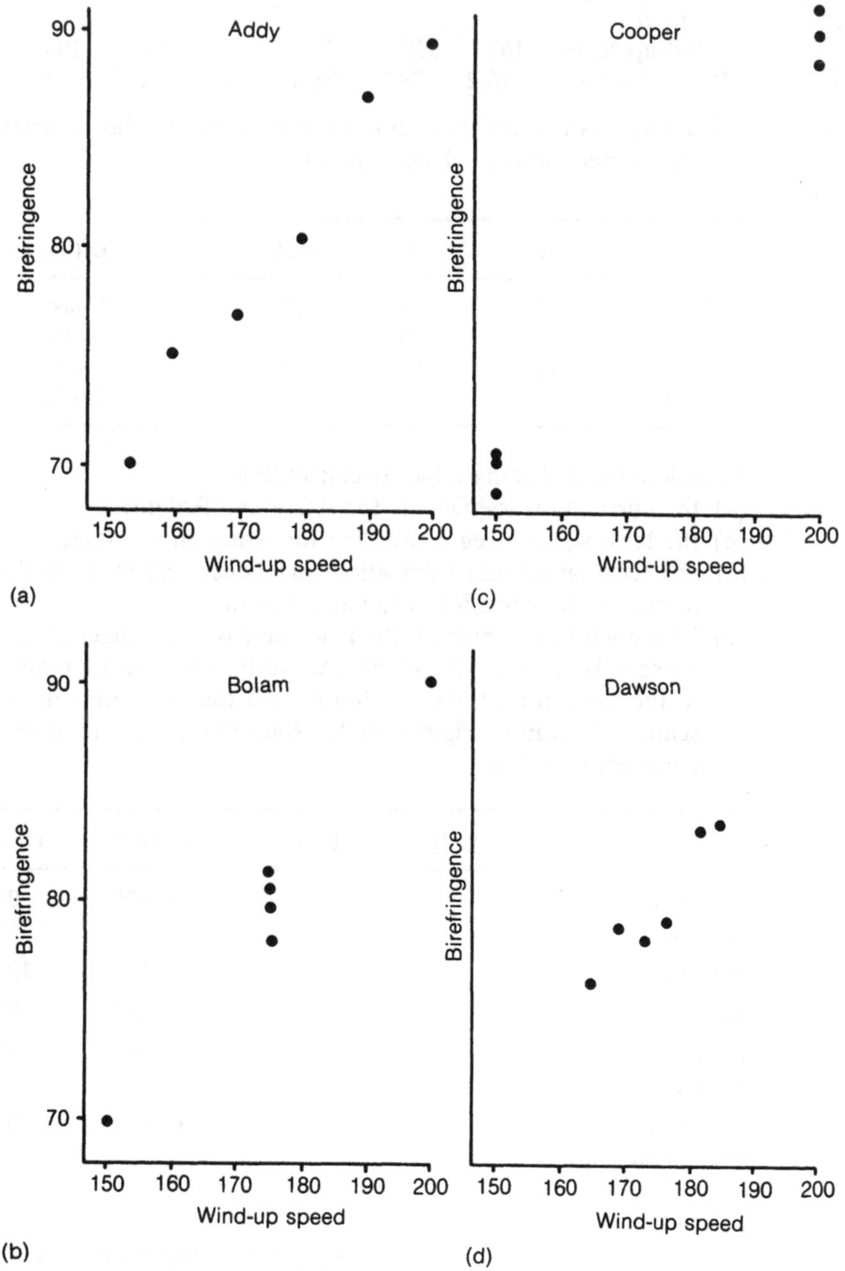

Figure 10.7 Birefringence at wind-up speed scatter diagrams

(e) Would you have expected the slope coefficients and the residual standard deviations to be less variable between the experiments?
(f) Why are the confidence intervals for the slope markedly different?
(g) The four correlation coefficients are different. Is this due only to sampling error?
(h) Why is there more curvature on the confidence band for Bolam than Cooper?
(i) Is there any danger in using Dawson's design?
(j) Is there any danger in using Cooper's design?

11
Why was the experiment not successful?

11.1 INTRODUCTION

The previous chapter was centred around a situation in which we attempted to explain why the percentage impurity in a pigment varied from batch to batch. On each of ten batches several variables had been measured and we selected the weight of special ingredient (x) as the independent variable which was most highly correlated with the dependent variable (percentage impurity). After fitting a simple regression equation we were able to make predictions of the impurity that could be expected in future batches if certain weights of special ingredient were used. Unfortunately the confidence intervals associated with these predictions were rather wide, which raised doubts about the usefulness of the regression equation.

We fitted a second simple equation in which the feedrate (z) was the independent variable. Whilst this new equation had a lower percentage fit (61% compared with 72%) and consequently even wider confidence intervals, it might nonetheless be more attractive to the plant manager on economic grounds. Perhaps it would be cheaper to increase the feedrate than to increase the weight of special ingredient in each batch.

In this chapter we will examine regression equations in which there are two or more independent variables. These equations will give a higher percentage fit than the simple equations discussed in Chapter 10 and we can therefore look forward to narrower confidence intervals. We will start by fitting a regression equation with two independent variables and it seems reasonable that we should choose feedrate (z) and weight of special ingredient (x) from the many possibilities open to us.

11.2 AN EQUATION WITH TWO INDEPENDENT VARIABLES

The two simple regression equations fitted in the previous chapter were:

$$y = 8.0 - x \qquad (72\% \text{ fit})$$

$$y = 8.6 - 0.90z \quad (61\% \text{ fit})$$

We will now fit an equation which contains both x and z as independent variables. This will have the form:

$$y = a + bx + cz$$

and values of a, b and c will be calculated using the y, x and z measurements in Table 10.1. It is reasonable to expect that this multiple regression equation will give a higher percentage fit than either of the simple equations. On the other hand we would be foolish to expect 61% + 72% (i.e. 133%) as this would exceed the obvious upper limit of 100%.

Fitting of a multiple regression equation involves rather tedious calculations and would usually be carried out on a computer, especially if the equation contained three or more independent variables. The y, x and z values for the ten batches are fed into a multiple regression program on a desktop computer. The coefficients of the regression equation are calculated and displayed as:

$$a = 7.0$$

$$b = -2.0$$

$$c = 1.0$$

The least squares multiple regression equation containing weight of special ingredient (x) and feedrate (z) as independent variables is, therefore:

$$y = 7.0 - 2.0x + 1.0z$$

What does this equation tell us about the values of x and z that the plant manager should use in order to reduce the impurity in future batches of pigment? The coefficient of x is -2.0 so we can expect that an increase of 1 unit in the weight of special ingredient (x) will give a 2% reduction in impurity. This differs from the 1% reduction offered by the simple equation $(y = 8.0 - x)$. The coefficient of z in the multiple regression equation suggests that a unit increase in the feedrate (z) will give a 1% increase in impurity. This is completely at odds with the 0.9% decrease promised by the simple equation $(y = 8.6 - 0.9z)$.

The three equations that we have fitted would seem to offer very conflicting recommendations concerning the operating conditions that should be used to reduce impurity. We are not at all sure just how large a decrease in impurity could be expected from an increase in the weight of special ingredient (x). As far as the other independent variable (z) is concerned we are not even sure whether an increase or a decrease in impurity is likely to result from an increase in feedrate.

198   Why was the experiment not successful?

It is possible that any one of the three equations may be offering a useful indication to the plant manager but it is obviously impossible to reconcile the conflicting advice which results when we put all three equations together.

It is not unreasonable to suggest, therefore, that the use of regression analysis with this set of data has been very unsatisfactory. In fairness to the statistical technique, however, it must be pointed out that the reason for our lack of success lies not in the technique, but in the data that we have attempted to analyse. Multiple regression analysis is, without doubt, a very useful technique but great care must be exercised because certain characteristics in a set of data can result in the drawing of invalid conclusions. Before we attempt to use multiple regression analysis we must examine the intercorrelation of the independent variables.

You may have imagined that the five independent variables, x, w, z, t and s vary from batch to batch independently of each other. This is certainly not so, as we can see from the correlation matrix of Table 11.1.

We see from Table 11.1 that the correlation between x and z is 0.97. This implies that those batches in which a high feedrate (z) was used also had a large quantity of special ingredient (x) and vice versa. Reference to Table 10.1 confirms that this is indeed the case and we see, for example, that the third batch had $x = 6$, $z = 7$, whilst at the other extreme the eighth batch had $x = 0$, $z = 1$.

This high correlation between the two independent variables, x and z, is telling us that the plant manager, in producing these ten batches, was carrying out a bad experiment. It is this same correlation which would have warned us of the danger of using regression analysis with this data had we first examined the correlation matrix. Furthermore it is this same correlation (0.97) which has caused the coefficients of x and z in the multiple regression equation to differ greatly from the coefficients in the two simple equations. Indeed, when using multiple regression analysis, it is wise to watch out for large changes in coefficients as new variables enter the equation. Such changes are an indication of correlation between independent variables.

Table 11.1 Correlation matrix

	x	w	z	t	s
x	1.00	−0.26	0.97	−0.07	0.00
w	−0.26	1.00	−0.20	−0.05	0.44
z	0.97	−0.20	1.00	0.07	−0.06
t	−0.07	−0.05	0.07	1.00	0.00
s	0.00	0.44	−0.06	0.00	1.00

Intercorrelation of the independent variables is entirely detrimental. We have seen the disadvantages which follow from the high correlation between feedrate and weight of special ingredient. There are no compensating advantages. The correlation between x and z even intrudes when we attempt to calculate the percentage fit of the multiple regression equation. When designing an experiment one objective is to reduce the correlation between pairs of independent variables.

In the previous chapter we calculated the percentage fit of a simple regression equation using:

$$\text{Percentage fit} = 100r^2 \qquad (11.1)$$

To obtain the percentage fit of an equation with two independent variables we use the more complex expression:

$$\text{Percentage fit} = 100(r_{xy}^2 + r_{zy}^2 - 2r_{xy}r_{xz})/(1 - r_{xz}^2) \qquad (11.2)$$

As there are three correlation coefficients involved we use subscripts to distinguish between them. Thus r_{xy} represents the correlation between x and y, etc. Substituting into this equation we get:

$$\begin{aligned}
\text{Percentage fit} &= 100[(-0.85)^2 + (-0.78)^2 - 2(-0.85)(-0.78)(0.97)]/ \\
&\qquad [1 - (0.97)^2] \\
&= 100(0.723 + 0.608 - 1.286)/(0.059) \\
&= 76\%
\end{aligned}$$

The percentage fit of the multiple regression equation (76%) is little higher than the percentage fit of the two simple equations (72% and 61%). Once again we can blame the correlation between x and z. The effect of inter-correlation between the independent variables is highlighted if we let r_{xz} equal 0 in equation (11.2), which then becomes:

$$\text{Percentage fit} = 100(r_{xy}^2 + r_{zy}^2) \qquad (11.3)$$

Equation (11.3), which is very similar in form to equation (11.1), would give us the percentage fit if x and z were not correlated. In other words equation (11.3) would be applicable if the plant manager had carried out an experiment in which feedrate (z) and weight of special ingredient (x) did not vary from batch to batch in sympathy with each other.

11.3 MULTIPLE REGRESSION ANALYSIS ON A COMPUTER

It was stated earlier that multiple regression analysis would not usually be attempted without the support of a computer together with a suitable computer program. Perhaps it has occurred to you that the use of com-

puting facilities could offer a safeguard against drawing erroneous conclusions from data in which the independent variables are intercorrelated.
Such faith in computers can be misplaced. Certainly some regression
analysis programs (or packages as they are often called) do print out
warnings of potential disasters, but many others do not. Even if you avail
yourself of an excellent package it can only warn you that the data are
suspect; it cannot turn a bad experiment into a good one. We will see in
later chapters that the plant manager's experiment can be greatly improved
but the improvement comes from producing several more batches and not
from improving the method of data analysis.

Just as computers come in many different sizes, ranging from micros
through minis to main frames, so multiple regression packages have different levels of complexity. The more sophisticated packages tend to be
located in the larger machines but recent advances in computer hardware
have made available very powerful multiple regression facilities on relatively
inexpensive desktop computers. Just what do these packages offer?

We will answer this question by describing how one particular regression
package would analyse the data from our ten batches of pigment and the
print-out from this package is reproduced in Appendix D. An important
feature, which this package shares with many others, is automatic variable
selection by means of which the program selects one independent variable,
then a second, then a third, etc. Thus a sequence of regression equations is
printed out until the 'best' equation is reached and then the sequence is
terminated. Each new variable is selected on purely statistical grounds, of
course, the program having no knowledge whatsoever of the underlying
science/technology. Using this statistical criterion the following equations
are fitted:

$$y = 8.0 - 1.0x \qquad 72\% \text{ fit}$$

$$y = 5.2 - 0.96x + 1.34t \quad 93.4\% \text{ fit}$$

When selecting the first independent variable the program chooses the one
which gives the greatest percentage fit. This procedure is equivalent to the
one we adopted in the previous chapter when we selected weight of special
ingredient (x) because it had the greatest correlation coefficient.

The inlet temperature (t) is selected as the second independent variable
because it offers the greatest increase in percentage fit. With x already in
the equation we have four variables $(z, t, s$ and $w)$ from which to choose.
The computer program carries out calculations which show that x and t
together will give a greater percentage fit than either x and z, x and s, or x
and w. We will not pursue these calculations, but the wisdom of the
decision can be seen if we examine Figure 11.1.

Figure 11.1 is very similar to Figure 10.1. The only difference between

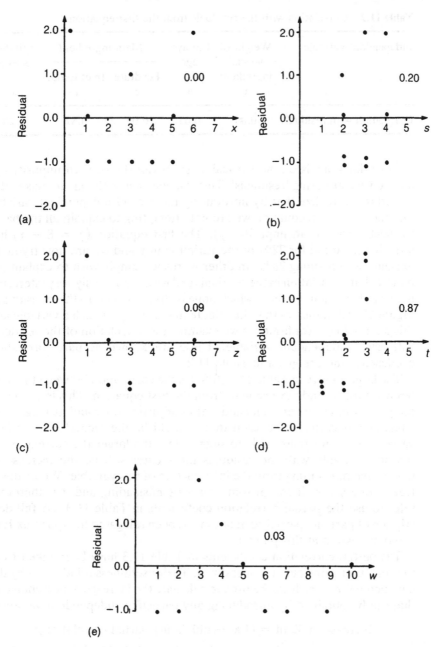

Figure 11.1   Residuals from $y = a + bx$ plotted against the independent variables

Table 11.2   Correlation with the residuals from the first equation

Independent variable	Weight of special ingredient	Catalyst age	Main ingredient		Agitation speed
			Feedrate	Inlet temp.	
	x	w	z	t	s
Correlation with residuals	0.00	0.03	0.09	0.87	0.20

the two diagrams is on the vertical axis; previously we used impurity (y) and now we are using 'residuals'. This is a reasonable change to make. It is important to realize that by introducing the second independent variable into the regression equation we are not attempting to explain all the batch to batch variation in impurity (y). The first equation ($y = 8 - x$) has already accounted for 72% of the variation in y and we are now trying to explain the remaining 28%. In other words we simply wish to explain the batch to batch variation in the residuals and we are obviously very interested in any independent variable which appears to be related to these residuals. Figure 11.1(d) indicates that the inlet temperature (t) is such a variable and this impression is confirmed if we examine the correlation of the residuals with each independent variable. These are known as part correlation coefficients and are given in Table 11.2.

The largest entry in Table 11.2 (0.87) is the correlation between the inlet temperature (t) and the residuals from the first equation. This is known as the part correlation between t and y after adjusting for x and its magnitude confirms our earlier impression that t should be the second independent variable. It seems reasonable to suggest that the larger the correlation of the new variable with the residuals the greater will be the increase in percentage fit resulting from the introduction of this variable. Whilst this is true in one sense, it can in some cases be misleading and it is therefore safer to use the partial correlation coefficients in Table 11.3. (A full description of part and partial correlation has been relegated to Appendix E to avoid digression at this point.)

The partial correlation coefficients in Table 11.3 have been taken from the computer print-out in Appendix D. Their usefulness is indicated by the equation below which can be used to calculate the increase in percentage fit that can be obtained by introducing any particular independent variable:

$$\text{Increase in \% fit} = (100 - \text{old \% fit})(\text{partial correlation})^2$$

By introducing inlet temperature (t) into the regression equation we can expect to get an increase in percentage fit equal to:

$$(100 - 72)(0.874)^2 = 21.4\%$$

Table 11.3 Partial correlation and increase in percentage fit

Independent variable	Weight of special ingredient	Catalyst age	Main ingredient		Agitation speed
			Feedrate	Inlet temp.	
	x	w	z	t	s
Partial correlation with the dependent variable (x being fixed)	0.000	0.030	0.377	0.874	0.203
Increase in % fit by introducing this variable	0.0%	0.02%	4.0%	21.4%	1.2%

After t has been introduced as the second independent variable the percentage fit should have increased, therefore, from 72% to 93.4%. This is indeed the case.

Clearly the increase in percentage fit must be statistically significant or the second regression equation would not have been printed out by the computer program. We can, however, check this significance by means of a t-test in which the test statistic is based on the increase in percentage fit.

Null hypothesis – impurity (y) is not dependent on inlet temperature (t).

Alternative hypothesis – impurity is dependent on inlet temperature (t).

Test statistic = $\sqrt{\{[($new % fit $-$ old % fit$)($degrees of freedom$)]/$
$(100\% -$ new % fit$)\}}$

(Degrees of freedom $= n - k - 1 = 10 - 2 - 1 = 7$, where k is the number of independent variables in the equation.)

$$= \sqrt{\left[\frac{(93.4\% - 72\%)(7)}{(100\% - 93.4\%)}\right]}$$

$$= 4.76$$

Critical values – from the two-sided t-table with 7 degrees of freedom:

2.36 at 5% significance level
3.50 at 1% significance level.

Decision – we reject the null hypothesis at the 1% level of significance.

Conclusion – we conclude that impurity (y) is related to the inlet temperature.

This significance test is based on the knowledge that a relationship between impurity (y) and weight of special ingredient (x) has already been estab-

Table 11.4   Residuals from the second regression equation

Weight of special ingredient	Inlet temp.	Actual impurity	Predicted impurity	Confidence interval for true impurity	Residual
x	t	y			
3	1	4	3.66	2.82 to 4.50	0.34
4	2	3	4.04	3.46 to 4.62	−1.04
6	3	4	3.45	2.25 to 4.65	0.55
3	3	6	6.34	5.50 to 7.18	−0.34
1	2	7	6.93	6.18 to 7.67	0.07
5	1	2	1.73	0.76 to 2.71	0.27
1	2	6	6.93	6.18 to 7.67	−0.93
0	3	10	9.23	8.09 to 10.36	0.77
2	1	5	4.62	3.73 to 5.52	0.38
5	2	3	3.07	2.33 to 3.82	−0.07

lished. We are now able to conclude that both the variation in inlet temperature (t) and the variation in weight of special ingredient (x) are contributing to the batch to batch variation in impurity. If the plant manager accepts this conclusion then any strategy put forward for the reduction of impurity in future batches should specify levels for the weight of special ingredient (x) and for the inlet temperature (t).

Also printed out by our computer program is a table containing residuals, predicted impurity values and confidence intervals for true impurity values. This very useful table is given as Table 11.4.

In Table 11.4 the column of residuals has a total of −0.04. Had there been no rounding errors in the individual residuals this total would have been zero. The sum of the squared residuals is equal to 3.30.

The confidence intervals in Table 11.4 vary in width. The narrowest interval has a width of 1.16 (the batch with $x = 4$ and $t = 2$) whilst the widest interval has a width of 2.4 (the batch with $x = 5$ and $t = 3$). This variation is consistent with what we found when we plotted confidence intervals from the first regression equation in Figure 10.6. If the values of the independent variables (x and t) are close to the means ($\bar{x} = 3$ and $\bar{t} = 2$) then the confidence interval will be narrower than it would be if the values of x and t were far removed from the means.

Before we leave the computer print-out in Appendix D let us summarize the conclusions that it suggests and record our reservations:

(a) Using a statistical criterion at each stage the computer program has given us the best equation:

$$y = 5.2 - 0.96x + 1.34t \quad 93.4\% \text{ fit}$$

(b) This equation gives predicted impurity values which have much narrower confidence intervals than those which were given by the first regression equation.

(c) The correlation matrix which is included in the print-out contains one very high value (0.97) which must not be ignored. This correlation between weight of special ingredient (x) and feedrate (z) warns us that it will not be possible to distinguish between changes in impurity due to changes in x and changes in impurity due to changes in z.

Though the multiple regression equation containing x and t has been declared the best equation on statistical grounds it may have occurred to you that there could be a multiple regression equation containing the feedrate (z) and other independent variables, which is almost as good on statistical grounds and more useful to the plant manager. We will now explore this possibility.

11.4 AN ALTERNATIVE MULTIPLE REGRESSION EQUATION

As we have already noted the computer program has a preference for weight of special ingredient (x) as the first independent variable. This selection is based on the correlation coefficients between the dependent variable (y) and the five independent variables (x, w, z, t and s) (Table 11.5).

So x is selected as the first independent variable and, as we have discovered, t is the second. The computer program stops at this point because the third independent variable (agitation speed) offers an increase in percentage fit which is not statistically significant. Had the program continued, however, we would have found that catalyst age (w) was the fourth variable and feedrate (z) was the last. With hindsight we can see

Table 11.5 Correlation with percentage impurity

Variable	Weight of special ingredient	Catalyst age	Main ingredient		Agitation speed
			Feedrate	Inlet temperature	
	x	w	z	t	s
Correlation	−0.85	0.23	−0.78	0.52	0.11

that when x enters the regression equation z is excluded. This occurs, of course, because of the very high correlation (0.97) between x and z.

We can ensure that the weight of special ingredient (x) does not dominate the regression equations in this way if we instruct the computer program to fit an equation with impurity (y) as the dependent variable and with only four independent variables, z, w, t and s, from which to choose. When we adopt this strategy we find that the first variable to enter the equation is feedrate (z) and the second variable is inlet temperature (t) with the other two independent variables proving to be statistically non-significant. The fitted equations are:

$$y = 8.58 - 0.895z \qquad 60.8\% \text{ fit}$$

$$y = 5.44 - 0.94z + 1.66t \quad 93.6\% \text{ fit}$$

We have already met the first of these two equations in the previous chapter and no further explanation is needed. The second equation is new to us and several points are worthy of note:

(a) The coefficient of z is -0.94 which implies that a unit increase in feed-rate (z) would result in an impurity reduction of 0.94%. As we have stated earlier this figure must be treated with great suspicion because of the very high correlation between the two independent variables x and z.

(b) The coefficient of t is $+1.66$ which implies that a unit decrease in the inlet temperature would give an impurity reduction of 1.66%. This figure is in fairly close agreement with the 1.34% reduction offered by the 'best' equation ($y = 5.2 - 0.96x - 1.34t$) fitted earlier. In the light of these two equations, therefore, it would be reasonable for the plant manager to conclude that a reduction in impurity of approximately 1.5% could be obtained by reducing the inlet temperature by 1 unit.

(c) The percentage fit of the most recent equation (93.6%) is actually a little higher than the 93.4% achieved by the 'best' equation. It was pointed out earlier that the automatic variable selection methods used in some multiple regression packages do not always produce satisfactory results. It is possible that a different method of selection would have given a different 'best' equation. In order to get the best out of the computing facilities it is important for the scientist/technologist to interact with the package, making full use of accumulated experience.

We now find ourselves in the rather unsatisfactory position of having two multiple regression equations:

$$y = 5.2 - 0.96x + 1.34t$$

$$y = 5.44 - 0.94z + 1.66t$$

Though both equations have a high percentage fit we are unable to decide which of the two independent variables, x or z, is important. This uncertainty arises because of high correlation between x and z. In later chapters we will attempt to improve our position by extending the plant manager's experiment. Before doing so let us consider how the first of the two equations can be represented graphically.

11.5 GRAPHICAL REPRESENTATION OF A MULTIPLE REGRESSION EQUATION

In the previous chapter we had no difficulty in representing the simple regression equations in graphical form. We drew two axes at right angles; one axis for the dependent variable and one for the independent variable. To represent a regression equation which contains two independent variables we need three axes at right angles to each other. Such a system of axes requires a three-dimensional space and cannot be accommodated on a sheet of paper.

The multiple regression equation, $y = a + bx + ct$, could be represented in two dimensions, however, if we eliminated one of the independent variables. This can be achieved by giving that variable a fixed value. To reduce our equation $y = 5.2 - 0.96x + 1.34t$ we will fix the value of t at three suitable levels ($t = 1$, $t = 2$ and $t = 3$) to obtain the equations:

$$y = 6.5 - 0.96x \quad \text{when } t = 1$$

$$y = 7.9 - 0.96x \quad \text{when } t = 2$$

$$y = 9.2 - 0.96x \quad \text{when } t = 3$$

These equations can be represented by the three straight lines in Figure 11.2(a). Alternatively we can eliminate the other independent variable (x) from the equation by giving it suitable values. As the weight of special ingredient (x) varies from 0 to 6 within the ten batches it is reasonable to let x equal 0, 3 and 6. Doing so gives three simple equations which can be represented by the three straight lines in Figure 11.2(b).

Either of the graphs, Figure 11.2(a) or (b), might be of use to the plant manager in his search for operating conditions which will give a tolerable level of impurity. If he has concluded that weight of special ingredient (x) and inlet temperature (t) are the only two variables requiring attention then Figure 11.2 may help him to compare the levels of impurity that are predicted for many different combinations of x and t.

You may have noticed that in Figure 11.2(a) and in Figure 11.2(b) we have a family of parallel straight lines and that, furthermore, the lines are equidistant. An equation of the form $y = a + bx + ct$ will always give a family of lines which are parallel and straight. Clearly such an equation is

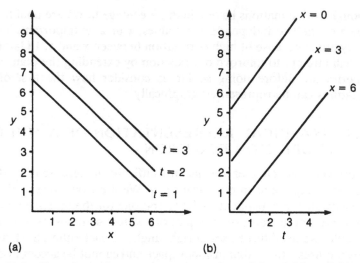

Figure 11.2   Two alternative representations of $y = 5.2 - 0.96x + 1.34t$

inadequate if you have a set of data which gives a graph resembling either of the scatter diagrams in Figure 11.3.

11.6 NON-LINEAR RELATIONSHIPS

Figure 11.3(a) and Figure 11.3(b) are alternative representations of the same hypothetical set of data. Clearly the relationship between x and y is

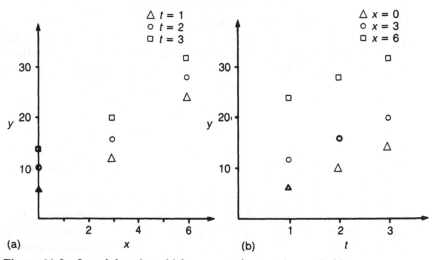

Figure 11.3   Set of data for which $y = a + bx + ct$ is unsuitable

non-linear for each of the three values of t. Though the relationship between t and y is linear we can see in Figure 11.3(b) that equidistant straight lines are inappropriate because of the curved relationship between x and y. The data which are plotted in Figure 11.3 require an equation of the form:

$$y = a + bx + ct + dx^2$$

To fit such an equation we require four columns of data. Three of these columns will already be available because observations have been made of y, x and t and these values were used in plotting the graphs in Figure 11.3. The values of x^2 could be calculated and then fed into the computer to give us the fourth column of data that we need. Fortunately this is rarely necessary because most multiple regression packages can generate new variables when asked to do so.

Returning to the data gathered by the plant manager in the previous chapter, we could ask the regression packages to generate a quadratic term corresponding to each of the five independent variables originally measured. We would then have a total of ten independent variables (x, w, z, t, s, x^2, w^2, z^2, t^2 and s^2) which could be included in regression equations.

It is possible to generate variables having higher powers such as x^3 or z^4 but this is rarely done. In some applications logarithms, exponentials or trigonometric functions are found to be useful but will not be discussed here. Far more important is the ability to produce equations which will fit reasonably to data like those in Figure 11.4.

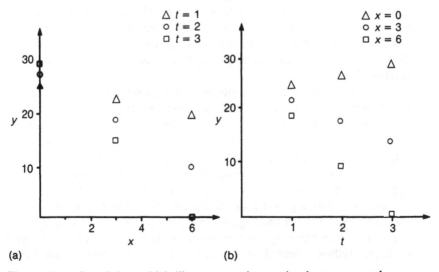

(a)   (b)

Figure 11.4 Set of data which illustrates an interaction between x and t

11.7 INTERACTIONS BETWEEN INDEPENDENT VARIABLES

The data displayed in Figure 11.4 do not exhibit curved relationships between the variables. In Figure 11.4(a) we can see that a linear relationship exists between x and y for each value of t, but the straight lines are not parallel. Three straight lines are needed but each must have a different slope. Figure 11.4(b) is an alternative representation of the data in Figure 11.4(a) and again we see the need for non-parallel lines; for each value of x there exists a linear relationship between t and y but it is a different relationship in each case. The statistician would summarize Figure 11.4 by saying that there appears to be an interaction between x and t. We say that there is an *interaction* between two independent variables when the effect of one (on the dependent variable) depends upon the value of the other.

The data represented in Figure 11.4 suggest very strongly that the relationship between x and y depends on the value of t or, from the alternative standpoint, the relationship between t and y depends on the value of x. Interactions between pairs of independent variables are extremely important in research and development work. They will be discussed further in later chapters.

Our immediate concern is to find a multiple regression equation which will fit the data in Figure 11.4. Such an equation must give non-parallel straight lines when plotted and this can be achieved by including a cross-product term (dxt), i.e.:

$$y = a + bx + ct + dxt$$

The new variable (xt) can be generated in the computer by multiplying each x value by the corresponding t value. The generation of such cross-product variables is even more important than the generation of quadratic variables in a multiple regression package for the simple reason that the detection of interactions is more important than the detection of curved relationships.

When we attempt to use a multiple regression package to find significant interactions between pairs of independent variables we find that there are many possibilities to investigate. With the data from ten batches of pigment we have only five independent variables (x, w, z, t and s) in the original data but these give rise to ten pairs of variables ($xw, xz, xt, xs, wz, wt, ws, zt, zs$ and ts) each of which can be used as a cross-product term in a regression equation. Add to these five quadratic terms (x^2, w^2, z^2, t^2 and s^2) and we have a grand total of 20 variables which can be included in the right-hand side of the regression equation. As we have only ten data points (i.e. ten batches) we are not in a good position to investigate all these possibilities, even if we ignore the crippling intercorrelation of x and

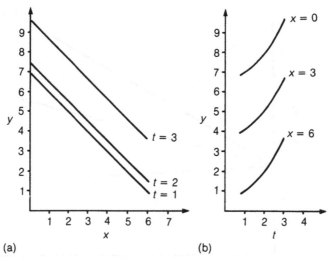

Figure 11.5 Two alternative representations of $y = 8.1 - 0.99x - 1.97t + 0.83t^2$

z, but the following equations have been fitted to illustrate what we might be able to achieve when the plant manager's experiment is extended in a later chapter:

$$y = 8.1 - 0.99x - 1.97t + 0.83t^2 \quad 96.7\% \text{ fit}$$
$$y = 3.5 - 0.37w + 0.321wt \quad\quad 46.7\% \text{ fit}$$

These two equations are represented in Figures 11.5 and 11.6. Figure 11.5 portrays the quadratic equation whilst Figure 11.6 portrays the equation containing the cross-product term (wt). The latter diagram illustrates an interaction between temperature (t) and catalyst age (w). We see that there is very little change in impurity with age if a low temperature is used but a substantial increase in impurity as the catalyst ages if a high temperature is used.

11.8 SUMMARY

In this chapter you have seen some of the power and versatility of multiple regression analysis as we have attempted to explain the batch to batch variation in pigment impurity. It is true that we have not been very successful but the following conclusions can now be drawn:

(a) The inlet temperature (t) of the main ingredient should be controlled in future batches and a reduction of one unit can be expected to give an impurity reduction of about 1.5%.

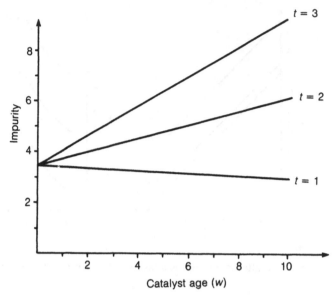

Figure 11.6   Graphical representation of $y = 3.5 - 0.37w + 0.321wt$

(b) The weight of special ingredient (x) and the feedrate (z) cannot be ignored, though we are unable to decide which of the two variables (or both) is important. We are unable to separate the effect of these variables because of the very bad design of the plant manager's experiment. The weakness of the experiment was revealed by inspecting the intercorrelations of the independent variables.

(c) Having found this deficiency in the experiment we must disregard the tentative conclusions drawn in Chapter 10.

(d) There is some indication that curved relationships might exist and that there might be an interaction between two of the independent variables. These will be explored later.

We must now set aside our techniques of statistical analysis whilst we consider the principles on which experiments should be based if we are to avoid some of the problems that we have encountered in this chapter.

PROBLEMS

11.1   A research chemist is investigating the factors which determine the tensile strength (y) of a man-made fibre. He has carried out 12 production runs on small scale spinning and drawing plant using various values of the following independent variables:

(i) drying time of the polymer (w);

(ii) spinning temperature (x);

(iii) draw ratio (z).

Values of the response (y) and the three independent variables (w, x and z) are fed into a stepwise multiple regression package which gives the following regression equations:

$$y = 12.65 + 9.731z \qquad\qquad 50.7\% \text{ fit}$$

$$y = 53.81 + 9.731z - 0.1136x \qquad 69.3\% \text{ fit}$$

$$y = 68.73 + 5.824z - 0.0527x + 9.113w \quad 73.8\% \text{ fit}$$

(a) Carry out *t*-tests to check the statistical significance of the three independent variables.

(b) By careful inspection of the equations and their percentage fits, what can you deduce about the correlations between the four variables?

11.2 If you wish to use a computer-based multiple regression package or to discuss a multivariable problem with a statistician then your success will depend, at least in part, on your understanding of the terminology used in Chapters 10 and 11. You can test the extent to which you have absorbed this terminology by attempting to insert the missing words in the passage below. Don't be surprised if you need to refer back to the text or even to peep at the answers occasionally.

When using regression analysis we must distinguish between the dependent variable (which is often referred to as the (1)
and appears in the left-hand side of the equation), on the one hand and the (2)   variables on the other. In a multiple regression equation we may have several (3)   variables but only one response. When choosing the independent variables to include in an equation we are not restricted to the variables that have been measured for we can also generate quadratic variables and (4)   variables. Inclusion of quadratic variables helps us to accommodate curved relationships whilst the use of cross product terms in the equation allows us to take account of (5)
            between pairs of independent variables. We say that there is an (6)   between two independent variables when the effect of one variable upon the (7)   depends upon the (8)   of the other. If, for example, we fitted the equation

$$y = a + bx + cxz + dz$$

and all three independent variables were found to be statistically significant then we would conclude that there was an interaction between (9)   and (10)   and it would be misleading to speak of the effect of x on y without mentioning (11)

To check the significance of each variable as it enters the equation we could carry out a (12)   with the test statistic being calculated from the new (13)   the old (14) and $(n - k - 1)$, the latter being known as the (15)   . If we put forward x, z and xz as independent variables and the fitted equations were

$$y = a + bx$$

then

$$y = c + dx + ez$$

and finally

$$y = f + gx + hz + ixz$$

we would know that:

(a) the correlation between y and (16)   would be greater than the correlation between y and z;
(b) at the second step the inclusion of z would offer a greater increase in (17)   than would inclusion of xz;
(c) the coefficient of x in the first equation (i.e. b) would only be equal to the coefficient of x in the second equation (i.e. d) if the correlation between x and z were equal to (18)   ;
(d) the percentage of fit of the first equation could be calculated using the correlation coefficient between (19)   and (20)
(e) the percentage fit of the second equation could be calculated using the three correlation coefficients r_{xy}, (21) and (22)

At each step in the stepwise regression procedure the equation is fitted by the method of least squares. When using this method the values of the coefficients are chosen so as to minimize the (23)   . As each new variable enters the equation we get a further reduction in the residual sum of squares, unless the correlation between:

(i) the new independent variable and
(ii) the (24)   from the previous equation

is equal to zero, in which case the old and new equations will have the same residual sum of squares and the same (25)

If we divide the residual sum of squares by its (26) and then take the square root we obtain the (27) which can be used in the calculation of confidence intervals for the true intercept, the true slopes and the true mean value of the (28)   variable for specified values of the independent variable(s). The residual standard deviation is a measure of the (29)   in the response variable that we would get with (30)   values of the independent variables that are included in the equation.

11.3   A research and development chemist wishes to investigate the effect upon the tensile strength of a synthetic yarn, of varying the drying time of the polymer and the spinning temperature. Five runs of the spinning process under experimental conditions yield the following results:

Run	Drying time (min)	Spinning temperature (°C)	Tensile strength (kg)
A	60	240	3.5
B	70	290	3.9
C	30	210	3.1
D	70	260	3.7
E	45	250	3.8

These data are fed into a multiple regression computer package using the symbols:

x for drying time
z for spinning temperature
y for tensile strength.

The print-out from this package contains many regression equations but no information about goodness of fit or the statistical significance of the relationships.

Basic analysis

	x	z	y
Mean	55.0	250.0	3.60
SD	17.321	29.155	0.3162
C of V	31.5%	11.7%	8.8%

Correlation matrix

	x	z	y
x	1.000	0.8416	0.7303
z	0.8416	1.000	0.9220
y	0.7303	0.9220	1.000

Simple regression equations

(1) $y = 1.100 + 0.0100z$
(2) $z = -56.000 + 85.0000y$
(3) $y = 2.867 + 0.0133x$
(4) $x = -89.000 + 40.0000y$
(5) $x = -70.000 + 0.5000z$
(6) $z = 172.083 + 1.4167x$

Multiple regression equations

(7) $y = 0.9000 - 0.00286x + 0.01143z$
(8) $x = -51.6667 + 0.66667z - 16.6666y$
(9) $z = -1.9643 + 0.60714x + 60.7143y$

(a) Which of the six simple regression equations offers the best explanation of the variation in tensile strength of the yarn?
(b) Calculate the percentage fit for each of the six simple regression equations.
(c) Calculate the partial correlation coefficient for y and x with z as the fixed variable.
(d) Use the partial correlation from (c) to calculate the increase in percentage fit you would expect to get by introducing x into the equation $y = a + bz$.
(e) Calculate the percentage fit of the first multiple regression equation using the simple correlation coefficients from the matrix.

Table 11.6

Run	Actual y	Actual z	Predicted y = 1.100 + 0.01z	Residual
A	3.5	240	3.5	0.0
B	3.9	290	4.0	-0.1
C	3.1	210	3.2	-0.1
D	3.7	260	3.7	0.0
E				

Table 11.7

Run	Actual x	Actual z	Predicted x $= -70.0 + 0.5z$	Residual
A	60	240	50	10
B	70	290	75	-5
C	30	210	35	-5
D	70	260	60	10
E				

(f) Complete Tables 11.6 and 11.7 to obtain the residuals for the two simple regression equations which have z as the independent variable.

(g) Calculate the simple correlation between the residuals in Table 11.6 and the residuals in Table 11.7. Note the similarity between this result and that obtained in (c).

12

Some simple but effective experiments

12.1 INTRODUCTION

In Chapters 10 and 11 we made use of regression analysis in an attempt to draw conclusions from a set of data. Despite the great power and versatility of the statistical technique our efforts were not very successful because of intercorrelation between two of the independent variables. This correlation was a feature of the experiment and would not have arisen if the plant manager had used different values of feedrate (z) and/or weight of special ingredient (x) in some of the ten batches.

Having criticized the efforts of the plant manager we will now attempt the much more difficult task of giving him some positive advice. We will suggest what he should have done in order to obtain a set of data from which valid conclusions could be drawn. Armed with this advice he would be in a much stronger position should he decide to abandon his original experiment and to make a fresh start.

We can, however, suggest an alternative strategy which will surely appeal more strongly to the plant manager. This strategy will spell out precisely what he should now do in order to salvage as much as possible from the work he has carried out so far. He will need to manufacture some more batches, but when he has done so he will be able to draw valid conclusions concerning the effect of the five independent variables on the pigment impurity.

The ambitious programme we have just outlined cannot be implemented in this chapter, however, for we must first consider the relative advantages and disadvantages of certain simple experimental designs. We will, therefore, set aside the data that have held our attention so far and consider a variety of problems.

12.2 THE CLASSICAL EXPERIMENT (ONE VARIABLE AT A TIME)

An industrial chemist, employed by Trisell, wishes to investigate a process in which it is suspected that the yield of a triacetate depends upon the

Table 12.1 Estimating the effect of temperature change

Temperature (°C)	Feedrate	Yield
60	40	76
70	40	72

temperature of the main ingredient at the start of the reaction and upon the feedrate of a secondary raw material. Past experience would suggest that temperatures between 55°C and 70°C might be suitable together with feedrates in the range 35 to 55. In order to find those values of temperature and feedrate which give maximum yield the researcher carries out an experiment in which he considers one variable at a time as follows:

Stage 1   To estimate the effect of changing temperature.

Using a feedrate of 40 units he carried out two trials using temperatures of 60°C and 70°C. The yields of triacetate obtained from the two trials are given in Table 12.1.

The results of the two trials in stage 1 indicate that the lower temperature (60°C) is giving the higher yield. We can make a quantitative estimate of the effect of temperature using:

$$\text{Temperature effect} = \text{yield at high temperature}$$
$$- \text{ yield at low temperature}$$
$$= 72 - 76 = -4$$

This estimate suggests that the effect of increasing temperature (by 10°C) is to reduce yield by 4 units.

Stage 2   To estimate the effect of changing feedrate.

Since, in stage 1, the lower temperature (60°C) gave the higher yield the researcher uses this temperature again in a third trial with a feedrate of 50. The yield from this trial is 70 which is included in Table 12.2 together

Table 12.2 Estimating the effect of feedrate change

Temperature (°C)	Feedrate	Yield
60	40	76
60	50	70

with the yield from the earlier trial, which was carried out at the same temperature.

The results of the two trials in Table 12.2 indicate that the lower feedrate (40) is giving the higher yield and a quantitative estimate of the effect of feedrate is given by:

$$\text{Feedrate effect} = \text{yield at high feedrate}$$
$$- \text{yield at low feedrate}$$
$$= 70 - 76 = -6$$

This estimate suggests that the effect of increasing feedrate (by 10 units) is to reduce yield by 6 units.

From the two stages of this classical experiment the researcher draws the conclusion that future batches should be produced using a starting temperature of 60°C and a feedrate of 40. His decision is supported by the simple graph of Figure 12.1.

One would hesitate, of course, to base an important decision on an experiment which consisted of only three trials, if it were possible to carry out a more extensive experiment. Putting aside this reservation let us question the usefulness of this type of experiment, in which one variable is changed at a time, by asking: is this experiment likely to lead the researcher to the operating conditions which give the maximum possible yield from the process? Whilst it is unwise to give a simple yes or no answer to this question, it can be said most emphatically that this classical approach to experimentation can lead the researcher badly astray.

It is quite possible that the relationship between temperature, feedrate and yield for a chemical process is so complex that it would take many

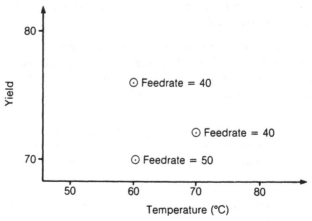

Figure 12.1   Data from Stages 1 and 2 of the experiment

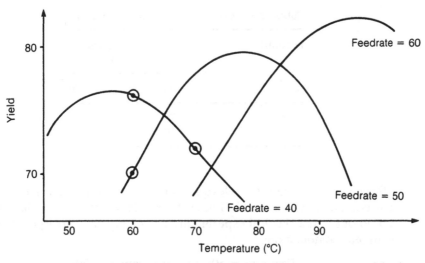

Figure 12.2 Hypothetical relationship between yield, temperature and feedrate

expensive experiments to unravel the true nature of this relationship. Suppose, for the sake of argument, that the true relationship is similar to that depicted in Figure 12.2.

The two striking features of Figure 12.2 are:

(a) For any value of feedrate the relationship between temperature and yield is represented by a curve, though the curves are fairly straight in parts. To investigate a curved relationship a variable must have more than two levels.
(b) There is an interaction between temperature and feedrate.

The concept of an interaction was introduced in Chapter 11. Its importance cannot be emphasized too strongly since there must be few, if any, industrial processes which do not have an interaction between two or more variables. A researcher who fails to recognize possible interactions when designing investigations or analysing data may well obtain results which are misleading.

12.3 FACTORIAL EXPERIMENTS

How might we improve the classical experiment which is clearly inadequate in any situation where an interaction exists? First we will highlight the deficiency of the classical design by bringing together the results of Table 12.1 and Table 12.2 into a two-way table Table 12.3.

Table 12.3 Results of classical experiment

Feedrate	Temperature (°C)	
	60	70
40	76	72
50	70	

Since each variable has two levels, Table 12.3 has four cells. Three of the cells contain values of yield. These three cells represent the three trials carried out in the classical experiment. The empty cell corresponds to the treatment combination:

$$\text{Feedrate} = 50, \quad \text{temperature} = 70$$

If we were to carry out a fourth trial using this treatment combination then the four trials together would constitute an experiment which is known as a 2^2 factorial experiment. It is so called because we have two variables each at two levels and we have used all of the possible treatment combinations in the $2^2 = 4$ trials. (Note the notation. A 3^4 factorial experiment, for example, would have four factors, each at three levels, and would contain 81 trials.)

If we had carried out the fourth trial to complete the 2^2 factorial experiment the full set of results indicated by the relationship in Figure 12.2 would be as in Table 12.4.

What benefits do we get from this enlarged experiment which we did not get from the classical design? One benefit is that the interaction, which could not be seen in Table 12.3, stands out very clearly in Table 12.4. By comparing the two rows of the table we see that:

(a) increase in temperature causes an increase in yield, if we are using a feedrate of 50 units; whereas

Table 12.4 Results of 2^2 factorial experiment

Feedrate	Temperature (°C)	
	60	70
40	76	72
50	70	78

(b) increase in temperature causes a decrease in yield, if we are using a feedrate of 40 units.

Alternatively, by comparing the two columns of Table 12.4 we see that:

(a) an increase in feedrate causes an increase in yield, if we are using a temperature of 70°C; whilst
(b) an increase in feedrate causes a decrease in yield, if we are using a temperature of 60°C.

Clearly the relationship between feedrate and yield depends upon the temperature used, or, alternatively, the relationship between temperature and yield depends upon the feedrate used. In a nutshell there is an interaction between feedrate and temperature. Perhaps the easiest way to spot an interaction between two factors is to compare the diagonals of a two-way table. In Table 12.4 the mean response for one diagonal is 77 whilst the mean response for the other diagonal is 71. Such a large difference (compared with some standard, which depends on residual variation) is an indication of an interaction.

A second benefit from carrying out the 2^2 factorial experiment rather than following the smaller classical design, is the increased accuracy with which we can estimate the effects of temperature and feedrate. We could of course use multiple regression to analyse the data from any experiment but with such simple designs the calculations can be carried out so easily that computing facilities are not needed. We will now consider how the effect estimates can be calculated.

12.4 ESTIMATION OF MAIN EFFECTS AND INTERACTION FROM THE RESULTS OF A 2^2 EXPERIMENT

We have already seen how the results of the classical experiment can be used to estimate the temperature effect and the feedrate effect. You may recall that we used

$$\text{Temperature effect} = \text{yield at high level of temperature}$$
$$- \text{yield at low level of temperature}$$
$$= 72 - 76 = -4$$

In the classical design we had three yield values but we were only able to use two of the three in the above calculation because the third yield value came from a trial in which a different level of feedrate (50) had been used. The classical design has an inherent lack of balance which is pointed out to us by the empty cell in Table 12.3. The 2^2 factorial design is balanced, in the sense that:

(a) we have two trials at each level of feedrate;
(b) we have two trials at each level of temperature.

To get the full benefit of the balanced design we change our method of calculation. At the same time we will refer to the effect of a variable as its main effect to distinguish it from interaction effects which we will also estimate.

> Main effect of a variable
>
> > = mean response at the high level of the variable
> >
> > − mean response at the low level of the variable

This formula allows us to use all four response values in calculating each of the main effects.

$$\text{Main effect of temperature} = \tfrac{1}{2}(72 + 78) - \tfrac{1}{2}(76 + 70)$$
$$= 2$$
$$\text{Main effect of feedrate} = \tfrac{1}{2}(70 + 78) - \tfrac{1}{2}(76 + 72)$$
$$= 0$$

These calculations could have been performed very easily using the row means and the column means of Table 12.4. These are included in Table 12.5 together with the diagonal means which can be used in the calculation of an estimate of the interaction.

It is clear from Table 12.5 that:

> Main effect of one variable = difference between row means
>
> Main effect of other variable = difference between column means

We can add to these formulae a third to be used for the estimation of the interaction effect.

Table 12.5 Calculation of effect estimates

Feedrate	Temperature (°C) 60	Temperature (°C) 70	Mean	
				----- 71
40	76	72	74	
50	70	78	74	
Mean	73	75	74	
				----- 77

$$\text{Interaction effect} = \text{difference between diagonal means}$$
$$= 77 - 71 = 6$$

The interaction effect could also be estimated using either:

$$\text{Interaction} = \tfrac{1}{2}(\text{effect of temperature at high level of feedrate}$$
$$- \text{ effect of temperature at low level of feedrate})$$

or

$$\text{Interaction} = \tfrac{1}{2}(\text{effect of feedrate at high level of temperature}$$
$$- \text{ effect of feedrate at low level of temperature})$$

All three formulae would give the same estimate of the interaction effect. Let us now bring together the results of our calculations which have given us three estimates:

$$\text{Temperature effect} = 2$$

$$\text{Feedrate effect} = 0$$

$$\text{Temperature} \times \text{feedrate interaction effect} = 6$$

We can interpret these estimates as follows:

(a) The effect of increasing temperature (by 10°C) is to increase the yield by 2 units, on average.
(b) The effect of increasing feedrate (by 10 units) is to cause no change in yield, on average.
(c) There is an interaction between feedrate and temperature. The existence of this interaction nullifies the value of statements (a) and (b) above. These statements are not incorrect, but they are worthless and potentially misleading. Whenever an interaction exists it is unwise to speak of either of the two main effects in isolation. The existence of the interaction tells us that the effect of temperature depends on the level of feedrate and that the effect of feedrate depends on the level of temperature. The relationship between the three variables, temperature, feedrate and yield is best described with reference to Table 12.5. This indicates that, for maximum yield, we should use either low values for both temperature and feedrate or high values for both independent variables.

12.5 DISTINGUISHING BETWEEN REAL AND CHANCE EFFECTS

From the results of the four trials in our 2^2 factorial experiment we have calculated three effect estimates. Had we carried out the four trials in a different order or on a different day, we would almost certainly have got different results and, consequently, different estimates. The estimated effects, therefore, are unlikely to be exactly equal to the true effects and we cannot dismiss the possibility that the true effects could be equal to zero. How are we to decide whether the independent variables really do influence the yield or not?

The reader will not be surprised to learn that we can use a significance test to distinguish between real and chance effects. The basis of this test will be clearer if we introduce a model to illustrate the relationship between the independent variables and the response. You will recall that, in earlier chapters, we used models such as

$$y = \alpha + \beta x + \text{error}$$

and

$$y = \alpha + \beta x + \gamma t + \text{error}$$

Such models were appropriate because we were fitting regression equations. For the method of analysis that we have used with the results of a 2^2 factorial experiment a rather different model is appropriate as in Table 12.6.

In this model y_1, y_2, y_3 and y_4 represent the observed yields whilst μ represents the true yield at the centre of the experimental region, i.e. with temperature equal to 65 and feedrate equal to 45. T represents the true temperature main effect, F represents the true feedrate main effect and I represents the true interaction effect. (It would have been appropriate to use Greek letters rather than T, F and I but the author is reluctant to introduce three more Greek letters at this point.) Each observed yield will differ from the true yield corresponding to the temperature and feedrate used in the manufacture of the batch. The differences between observed

Table 12.6 2^2 factorial experiments: a model

Feedrate	Temperature (°C) 60	70
40	$y_1 = \mu - \dfrac{T}{2} - \dfrac{F}{2} + \dfrac{I}{2} + e_1$	$y_2 = \mu + \dfrac{T}{2} - \dfrac{F}{2} - \dfrac{I}{2} + e_2$
50	$y_3 = \mu - \dfrac{T}{2} + \dfrac{F}{2} - \dfrac{I}{2} + e_3$	$y_4 = \mu + \dfrac{T}{2} + \dfrac{F}{2} + \dfrac{I}{2} + e_4$

and true yields are represented by the four errors (e_1, e_2, e_3 and e_4) in the model.

We have already calculated estimates for the two main effects and the interaction (i.e. we have estimated that $T = 2$, $F = 0$ and $I = 6$). Using the mean yield of all four cells we can estimate that μ is equal to 74.

Having estimated T, F, I and μ we can use the equations in Table 12.6 to calculate a predicted yield for each of the four cells:

$$\text{Predicted } y_1 = \mu - (T/2) - (F/2) + (I/2)$$
$$= 74 - (2/2) - (0/2) + (6/2) = 76$$

$$\text{Predicted } y_2 = \mu + (T/2) - (F/2) - (I/2)$$
$$= 74 + (2/2) - (0/2) - (6/2) = 72$$

$$\text{Predicted } y_3 = \mu - (T/2) + (F/2) - (I/2)$$
$$= 74 - (2/2) + (0/2) - (6/2) = 70$$

$$\text{Predicted } y_4 = \mu + (T/2) + (F/2) + (I/2)$$
$$= 74 + (2/2) + (0/2) + (6/2) = 78$$

If you compare these predicted yields with the actual yields in Table 12.5 you will see that they are identical. Our model fits the data perfectly. You will realize, on reflection, that this is no great achievement. If we had two points on a scatter diagram you would not be surprised to find that a straight line, $y = a + bx$, gives 100% fit. Similarly, with three points we would get 100% fit with $y = a + bx + cx^2$ or with $y = a + bx + cz$.

The simple truth is that we have extracted the maximum from the data whilst estimating μ, T, F and I. We cannot now estimate the four errors e_1, e_2, e_3, and e_4.

You could argue that the four errors are of little or no interest because each error is random and, therefore, never likely to be repeated. It is true that the individual errors are not important in themselves, but it is also true that a knowledge of the whole population of errors would be useful. If we knew, for example, that the errors came from a population which had a normal distribution with a mean of zero and a standard deviation of 2.5, say, then we would be able to decide whether the effect estimates were likely to have arisen simply by chance.

Unfortunately we cannot obtain an estimate of the error standard deviation from our 2^2 factorial experiment. To calculate such an estimate we would need the results of at least two trials carried out under identical conditions. As each of the four batches in our experiment was manufactured with a unique combination of temperature and feedrate we do not have a means of estimating the error standard deviation, or residual standard deviation as it is often called. A 2^n factorial experiment is a very

Table 12.7 Two replicates of 2^2 factorial experiment: a model

Feedrate	Temperature (°C)	
	60	70
40	$y_1 = \mu - \dfrac{T}{2} - \dfrac{F}{2} + \dfrac{I}{2} + e_1$	$y_2 = \mu + \dfrac{T}{2} - \dfrac{F}{2} - \dfrac{I}{2} + e_2$
	$y_5 = \mu - \dfrac{T}{2} - \dfrac{F}{2} + \dfrac{I}{2} + e_5$	$y_6 = \mu + \dfrac{T}{2} - \dfrac{F}{2} - \dfrac{I}{2} + e_6$
50	$y_3 = \mu - \dfrac{T}{2} + \dfrac{F}{2} - \dfrac{I}{2} + e_3$	$y_4 = \mu + \dfrac{T}{2} + \dfrac{F}{2} + \dfrac{I}{2} + e_4$
	$y_7 = \mu - \dfrac{T}{2} + \dfrac{F}{2} - \dfrac{I}{2} + e_7$	$y_8 = \mu + \dfrac{T}{2} + \dfrac{F}{2} + \dfrac{I}{2} + e_8$

efficient means of estimating main effects and interactions but it does not give an estimate of residual variation. There are several ways in which we could obtain an estimate of the residual standard deviation. One possibility is to repeat the whole of the 2^2 factorial experiment. The yield values from the four new batches could be combined with the yield values of the four original batches; then, from the enlarged set of data, we would be able to estimate the two main effects (F and T) and the interaction (I) plus the residual standard deviation. (It would be better to mingle the two factorial experiments and carry out the eight trials in random order, as we shall see later.) The benefits to be gained by carrying out two replicates of a 2^2 factorial experiment may be clearer if we examine the model in Table 12.7.

It should be clear from Table 12.7 that if the two yield values in any particular cell differ from each other, then this difference can only be due to random error and is not caused by changes of temperature or feedrate. We could, therefore, use the differences within the four cells to calculate the residual standard deviation. An alternative approach, which is more consistent with procedures we have used earlier, starts with the calculation of a mean and standard deviation for each of the four cells in Table 12.7. The cell means are then used to calculate the effect estimates whilst the cell standard deviations are combined to give an estimate of the residual standard deviation. You may recall that we combined several standard deviations in Chapter 7 using:

$$\text{Combined standard deviation} = \sqrt{\left\{\dfrac{\Sigma[(\text{d.f.})(\text{SD}^2)]}{\Sigma(\text{d.f.})}\right\}}$$

As each of our four standard deviations has 1 degree of freedom this equation becomes:

$$\text{Residual standard deviation (RSD)} = \sqrt{\frac{s_1{}^2 + s_2{}^2 + s_3{}^2 + s_4{}^2}{4}}$$

To illustrate this approach let us assume that we have carried out two replicates of the 2^2 factorial experiment and the results of the eight trials are the yield values in Table 12.8. For each of the four cells in Table 12.8 the mean and SD have been calculated. They are displayed in Table 12.9.

The four cell standard deviations can be combined as follows:

Residual standard deviation (RSD)

$$= \sqrt{\left[\frac{(1.414)^2 + (1.414)^2 + (4.243)^2 + (2.828)^2}{4}\right]}$$

$$= 2.739$$

If we were to manufacture a series of batches under identical conditions then we would expect the standard deviation of the batch yields to be approximately 2.739. This is not a very good estimate of batch to batch variability, of course, as it is based on only four degrees of freedom, but it is better than no estimate at all.

The four cell means in Table 12.8 are exactly equal to the four yield values in Table 12.5. This larger experiment will, therefore, give exactly the same effect estimates that we obtained from the original 2^2 factorial experiment (i.e. $T = 2$, $F = 0$ and $I = 6$). It is important to note, however, that we can place more confidence in these new estimates because each is based on eight yield values, whereas each of the original effect estimates was based on only four. This point is taken into account when we carry out

Table 12.8   Yields of eight batches in a 2×2^2 factorial experiment

Feedrate	Temperature (°C)	
	60	70
40	$y_1 = 77$ $y_5 = 55$	$y_2 = 73$ $y_6 = 71$
50	$y_3 = 67$ $y_7 = 73$	$y_4 = 80$ $y_8 = 76$

Table 12.9 Cell means and standard deviation for the data in Table 12.8

Feedrate	Temperature (°C)	
	60	70
40	$\bar{y}_1 = 76.0$ $s_1 = 1.414$	$\bar{y}_2 = 72.0$ $s_2 = 1.414$
50	$\bar{y}_3 = 70.0$ $s_3 = 4.243$	$\bar{y}_4 = 78.0$ $s_4 = 2.828$

a t-test to check the statistical significance of the effect estimates. The test statistic for the t-test is calculated as follows:

(effect estimate)/RSD   for a 2^2 factorial experiment
(effect estimate)/(RSD/$\sqrt{2}$)   for two replicates of a 2^2 factorial experiment.

When carrying out a t-test on the effect estimates obtained from p replicates of a 2^n factorial experiment the test statistic can be calculated from:

$$(\text{effect estimate})/[\text{RSD}/\sqrt{(p2^{n-2})}]$$

where RSD is the residual standard deviation. For a discussion of the above formula the reader should refer to Appendix F. We will make use of the formula to carry out t-tests on the effect estimates and we will start with the largest of the three.

Null hypothesis – there is no interaction between temperature and feedrate (i.e. $I = 0$).

Alternative hypothesis – there is an interaction between temperature and feedrate (i.e. $I \neq 0$).

$$\text{Test statistic} = \frac{|\text{effect estimate}|}{\text{RSD}/\sqrt{2}}$$

$$= \frac{6}{2.739/\sqrt{2}}$$

$$= 3.10$$

Critical values – from the two-sided t-table with 4 degrees of freedom:

2.78 at the 5% significance level
4.60 at the 1% significance level.

Decision – we reject the null hypothesis at the 5% level of significance.

Conclusion – we conclude that there is an interaction between temperature and feedrate.

Carrying out similar t-tests on the two main effect estimates would give inconclusive results in that we would be unable to reject the null hypothesis in either case. If, however, we had a better estimate of the residual standard deviation we might have found the temperature main effect to be significant also. A better estimate of the residual standard deviation could be obtained by either:

(a) carrying out further replicates of the 2^2 factorial experiment;
(b) manufacturing several batches of triacetate under fixed conditions (n batches would give an estimate of RSD with ($n - 1$) degrees of freedom);
(c) referring to the plant records to extract the yields of several batches which were produced during a period when conditions were stable. From these yields an estimate of the residual standard deviation could be calculated.

You might imagine that the third alternative would be dangerous. If we obtained an estimate of residual standard deviation that was much smaller than the true value then there would be an increased risk of declaring non-existent effects to be significant. In reality the reverse is more likely to occur. An estimate of residual standard deviation obtained from routine plant data is likely to be inflated because of unintended variation in operating conditions. Thus we would incur a greater risk of failing to detect real effects by using such an estimate.

Before we progress to consider a factorial experiment with three independent variables we will return to the results of the 2^2 factorial experiment for a second analysis. This time we will use multiple regression analysis.

12.6 THE USE OF MULTIPLE REGRESSION WITH A 2^2 EXPERIMENT

The results of the 2^2 factorial experiment are set out in Table 12.4. This style of tabulation proved to be very useful when we needed to calculate the row, column and diagonal means but it is not the type of table that we used in earlier chapters. Before we carry out a regression analysis we will

Table 12.10 Data from the 2^2 factorial experiment

	Temperature x	Feedrate z	Yield y
	60	40	76
	70	40	72
	60	50	70
	70	50	78
Mean	65.0	45.0	74.0
SD	5.77	5.77	3.65

set out the data in the manner (Table 12.10) that would be expected by a multiple regression package.

If we wish to analyse the data in Table 12.10 by means of our multiple regression package we must specify that y is the dependent variable whilst x and z are independent variables. In addition we will specify that the program should add the independent variables, one by one, to the equation without checking the statistical significance. In this way we can be sure that the final equation will contain all of the independent variables, including those which are significant and those which are not. The program responds by printing out the following equations:

$$y = 61 + 0.2x \qquad 10\% \text{ fit}$$

$$y = 61 + 0.2x + 0.0z \qquad 10\% \text{ fit}$$

At first sight these equations may appear to have no connection with the analysis carried out earlier on these same data. You will recall that we calculated the three effect estimates to be:

$$\text{temperature effect} = 2$$

$$\text{feedrate effect} = 0$$

$$\text{temperature} \times \text{feedrate interaction} = 6$$

On further inspection of the two alternative analyses we find a certain measure of agreement between them, the following points being worthy of note:

(a) The coefficient of x (i.e. $+0.2$) in the regression equations indicates a yield increase of 0.2 for a unit increase in temperature. This is in perfect agreement with our earlier estimate of $+2$ for the temperature effect since this latter figure relates to a total temperature change of 10 degrees.

Table 12.11 Correlation matrix for 2^2 experiment

	x	z	y
x	1.00	0.00	0.316
z		1.00	0.00
y			1.00

(b) The coefficient of z (i.e. 0.0) in the regression equations is in agreement with our earlier estimate of the feedrate effect (i.e. 0).
(c) The regression analysis has not embraced the temperature × feedrate interaction, but this can be rectified by introducing a cross-product term.

Perhaps our strongest memory of the previous chapter is the trouble that was caused by the intercorrelation of two independent variables. In fact the main reason for considering experimental designs was to obtain data in which such things did not arise.

To see how the 2^2 factorial design matches up to our requirements we can examine the correlation matrix printed out by the regression package in Table 12.11. We see that the correlation between the independent variables, x and z, is equal to zero. This is highly satisfactory.

Let us now extend our regression equation to include a cross-product term in order to assess any interaction between temperature and feedrate. To do this we must generate a new variable so that our complete set of data is that in Table 12.12.

If the data of Table 12.12 are fed into the multiple regression package they yield the following equations:

$$y = 61 + 0.2x \qquad 10\% \text{ fit}$$
$$y = 61 + 0.168235x + 0.000706xz \qquad 10.53\% \text{ fit}$$
$$y = 412 - 5.12x + 0.12xz - 7.8z \qquad 100\% \text{ fit}$$

The first equation is already familiar to us. It is identical with the first equation printed out earlier and it has the same percentage fit (10%).

You may have expected, however, that the cross-product variable (xz) would have been the first to enter the equation. After all, it was the interaction between feedrate and temperature which was earlier found to be significant whilst the two main effects were not.

When the cross-product term does enter the regression equation the percentage fit increases from 10% to 10.53%. This minute increase is in

Table 12.12 Extended data for the 2^2 experiment

	Temperature	Feedrate	Temperature × feedrate	Yield
	x	z	xz	y
	60	40	2400	76
	70	40	2800	72
	60	50	3000	70
	70	50	3500	78
Mean	65.0	45.0	2925.0	74.0
SD	5.77	5.77	457.35	3.65

stark contrast to the massive jump which occurs when the feedrate (z) finally enters the equation. When you recall that the feedrate effect was earlier estimated to be zero you will realize that multiple regression analysis has again revealed its darker side.

Once again the trouble was caused by intercorrelation of the independent variables. This is revealed if we examine the correlation matrix printed out by the regression package and reproduced in Table 12.13. Though there is no correlation between the two original variables, x and z, each one of these is correlated with the newly introduced variable (xz).

The two offending correlations are 0.568 and 0.821. Neither of these is as large as the 0.97 that we found between two independent variables when analysing a different set of data in the previous chapter. Nonetheless their combined effect is to give a series of equations which could be misleading and they serve to remind us yet again that the use of multiple regression analysis can be dangerous.

Fortunately, with data from a 2^2 factorial experiment the intercorrelation of the independent variables can be eliminated before equations are fitted. This is achieved by standardizing (or scaling or transforming) the values of the two independent variables x and z as follows:

Table 12.13 Correlation matrix

	x	z	xz	y
x	1.00	0.00	0.568	0.316
z		1.00	0.821	0.00
xz			1.00	0.240
y				1.00

Table 12.14   Transformed data

	X	Z	XZ	y
	−1	−1	+1	76
	+1	−1	−1	72
	−1	+1	−1	70
	+1	+1	+1	78
Mean	0.00	0.00	0.00	74.00
SD	1.00	1.00	1.00	3.65

$$X = (x - \bar{x})/(\text{half range of } x)$$
$$Z = (z - \bar{z})/(\text{half range of } z)$$

i.e.

$$X = (x - 65.0)/5.00$$
$$Z = (z - 45.0)/5.00$$

After carrying out this transformation the data from the 2^2 factorial experiment are as given in Table 12.14.

To see what effect this scaling operation has had let us re-examine the correlation matrix. The correlation coefficients for the scaled data are given in Table 12.15. Comparing Table 12.15 with Table 12.13 we see that the scaling operation has had two beneficial effects. The intercorrelation of the independent variables has been removed. The correlations in the y column of Table 12.15 now reflect the relative statistical importance of the three independent variables, as revealed by our earlier analysis. Submitting the transformed data to the multiple regression package results in the following equations being printed:

Table 12.15   Correlation matrix for transformed data

	X	Z	XZ	y
X	1.00	0.00	0.00	0.316
Z		1.00	0.00	0.000
XZ			1.00	0.949
y				1.00

$$y = 74.0 + 3.0XZ \qquad\qquad\qquad 90\% \text{ fit}$$

$$y = 74.0 + 3.0XZ + 1.0X \qquad\qquad 100\% \text{ fit}$$

$$y = 74.0 + 3.0XZ + 1.0X + 0.0Z \qquad 100\% \text{ fit}$$

This set of equations is much more reasonable. The first variable to enter the equation is XZ which represents the interaction effect. The percentage fit of this first equation is 90%. The second independent variable to enter the equation is X which adds a further 10% to the percentage fit whilst the third variable Z adds nothing.

The coefficients of the three variables in the third equation (3.0, 1.0, 0.0) are also in agreement with the effect estimates calculated earlier (6.0, 2.0, 0.0) if a factor of 2 is taken into account. This 2 arises because X, Z and XZ are varying from -1 to $+1$ which is a total change of 2 units.

By the use of variable transformation (or scaling as it is often called) we have managed to obtain regression equations which would lead us to conclusions rather similar to those suggested by the effect estimates calculated from the two-way table (Table 12.5). Without the use of scaling the multiple regression package gives us a sequence of equations which could lead someone to very different conclusions. It is possible, for example, that an inexperienced data analyst might conclude that the least important of the independent variables (i.e. z) was actually the most important.

In fairness to the statistical technique and in summary of this discussion the following points should be noted:

(a) The use of multiple regression analysis to analyse the data from a 2^2 factorial experiment is neither necessary nor desirable. As we have no way of feeding into the program an estimate of residual variance, no sensible t-tests can be carried out.

(b) As the 2^2 factorial experiment gives only four data points we are certain to get 100% fit when we include the three independent variables (x, z and xz) in the equation.

(c) By scaling (or transforming) the independent variables we get a much more meaningful set of equations. The need for scaling arises when we introduce the cross-product variable (xz) which is correlated with x and with z unless scaling is used.

(d) The use of scaling would not have eliminated the intercorrelation that was so troublesome in the previous chapter. That particular correlation was between two of the measured variables. It resulted from bad design and not from the introduction of extra variables.

A warning

Whenever cross-product variables and/or quadratic variables are introduced into a regression equation the independent variables should be scaled.

12.7 A 2^3 FACTORIAL EXPERIMENT

A research and development chemist is attempting to increase the tensile strength of a particular type of rubber so that it will be suitable for a new application. He realizes that several investigations may be required before a worthwhile increase in tensile strength is achieved and he limits the first investigation to examining the effects of certain changes in the formulation. He has selected three variables for inclusion in a 2^3 factorial experiment:

(a) the carbon black content;
(b) the type of accelerator;
(c) the percentage of natural rubber in the polymer.

A list of the eight trials is drawn up and random numbers are used to put the trials into random order. This is the order in which the trials are carried out and the results are listed in Table 12.16. The use of randomization in the execution of experiments is a means of avoiding unsuspected systematic errors. For a full discussion of randomization see Cox (1958).

The chemist does not expect to draw final conclusions from this experiment. If he had had such expectations he would have used a different design which will be discussed in a later chapter. He simply hopes that an analysis of the variation in tensile strength in Table 12.6 will enable him to decide what further trials need to be carried out. Note that in Table 12.16 the variables have been labelled x, z and w. In addition to renaming the variables we will also refer to the high level and the low level of each variable as in Table 12.17.

To facilitate the analysis of the results of this experiment we subtract 200 from each response value in Table 12.16 and transfer the reduced figure to a standard three-way table, Table 12.18.

Table 12.16 Results of a 2^3 factorial experiment

Carbon black (parts per hundred) x	Type of accelerator z	% natural rubber in polymer w	Tensile strength (kg/cm^2) y
20	HLX	50	236
40	HLX	75	284
40	MD	75	272
20	MD	75	232
40	MD	50	280
40	HLX	50	264
20	HLX	75	256
20	MD	50	248

Table 12.17 Levels of the independent variables

Variable	Low level	High level
x (carbon black)	20	40
z (type of accelerator)	MD	HLX
w (% natural rubber)	50	75

Table 12.18 Three-way table of response values

			z		
		Low		High	
		w		w	
x		Low	High	Low	High
Low		48	32	36	56
High		80	72	64	84

From Table 12.18 we could calculate estimates of the three main effects using the formula that we used with a 2^2 design:

Main effect of a variable

= mean response at the high level of the variable

− mean response at the low level of the variable

It would not, however, be obvious how we could calculate estimates of the interaction effects from this table, so we will first break down the three-way table into three two-way tables. This is achieved by averaging the response values in pairs to give Table 12.19.

Table 12.19 Two-way tables of mean response values

	(a)			(b)			(c)	
	z			w			w	
x	Low	High	x	Low	High	z	Low	High
Low	40	46	Low	42	44	Low	64	52
High	76	74	High	72	78	High	50	70

Some care is needed in completing the two-way tables. The entry of 40 in Table 12.19(a) is the mean of 48 and 32 from Table 12.18, whilst the entry 52 in Table 12.19(c) is the mean of 32 and 72 from Table 12.18. From each two-way table we can estimate two main effects and a two factor inter-action using the row means, column means and diagonal means as follows:

From Table 12.19(a) we can calculate estimates of:

$$\text{Main effect } x = \tfrac{1}{2}(74 + 76) - \tfrac{1}{2}(46 + 40) = 32$$

$$\text{Main effect } z = \tfrac{1}{2}(74 + 46) - \tfrac{1}{2}(76 + 40) = 2$$

$$\text{Interaction } xz = \tfrac{1}{2}(74 + 40) - \tfrac{1}{2}(76 + 46) = -4$$

From Table 12.19(b) we can calculate estimates of:

$$\text{Main effect } x = \tfrac{1}{2}(78 + 72) - \tfrac{1}{2}(44 + 42) = 32$$

$$\text{Main effect } w = \tfrac{1}{2}(78 + 44) - \tfrac{1}{2}(72 + 42) = 4$$

$$\text{Interaction } xw = \tfrac{1}{2}(78 + 42) - \tfrac{1}{2}(72 + 44) = 2$$

From Table 12.19(c) we can calculate estimates of:

$$\text{Main effect } z = \tfrac{1}{2}(50 + 70) - \tfrac{1}{2}(52 + 64) = 2$$

$$\text{Main effect } w = \tfrac{1}{2}(70 + 52) - \tfrac{1}{2}(64 + 50) = 4$$

$$\text{Interaction } zw = \tfrac{1}{2}(70 + 64) - \tfrac{1}{2}(50 + 52) = 16$$

Each of the three main effect estimates can be calculated from either of two tables. The estimate is of course the same whichever table is used. Since our 2^3 experiment contains three variables we can also estimate the three-variable interaction using the following calculation which is based on Table 12.18:

$$\begin{aligned}
\text{Interaction } xzw = \tfrac{1}{2}\{&\text{interaction } xw \text{ with } z \text{ at high level} \\
&- \text{interaction } xw \text{ with } z \text{ at low level}\} \\
= \tfrac{1}{2}\{&[\tfrac{1}{2}(84 + 36) - \tfrac{1}{2}(64 + 56)] \\
&- [\tfrac{1}{2}(48 + 72) - \tfrac{1}{2}(80 + 32)]\} \\
= &-2
\end{aligned}$$

In many situations it is difficult to attach any meaning to a three-variable interaction. The interaction xzw tells us 'how the interaction xw depends upon the level of z', or, alternatively, 'how the interaction xz depends upon the level of w', or, alternatively, 'how the interaction zw depends on the level of x'. Taking the first of these three statements and translating it into the language of our experiment the three-factor interaction tells us how the interaction between the carbon black content and the percentage of natural rubber depends upon the type of accelerator used. Whether or not it is possible for such an interaction to exist can only be decided on chemical/ physical grounds. We will return to this point later.

Drawing together the results of our calculations we have the seven effect estimates listed below:

Main effect x	$= 32$
Main effect z	$= 2$
Main effect w	$= 4$
Interaction xz	$= -4$
Interaction xw	$= 2$
Interaction zw	$= 16$
Interaction xzw	$= -2$

These numerical estimates will be more meaningful if we return to the language of the chemist, from the language of the statistician in which the list is written. The value of 32 for our estimate of main effect x implies that the effect of increasing carbon black from 20 parts/hundred to 40 parts/ hundred is to increase the tensile strength by 32 kg/cm^2. The second entry in the list implies that the effect of changing the accelerator type from MD to HLX increases the tensile strength by 2 kg/cm^2.

Interpreting a two-factor interaction effect is best achieved by reference to a two-way table. The nature of interaction zw, for which we have such a large effect estimate, can be illustrated by Table 12.19(c), which tells us that a high tensile strength will result from using the HLX accelerator and 75% natural rubber in the polymer blend, whilst the second best combination of these two factors is to use the MD accelerator with 50% natural rubber. Clearly the effect of increasing % natural rubber (w) is to increase the tensile strength when we use the HLX accelerator but to decrease tensile strength if we use the MD accelerator.

Having calculated the effect estimates it would be wise to check their statistical significance before acting upon the conclusions they appear to imply. Unfortunately we cannot carry out a t-test upon the estimates without a suitable standard deviation. We have already noted that a 2^n factorial experiment does not yield such an estimate. Let us see what we can achieve with multiple regression analysis.

12.8 THE USE OF MULTIPLE REGRESSION WITH A 2^3 EXPERIMENT

Before we explore various methods of obtaining an estimate of residual variance we will make use of our multiple regression package to analyse

Table 12.20 2^3 experiment – data scaled for regression analysis

	X	Z	W	XZ	XW	ZW	XZW	y
	−1	+1	−1	−1	+1	−1	+1	236
	+1	+1	+1	+1	+1	+1	+1	284
	+1	−1	+1	−1	+1	−1	−1	272
	−1	−1	+1	+1	−1	−1	+1	232
	+1	−1	−1	−1	−1	+1	+1	280
	+1	+1	−1	+1	−1	−1	−1	264
	−1	+1	+1	−1	−1	+1	−1	256
	−1	−1	−1	+1	+1	+1	−1	248
Mean	0.0	0.0	0.0	0.0	0.0	0.0	0.0	259.0
SD	1.00	1.00	1.00	1.00	1.00	1.00	1.00	18.19

the data from the 2^3 factorial experiment. A prerequisite for this course of action is to quantify the qualitative variable, 'type of accelerator (z)'. If we replace MD with −1 and replace HLX with +1 there will be no necessity for scaling of this variable. In the light of our earlier experience we will, of course, scale the other two measured variables using:

$$X = (x - 30.0)/10.00 \quad \text{and} \quad W = (w - 62.5)/12.50$$

Introduction of cross-product variables gives us the complete set of data in Table 12.20.

If we now specify that the dependent variable is y and the independent variables are X, Z, W, XZ, XW, ZW and XZW the regression package prints out the correlation matrix (Table 12.21) and the following equations:

Table 12.21 Correlation matrix for 2^3 experiment

	X	Z	W	XZ	XW	ZW	XZW	y
X	1.00	0.00	0.00	0.00	0.00	0.00	0.00	0.879
Z		1.00	0.00	0.00	0.00	0.00	0.00	0.055
W			1.00	0.00	0.00	0.00	0.00	0.100
XZ				1.00	0.00	0.00	0.00	−0.110
XW					1.00	0.00	0.00	0.055
ZW						1.00	0.00	0.440
XZW							1.00	−0.055
y								1.00

$$y = 259.0 + 16.0X \qquad\qquad 77.34\% \text{ fit}$$

$$y = 259.0 + 16.0X + 8.0ZW \qquad\qquad 96.68\% \text{ fit}$$

$$y = 259.0 + 16.0X + 8.0ZW - 2.0XZ \qquad\qquad 97.89\% \text{ fit}$$

$$y = 259.0 + 16.0X + 8.0ZW - 2.0XZ + 2.0W \qquad 99.09\% \text{ fit}$$

$$y = 259.0 + 16.0X + 8.0ZW - 2.0XZ + 2.0W$$
$$- 1.0XZW \qquad\qquad 99.40\% \text{ fit}$$

$$y = 259.0 + 16.0X + 8.0ZW - 2.0XZ + 2.0W$$
$$- 1.0XZW + 1.0XW \qquad\qquad 99.70\% \text{ fit}$$

$$y = 259.0 + 16.0X + 8.0ZW - 2.0XZ + 2.0W$$
$$- 1.0XZW + 1.0XW + 1.0Z \qquad 100.00\% \text{ fit}$$

There is substantial agreement between the regression analysis above and the effect estimates calculated earlier. The following points should be noted:

(a) The order in which the independent variables enter the regression equation corresponds exactly with the size of the effect estimates. For example, the largest estimate (32) indicates that carbon black (x) is the most important variable and we see that X is the first variable to enter the equation.

(b) The coefficients of the variables already in the regression equation do not change as new variables enter. This is to be expected when the intercorrelations amongst the independent variables are all equal to zero as we see in Table 12.21.

Do not forget that the automatic significance testing was suppressed whilst the regression package produced the succession of equations which culminated in 100% fit for the seventh equation. If we were to carry out the significance tests we would find that the first two independent variables, X and ZW, were significant but the third was not. This would lead us to the conclusion that the 'carbon black content (x)' and the 'interaction between type of accelerator and % natural rubber' were important. The significance tests are not based on a 'good' estimate of the residual variance because, as we noted earlier, no two trials were given the same treatment combination. Accepting the results of the significance tests is tantamount to assuming that the other effects (z, w, xz, xw and xzw) do not exist. Scientists and technologists are very reluctant to make such assumptions when analysing data by hand but they are often willing to let a computer program make such assumptions on their behalf.

We have returned to the point made earlier that a 2^n factorial experiment does not give us a reliable estimate of residual variance. In practice there are three ways in which this problem can be overcome:

(a) An estimate of residual variance may have been available before the experiment was carried out. We will assume that the research and development chemist carrying out this experiment had no such estimate.
(b) Further trials can be carried out using one or more of the eight treatment combinations or even using a new treatment combination that was not included in the original experiment. One obvious possibility is to repeat (or replicate) the whole of the 2^3 factorial experiment and we will investigate the usefulness of this strategy in the next section.
(c) An assumption can be made that one or more of the interactions could not possibly exist. If, for example we assume that the three factor interaction (i.e. interaction xzw) does not exist then it would be possible to calculate an estimate of the residual standard deviation that had 1 degree of freedom. Two points need to be stressed immediately:

   (i) An estimate of residual variance which is based on only 1 degree of freedom is of doubtful value (the critical value from the t-table using 1 degree of freedom is 12.71 at 5% significance).
   (ii) Any assumption concerning the non-existence of an interaction is clearly very dangerous and should certainly be based on chemical rather than statistical reasoning. Note, however, that such an assumption is conservative in the sense that, if the assumption is invalid, the residual will be inflated and we may fail to detect an effect which actually exists.

12.9 TWO REPLICATES OF A 2^3 FACTORIAL EXPERIMENT

The research and development chemist decides to replicate the 2^3 factorial experiment. He or she takes great care to use the same levels of the variables which were used in the first experiment and goes to considerable lengths to ensure that extraneous variables which were controlled in the first experiment are similarly controlled throughout the eight trials of this second experiment. He or she ensures, for example, that the natural rubber and the carbon black come from the same batch that was used previously and that the milling of the rubber is carried out at the same speed and for the same duration. The order of the eight trials is randomized and the response values are included with those from the first experiment in Table 12.22.

The two response values in any cell of Table 12.22 are not simply repeat determinations of tensile strength on the same batch but are single deter-

Table 12.22   Two replicates of a 2^3 factorial experiment

	z			
	Low		High	
	w		w	
x	Low	High	Low	High
Low	48	32	36	56
	52	32	32	48
High	80	72	64	84
	76	80	56	80

Table 12.23   Means and standard deviations of data in Table 12.22

	z			
	Low		High	
	w		w	
x	Low	High	Low	High
Low	50.0	32.0	34.0	52.0
	2.828	0.000	2.828	5.657
High	78.0	76.0	60.0	82.0
	2.828	5.657	5.657	2.828

minations on two batches produced with the same formulation. The variability within any cell is not caused by changes in the independent variables but is a manifestation of unassignable variation or error. The first step in our analysis of the above data is to compute a table of cell means and cell standard deviations, as in Table 12.23.

The cell standard deviations can now be combined to obtain an estimate of the residual standard deviation with 8 degrees of freedom.

$$\text{Estimate of RSD} = \sqrt{\{[(2.828)^2 + (0.000)^2 + (2.828)^2 + (5.657)^2 \\ + (2.828)^2 + (5.657)^2 + (5.657)^2 + (2.828)^2]/8\}} \\ = 4.000$$

Table 12.24 Two-way tables of mean response values

(a)			(b)			(c)		
	z			w			w	
x	Low	High	x	Low	High	z	Low	High
Low	41	43	Low	42	42	Low	64	54
High	77	71	High	69	79	High	47	67

Had all 16 batches been produced using the same formulation we could have expected a standard deviation of approximately 4.00 for the tensile strength measurements. The standard deviation of the measurements in Table 12.22 is, of course, much greater than 4.00 because of the variation due to the changes in the independent variables. We will now calculate effect estimates in an attempt to quantify the effects of the three independent variables.

The three-way table can, of course, be presented as three two-way tables (Table 12.24) to facilitate the calculation of effect estimates. Each entry in the two-way tables is the mean of two cell means from Table 12.23.

Using the row means, the column means and the diagonal means we can calculate the effect estimates as follows:

$$\text{Main effect } x = \tfrac{1}{2}(77 + 71) - \tfrac{1}{2}(41 + 43) = 32$$

$$\text{Main effect } z = \tfrac{1}{2}(43 + 71) - \tfrac{1}{2}(41 + 77) = -2$$

$$\text{Main effect } w = \tfrac{1}{2}(54 + 67) - \tfrac{1}{2}(64 + 47) = 5$$

$$\text{Interaction } xz = \tfrac{1}{2}(41 + 71) - \tfrac{1}{2}(43 + 77) = -4$$

$$\text{Interaction } xw = \tfrac{1}{2}(42 + 79) - \tfrac{1}{2}(42 + 69) = 5$$

$$\text{Interaction } zw = \tfrac{1}{2}(64 + 67) - \tfrac{1}{2}(54 + 47) = 15$$

Returning to the three-way table, Table 12.22, we can also calculate the three-variable interaction as:

$$\begin{aligned}\text{Interaction } xzw = &\tfrac{1}{2}\{[\tfrac{1}{2}(34 + 82) - \tfrac{1}{2}(60 + 52)] \\ &- [\tfrac{1}{2}(50 + 76) - \tfrac{1}{2}(78 + 32)]\} \\ = &-3\end{aligned}$$

The statistical significance of each effect estimate can be assessed by means of a t-test. We will first test the largest estimate, main effect x.

Null hypothesis – tensile strength of the rubber is not dependent on the carbon black content.

Alternative hypothesis – tensile strength of the rubber is dependent on the carbon black content.

$$\text{Test statistic} = \frac{\text{effect estimate}}{RSD/\sqrt{(p2^{n-2})}} \quad (\text{where } p = 2 \text{ and } n = 3)$$

$$= \frac{32.0}{4.00/\sqrt{4}}$$

$$= 16.0$$

Critical values – from the two-sided t-table with 8 degrees of freedom:

2.31 at the 5% significance level
3.36 at the 1% significance level
5.04 at the 0.1% significance level.

Decision – we reject the null hypothesis at the 0.1% level of significance.

Conclusion – we conclude that the tensile strength of the rubber is dependent on the carbon black content.

Each of the seven effect estimates can be subjected to the t-test. Continuing down the list we would conclude that:

(a) interaction zw is very highly significant (i.e. at 0.1%);
(b) main effect w is significant (i.e. at 5%);
(c) interaction xw is significant (i.e. at 5%);
(d) main effect z, interaction xz and interaction xzw are not significant.

In carrying out the above tests we are concerned with statistical significance rather than practical importance. We must now return to the world of reality and express our conclusions in the language of the research and development chemist. In the statement of the problem it was suggested that this experiment was a preliminary investigation which would provide information on which to base further experiments. Are we now in a position to specify treatment combinations which are worthy of investigation in our quest for increased tensile strength?

Starting with the main effects we see that x and w are both significant and that both of the effect estimates are positive. To increase tensile strength, therefore, we should adopt the high level of variable x and the high level of variable w. In practical terms we should use 40 parts per hundred of carbon black and we should use 75% natural rubber in the formulation. The significance of interaction xw directs our attention to Table 12.24(b), where we see that the high levels of both factors give the greatest value of tensile strength. This confirms what we have already established. The significance of interaction zw directs our attention to Table 12.24(c) which indicates that maximum tensile strength results from:

(a) either low level of variable z with low level of variable w;
(b) or high level of variable z with high level of variable w.

Since we have already decided that the high level of variable w is desirable, we will now adopt the high level of variable z, i.e. we will use the HLX accelerator.

In the light of the above conclusions, the research and development chemist might advance his investigation one stage further by carrying out a 2^2 factorial experiment using:

(a) 40 parts per hundred and 50 parts per hundred of carbon black;
(b) 75% and 85% of natural rubber;
(c) the HLX accelerator in all four trials.

This proposed experiment is, of course, just one of the many possibilities open to the chemist at this point. Whether he or she carries out this 2^2 factorial experiment or some alternative, he will be conscious of many constraints on his freedom to explore. Most of these constraints have not even been mentioned in this discussion.

12.10 MORE REGRESSION ANALYSIS

Let us use our regression package yet again. The data in Table 12.22 will be reanalysed so that we can compare the conclusions suggested by multiple regression analysis with the conclusions reached after using the t-test. Will we again decide that x, zw, xw and z are statistically significant?

To make the two analyses more comparable we will not specify a significance level for the regression package. The result will be that all of the independent variables are brought into the equation, starting with the most significant and finishing with the least significant. This is the strategy we have adopted in all of the regression analyses carried out in this chapter.

The data fed into the computer contain values of the four measured variables:

(a) tensile strength (y);
(b) carbon black content (x);
(c) type of accelerator (z);
(d) % natural rubber (w).

The three independent variables (x, z, w) are first scaled and then four cross-product variables are generated to give the full set of data in Table 12.25.

Table 12.25   Two replicates of a 2^3 experiment

X	Z	W	XZ	XW	ZW	XZW	y
−1	−1	−1	+1	+1	+1	−1	248
−1	−1	−1	+1	+1	+1	−1	252
−1	−1	+1	+1	−1	−1	+1	232
−1	−1	+1	+1	−1	−1	+1	232
−1	+1	−1	−1	+1	−1	+1	236
−1	+1	−1	−1	+1	−1	+1	232
−1	+1	+1	−1	−1	+1	−1	256
−1	+1	+1	−1	−1	+1	−1	248
+1	−1	−1	−1	−1	+1	+1	280
+1	−1	−1	−1	−1	+1	+1	276
+1	−1	+1	−1	+1	−1	−1	272
+1	−1	+1	−1	+1	−1	−1	280
+1	+1	−1	+1	−1	−1	−1	264
+1	+1	−1	+1	−1	−1	−1	256
+1	+1	+1	+1	+1	+1	+1	284
+1	+1	+1	+1	+1	+1	+1	280

The correlation matrix for the data in Table 12.25 is largely predictable and contains the highly desirable zero intercorrelations between the seven independent variables. As we would expect with such a matrix the sequence of regression equations shows no changes in the coefficients as variables are introduced. The equations are:

$y = 258.0 + 16.0X$ 75.29% fit

$y = 258.0 + 16.0X + 7.5ZW$ 91.84% fit

$y = 258.0 + 16.0X + 7.5ZW + 2.5XW$ 93.68% fit

$y = 258.0 + 16.0X + 7.5ZW + 2.5XW + 2.5W$ 95.51% fit

$y = 258.0 + 16.0X + 7.5ZW + 2.5XW + 2.5W - 2.0XZ$ 96.69% fit

etc.   etc.

If we compare these equations with the earlier analysis we find substantial agreement. The coefficient of X (16.0) in the first regression equation is equal to exactly half of the effect estimate (32.0) calculated earlier. The former tells us that an increase in tensile strength of 16 units can be expected from a carbon black increase of 10 units. The latter tells us that a tensile strength increase of 32 units can be expected from the total change of 20 units of carbon black used in the experiment. Clearly it is not difficult to misinterpret the effect estimate from either of the two methods of

Table 12.26 *t*-tests on the multiple regression equation

Independent variables in the equation	Test statistic	Critical values from *t*-table	
		5%	1%
X	6.53	2.14	2.98
X, ZW	5.13	2.16	2.01
X, ZW, XW	1.86	2.18	3.05
X, ZW, XW, W	2.18	2.20	3.10
X, ZW, XW, W, XZ	1.89	2.23	3.17

analysis. Fortunately it is easier to interpret the regression equation after the independent variables (*X*, *W*, etc.) have been descaled. Before the descaling is carried out, however, we must decide which of the terms in the regression equation are statistically significant. This can be achieved by means of a sequence of *t*-tests based on the changes in percentage fit. You may recall that, in the previous chapter, we calculated the test statistic for this test using:

Test statistic
$$= \sqrt{\{[(\text{new \% fit} - \text{old \% fit})(n - k - 1)]/(100 - \text{new \% fit})\}}$$

where *k* is the number of independent variables in the regression equation.

Using this formula again we get the test statistics in Table 12.26. From this table we see that the first independent variable (*X*) and the second independent variable (*XW*) are statistically significant at the 1% level of significance. The third and subsequent variables are not significant, even at the 5% level. Unfortunately this is not in agreement with the conclusion suggested by the *t*-tests which indicated that *x*, *zw*, *xw* and *z* were significant. What is the reason for this discrepancy?

Consider the regression analysis after the third variable has entered the equation. The percentage fit is now 93.68%. Subtracting this figure from 100% we find that 6.32% of the variation in tensile strength remains unexplained. This 6.32% appears in the denominator of the test statistic where it is being used as a yardstick against which to compare the increase in percentage fit. It could be argued that the use of this yardstick does not give the third variable a fair chance of getting into the equation. Whilst part of the 6.32% is undoubtedly due to 'error' it is also possible that part is due to the effect of an independent variable which is not yet in the equation.

Our regression analysis, therefore, is more conservative than the analysis based on t-tests since the increase in percentage fit is being tested against an inflated residual. Greater comparability of the two approaches can be obtained if we use a regression package which is based on 'variable rejection' rather than 'variable selection'. Such a package would first fit an equation containing all seven of the independent variables (X, Z, W, XZ, XW, ZW and XZW) and the residual standard deviation from this equation would actually be equal to that calculated from Table 12.23 (i.e. 4.00). The least significant variable would then be tested against this residual standard deviation and rejected from the equation if the test statistic was less than the critical values. If this variable is rejected from the equation its sum of squares is included in the residual sum of squares before the next variable is tested. After a succession of rejections a final equation is reached and with some sets of data this equation would include a larger number of independent variables than the equation arrived at by a 'variable selection' method. In other cases the two methods would lead us to exactly the same conclusion.

12.11   FACTORIAL EXPERIMENTS MADE SIMPLE

This is a very long chapter. I hope you have not got bogged down in the detail and lost sight of the essential simplicity of 2^n factorial designs. For those who have, I will attempt to restore confidence by offering an alternative view of these very useful experiments.

First, to reduce the quantity and complexity, you could ignore all the sections in this chapter which focus on regression analysis. These were included to give you a deeper understanding of experimental design and data analysis. If they obscure your view of the essentials, disregard them. Secondly, you could ignore all mention of significance testing. Of course, it is very important to check that any effect estimate is a reflection of a true effect, and not just the result of random variation; but this can be achieved using confidence limits rather than a t-test. When we calculate an effect estimate we are simply taking the difference between two means. You may recall that, in Chapter 5, we calculated confidence limits for the difference between two population means using the formula

$$|\bar{x}_1 - \bar{x}_2| \pm ts\sqrt{[(1/n_1 + (1/n_2)]}$$

With a two-level factorial experiment, n_1 and n_2 will both be equal to half of 2^n. If we have p replicates n_1 and n_2 will be half of $p2^n$ and the formula becomes

$$\text{Effect estimate} \pm ts/\sqrt{(p2^{n-2})}$$

In section 12.9 we had two replicates of a 2^3 factorial experiment, so p was equal to 2 and n was equal to 3. Thus the formula becomes

$$\text{Effect estimate} \pm ts/2$$

From the results of this experiment we calculated seven effect estimates and a residual standard deviation, 4.000, which had 8 degrees of freedom. Confidence intervals for the true effects are

$$
\begin{aligned}
\text{Main effect } x &= 32 \pm 4.62 = 27.38 \text{ to } 36.62 \\
\text{Main effect } z &= -2 \pm 4.62 = -6.62 \text{ to } 2.62 \\
\text{Main effect } w &= 5 \pm 4.62 = 0.38 \text{ to } 9.62 \\
\text{Interaction } xz &= -4 \pm 4.62 = -8.62 \text{ to } 0.62 \\
\text{Interaction } xw &= 5 \pm 4.62 = 0.38 \text{ to } 9.62 \\
\text{Interaction } zw &= 15 \pm 4.62 = 10.38 \text{ to } 19.62 \\
\text{Interaction } xwz &= -3 \pm 4.62 = -7.62 \text{ to } 1.62
\end{aligned}
$$

The confidence intervals for x, w, xw and zw do not include zero. Thus we can conclude that these main effects and interactions are statistically significant. On the other hand, the confidence intervals for z, xz and xzw do include zero, so we are unable to conclude that these effects are significant. These conclusions coincide exactly with those reached in section 12.9 when we used the t-test.

You will have noticed, I am sure, that all seven confidence intervals are the same width. As the width of the intervals is ± 4.62 it is clear that an effect estimate will only be significant if its magnitude exceeds 4.62. Thus we could refer to the 4.62 as the 'least significant effect'. Indeed we could have saved ourselves a lot of effort if we had calculated this least significant effect rather than carrying out t-tests or producing confidence intervals.

With p replicates of a 2^n factorial experiment the least significant effect is equal to $ts/\sqrt{(p2^{n-2})}$.

Let us now examine the efficiency of 2^n factorial experiments. Perhaps I could remind you that an effect estimate calculated from the results of such an experiment is simply the difference between two means. Thus it is appropriate to recall another formula from Chapter 5, which told us that the sample sizes needed to estimate the difference between two population means to within $\pm c$, are

$$n_1 = n_2 = 2(ts/c)^2$$

If we know the population standard deviation we could replace s by σ, and t would then be equal to 1.96 which we could replace by 2. The formula would then become

$$n_1 = n_2 = 8(\sigma/c)^2$$

If we now put c equal to σ we get $n_1 = n_2 = 8$. Thus with both sample sizes equal to 8 we can expect a confidence interval of plus or minus one standard deviation for the difference between the population means. This width of interval is often acceptable in practice. Thus the researcher might be quite happy, for example, to produce 8 batches of product at one temperature and 8 batches at a second temperature to estimate the effect of the temperature change. In the language of this chapter, he or she is carrying out 16 trials to estimate one main effect. This is very inefficient. In 16 trials we can carry out a 2^4 factorial experiment, which will allow us to estimate 4 main effects, 6 two-factor interactions, 4 three-factor interactions and 1 four-factor interaction.

It is true that the 2^4 factorial experiment would not give us an estimate of the residual standard deviation, whereas the simple experiment would yield an estimate with 14 degrees of freedom. However, as we have seen earlier, there are ways of coping with this problem. For example, we could assume that the three-factor and four-factor interactions did not exist and then use their effect estimates to obtain an estimate of the residual standard deviation with 5 degrees of freedom. Perhaps you do not like to make such assumptions. Alternatively, we could examine the 15 effect estimates and combine the small ones, to get a standard deviation against which to assess the large ones. Perhaps that appeals to you even less. Before you reject the suggestion, however, let me explain the subtle way in which it can be done, using the normal probability graph paper introduced in Chapter 3.

Suppose we carried out a 2^4 factorial experiment in a situation where the four independent variables had no effect whatsoever on the response variable. The true main effects would be zero and the true interaction effects would be zero. Of course, the calculated effects would not be zero, because the random variation in the response variable would work its way through the means into the estimates. You may wonder what values these 15 estimates would be likely to have. It can be shown that the effect estimates would resemble a random sample from a normal distribution with a mean of zero and a standard deviation of $\sigma/\sqrt{2^{n-2}}$, where σ is the standard deviation of the random variation in the response. Thus the effect estimates should lie roughly on a straight line when plotted on normal probability paper.

If, on the other hand, a small number of the effects really did exist, we would expect a small number of the points to deviate from the straight line on which the other points lay. Thus, if we carry out a 2^4 factorial experiment, then plot the 15 effect estimates on normal paper and find that 3 of the 15 points deviate considerably from the straight line pattern of the other 12, we can reasonably conclude that these 3 points represent statistically significant effects.

In practice we prefer to plot the effect estimates on half-normal paper rather than the normal probability paper we used in Chapter 3. The statistical significance of an effect estimate depends simply upon its magnitude and not upon its sign. Thus we ignore any minus signs as we put the estimates in order of magnitude, and then we plot them on half-normal graph paper.

The percentages used in the plotting are 3.3, 10.0, 16.7 ... 96.7%. These are calculated from $100(i - 0.5)/15$ for i equals 1 to 15. With half-normal paper the straight line should pass through the origin.

We can see in Figure 12.3 that three of the effect estimates do not lie close to the straight line that was drawn through the other 12. Thus we can reasonably conclude that main effect X, main effect Z and interaction XW are statistically significant. We can also use the straight line to estimate the residual standard deviation. We simply find the value on the effect axis which corresponds to 68% on the percentage axis. From Figure 12.3 we get a residual standard deviation of approximately 13.5.

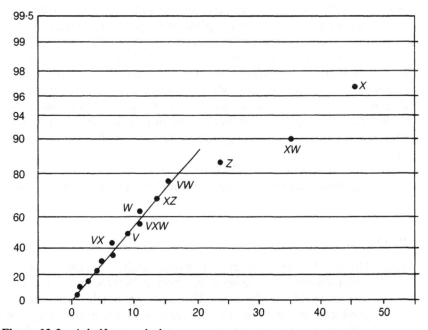

Figure 12.3   A half-normal plot

It is very easy to interpret Figure 12.3 because a large number of points can be fitted very well by a straight line through the origin, whilst the three other points are distinctly divergent from this pattern. In some cases the interpretation would not be so simple. If, for example, many of the points lay on a curve we might suspect that the response variable had a very skewed distribution. If many points lay on a straight line but this did not pass through the origin, we would be wise to explore the possibility that one of the response values was an outlier.

Clearly the half-normal plot offers us a very useful method for analysing the results of a 2^4 factorial experiment. It is even better with a 2^5 design. You should give serious consideration to including four or more independent variables in any industrial experiment. Before doing so, however, you should study the next chapter which might help you to reduce the size of your experiment.

12.12 SUMMARY

In this chapter we have examined 2^2 and 2^3 factorial experiments. By means of such experiments we can estimate the main effects of the variables and the interactions between the variables. These estimates are independent of each other as the independent variables are not correlated with each other in a 2^n factorial experiment. We can therefore reach more positive conclusions than we were able to reach in the previous chapter when analysing the results of the plant manager's experiment.

We have seen that regression analysis can be used to analyse the data from a factorial experiment but its indiscriminate use can lead us astray especially if the independent variables have not been scaled. When a 2^n factorial experiment has been designed and carried out according to plan then it is probably better to avoid the use of regression analysis. If, on the other hand, the experiment is incomplete or does not match up to the plan for some reason, the use of a regression package might help us to salvage a great deal.

In the next chapter we will consider ways in which the size of a factorial experiment can be reduced without sacrificing the most important estimates.

PROBLEMS

12.1   Dullness has been observed in recent batches of a particular dyestuff and a laboratory experiment has been called for in order to discover the cause of the dullness. The last ten batches that have been produced on the plant have brightness values in the range -4 to 0 with an average brightness of -2. It is important that the plant operating conditions be changed as soon as possible to obtain an average brightness of 0. (Brightness is measured on a scale from −7 to +7.)

It is decided that three variables will be included in the laboratory experiment the details of which are tabulated below:

Trial number	Speed of agitation	Reaction temperature (°C)	pH	Brightness
1	Fast	80	6	−2
2	Fast	90	8	0
3	Slow	90	6	−2
4	Fast	80	8	2
5	Slow	80	8	−4
6	Slow	80	6	0
7	Fast	90	6	−6
8	Slow	90	8	−4

(a) What is the name given to this type of experiment?
(b) What would you be able to estimate from the results of this experiment?
(c) Add 6 to the brightness values to eliminate the negative values.
(d) Put the results into a three-way table.
(e) Produce three two-way tables from the three-way table.
(f) Calculate estimates of the main effects and interaction effects from the two-way tables.
(g) Four earlier trials each with medium speed, 85°C and a pH of 7 were found to have a standard deviation of brightness equal to 1.15. Use this as an estimate of the residual standard deviation to test the significance of the effect estimates.

12.2 The data from Problem 1 were analysed using a multiple regression program. The actual values of temperature and pH were used and the speed was quantified as $0 =$ slow and $1 =$ fast. The interactions were generated by the program without coding the data. The following output was obtained, where s is speed, t is temperature, p is pH and y is brightness:

Correlation coefficients

	s	t	p	$s \times t$	$s \times p$	$t \times p$	$s \times t \times p$
y	0.20	−0.41	0.20	0.17	0.34	0.04	0.31

Correlation matrix

	s	t	p	$s \times t$	$s \times p$	$t \times p$	$s \times t \times p$
s	1.00	0.00	0.00	0.99	0.98	0.00	0.98
t		1.00	0.00	0.06	0.00	0.38	0.06
p			1.00	0.00	0.14	0.92	0.14
$s \times t$				1.00	0.98	0.02	0.98
$s \times p$					1.00	0.13	0.99
$t \times p$						1.00	0.15
$s \times t \times p$							1.00

Regression equations

$y = 15.00 - 0.20t$ 16.67%

$y = 14.2 - 0.20t + 0.23(s \times p)$ 28.43%

$y = 6.2 - 0.10t - 0.19(s \times t) + 2.50(s \times p)$ 77.08%

$y = 11.5 - 0.05t - 1.40p - 0.30(s \times t) + 3.80(s \times p)$ 94.60%

$y = 71.0 - 0.75t - 9.90p - 0.30(s \times t) + 3.80(s \times p)$
$\quad + 0.1(t \times p)$ 98.77%

$y = 76.0 - 10.0s - 0.8t - 10.0p - 0.2(s \times t)$
$\quad + 4.0(s \times p) + 0.1(t \times p)$ 100.0%

(a) Complete the following table which refers to values used by the computer in the multiple regression analysis:

Trial	s	t	p	$s \times t$	$s \times p$	$t \times p$	$s \times t \times p$
1	1	80	6	80	6	480	480
2		90	8				
3	0	90	6				
4		80	8				
5		80	8				
6		80	6				
7		90	6				
8		90	8				

(b) Examine the correlation matrix. Why are there so many high correlations?

(c) If the data were coded (-1 and $+1$), what correlation coefficient would be given by s and $(s \times t)$?

(d) Using the percentage fit statistic, carry out a significance test at each stage of the regression.

(e) If a 5% significance level had been stipulated as a criterion for cessation of the stepwise regression analysis, what conclusions would have been reached?

(f) Using the final multiple regression equation, calculate a predicted value for brightness with a fast agitator speed, a temperature of 90 and a pH of 8.

(g) Why was a 100% fit obtained without the inclusion of the three-variable interaction?

13

Adapting the simple experiments

13.1 INTRODUCTION

In the previous chapter we examined the 2^2 and 2^3 factorial experiments. These experiments are useful if we have two or three independent variables and we wish to explore only two values of each variable. You may recall that the plant manager, whose problems were discussed in Chapters 10 and 11, included five independent variables in his experiment and, furthermore, he used at least three values for each variable. As the plant manager only carried out ten trials (i.e. he produced ten batches) it is small wonder that his experiment lacked the admirable qualities that we find in a factorial experiment. On the other hand it would be out of the question to carry out an experiment involving 243 batches, which is the number required for a 3^5 factorial experiment.

In this chapter we will explore the possibility of getting some of the advantages of a factorial experiment without undertaking the whole of the experiment.

13.2 THE DESIGN MATRIX

In the previous chapter we analysed the results from several factorial experiments and for each experiment we used two methods of analysis. The first was based on two-way tables and the second made use of a multiple regression package. Both methods have their advantages but there are other methods which are found to be even more useful when dealing with parts of factorial experiments as we shall see later in this chapter.

Prominent amongst these other methods of analysis is the very popular Yates' technique. Unfortunately the Yates' approach is expressed in a mathematical notation which has to be mastered before analysis can begin. We will not, therefore, discuss Yates' technique in this book, preferring

Table 13.1 2^3 experiment – design matrix and response vector

X	Z	W	XZ	XW	ZW	XZW		y
−1	+1	−1	−1	+1	−1	+1		236
+1	+1	+1	+1	+1	+1	+1		284
+1	−1	+1	−1	+1	−1	−1		272
−1	−1	+1	+1	−1	−1	+1		232
+1	−1	−1	−1	−1	+1	+1		280
+1	+1	−1	+1	−1	−1	−1		264
−1	+1	+1	−1	−1	+1	−1		256
−1	−1	−1	+1	+1	+1	−1		248

instead to make use of what is known as the design matrix. This is basically very simple and involves very little notation; in fact we have already used several design matrices in the previous chapter without referring to them as such. One is reproduced as Table 13.1.

Table 13.1 above is very similar to Table 12.20. The main difference between the two is the separation of the y column from the other seven columns. By making this split we are distinguishing between the dependent variable (y) and the independent variables (X, Z, W, XZ, XW, ZW and XZW). To emphasize this split we refer to the y column as the **response vector** and to the other columns together as the **design matrix**. These names are very suitable as the design matrix is determined entirely by the design of the experiment and could be written down before the experiment is carried out, whereas the response vector contains the results of the experiment and is therefore not available until after the experiment is completed.

If we are contemplating carrying out a particular experiment we can set out the design matrix in advance and by examining its peculiarities we can determine whether or not the design meets our objectives. We will use the design matrix in this way later in the chapter but first we will make use of Table 13.1 to calculate the effect estimates which were obtained by two-way tables in the previous chapter.

To calculate the main effect of variable x we use the response vector and the X column of the design matrix as follows:

Main effect $x = (−236 + 284 + 272 − 232 + 280 + 264 − 256 − 248)/4$

$\qquad = 32.0$

In this calculation the numbers are taken from the response vector and the signs are taken from the X column of the design matrix. The other six columns of the matrix can be used in a similar manner. For example, the Z column of the design matrix together with the response vector can be used to calculate an estimate of main effect z:

Main effect $z = (+236 + 284 - 272 - 232 - 280 + 264 + 256 - 248)/4$
$$= 2.0$$

Clearly we can also calculate estimates of the other main effect (w), the two-variable interactions (xz, xw and zw) and the three-variable interaction (xzw). In each case we would obtain exactly the same value as that calculated from the two-way tables in the previous chapter. The design matrix offers us a simple way of analysing the results of a factorial experiment but the main use to which we will put the design matrix is to compare alternative experiments before they have been carried out. We will pursue this in the next section.

13.3   HALF REPLICATES OF A 2^n FACTORIAL DESIGN

In many situations it may not be possible to carry out a factorial experiment to investigate the factors in which we are interested. The prohibiting constraint may be the time or cost if we wish to examine several factors. It is unfortunately true that a 2^n factorial experiment requires a large number of trials if n is large, as we see in Table 13.2.

Another criticism of the factorial experiment is that it may give us estimates that we do not require. This is not true of the 2^2 and 2^3 experiments we have considered so far but would certainly be true of a 2^6 factorial experiment. Imagine that a researcher wished to investigate the effect of six factors on the impurity in a synthesized product. If he was prepared to limit each factor to only two levels the 2^6 factorial experiment would involve 64 trials which could be very time-consuming and/or very expensive. To compensate the researcher for his or her labours what estimates would be obtained from the experiment? They are rather numerous as the list below shows:

  6 main effects
15 two-factor interactions
20 three-factor interactions
15 four-factor interactions
  6 five-factor interactions
  1 six-factor interaction.

Table 13.2   The number of trials in 2^n factorial experiments

Number of factors (n)	2	3	4	5	6	7
Number of trials (2^n)	4	8	16	32	64	128

It is very doubtful if all of these estimates would be required. Indeed it might be very difficult to find any meaning for a four-factor interaction if it was shown to be significant.

Since the 2^6 experiment is, on the one hand, too large and on the other hand, too productive of effect estimates it is natural to ask 'Can I carry out part of the 2^6 factorial experiment in order to get the estimates I require whilst losing those which I neither need nor understand?' The answer is 'Yes', and we can use the design matrix to help the researcher to select which part of the 2^6 factorial experiment he or she will carry out.

To illustrate the usefulness of the design matrix for this purpose let us return to the 2^3 factorial experiment first described in the previous chapter. Suppose that the research and development chemist carrying out the experiment had been forced to call a halt after the first four trials had been completed. Since the full series of eight trials constitutes a 2^3 factorial experiment we refer to these four trials as a half replicate of a 2^3 factorial experiment. (Similarly we would call any two of the eight trials, a quarter replicate.)

What can the research and development chemist salvage from the half replicate? Can he or she calculate estimates of the three main effects and the four interactions? The answers to these questions must lie in the top four rows of the design matrix in Table 13.1, which are reproduced in Tables 13.3 and 13.4 with the corresponding response values.

Table 13.3 Design matrix for a half replicate of the 2^3 experiment

$$
\begin{array}{ccccccc}
X & Z & W & XZ & XW & ZW & XZW \\
\left[\begin{array}{ccccccc}
- & + & - & - & + & - & + \\
+ & + & + & + & + & + & + \\
+ & - & + & - & + & - & - \\
- & - & + & + & - & - & +
\end{array}\right]
\end{array}
\qquad
\begin{array}{c}
y \\
\left[\begin{array}{c}
236 \\
284 \\
272 \\
232
\end{array}\right]
\end{array}
$$

Table 13.4 Three-way table for a half replicate of the 2^3 experiment

		Z		
	Low		High	
	W		W	
X	Low	High	Low	High
Low		232	236	
High		272		284

Tables 13.3 and 13.4 present the same information in alternative forms and either the design matrix or the three-way table could be used to calculate effect estimates. Using the X column of the design matrix we get:

$$\text{Main effect } x = (-236 + 284 + 272 - 232) \div 2$$
$$= 44$$

and, using the Z column, we can estimate the second main effect:

$$\text{Main effect } z = (+236 + 284 - 272 - 232) \div 2$$
$$= 8$$

When we come to estimate main effect w we notice that the W column contains three plusses and only one minus. Clearly this particular half replicate is not balanced as was the full 2^3 factorial experiment from which it was taken. Whereas each of the seven columns in Table 13.1 contains four pluses and four minuses there is no such balance of plus and minus signs in Table 13.3. This does not prevent us from calculating effect estimates but it does mean that we cannot use simplified formulae similar to those which are applicable to the full factorial experiment. With an unbalanced design we would have to work from first principles in calculating effect estimates and sums of squares. Furthermore we would get less precise estimates than could be obtained from a balanced design of the same size and these estimates would not be independent because of inter-correlation of the columns in the design matrix.

If this particular half replicate is lacking in the fine qualities that we found in the full factorial experiment, can we find a different half replicate which is more desirable? Thinking in terms of a balanced design, can we find four rows in Table 13.1 such that each of the seven columns will contain two plus and two minus signs? Unfortunately this is not possible. Indeed, it would be rather optimistic to expect all the benefits of a 2^3 design from only four trials. Perhaps we should concentrate on getting 'balanced' estimates of the three main effects. We can see in Table 13.4 the nature of the unbalance of this particular half replicate. Whilst the four response values are equally shared between the two rows they are not equally shared between the four columns. If we changed one of the four trials a more

Table 13.5 Design matrix for a second half replicate

X	Z	W	XZ	XW	ZW	XZW	y
−	+	−	−	+	−	+	236
+	+	+	+	+	+	+	284
+	−	+	−	+	−	−	272
−	−	−	+	+	+	−	248

Table 13.6   Three-way table for a second half replicate

	Z			
	Low		High	
	W		W	
X	Low	High	Low	High
Low	248		236	
High		272		284

balanced design would result. The half replicate in Tables 13.5 and 13.6 is certainly better balanced than the one we have just considered.

From this half replicate we can calculate the following estimates, using either the design matrix or the three-way table:

$$\text{Main effect } x = 36$$

$$\text{Main effect } z = 0$$

$$\text{Main effect } w = 36$$

$$\text{Interaction } xz = 12$$

$$\text{Interaction } xw = ??$$

$$\text{Interaction } zw = 12$$

$$\text{Interaction } xzw = 0$$

You will notice in the above list that there is no estimate for interaction xw. It is not possible to calculate an estimate since the XW column of the design matrix contains four plus signs. You will also notice that the six estimates we have calculated fall into three pairs, i.e. we have the same value for main effect x and for main effect w, etc. This is not just a coincidence, as Table 13.5 shows. In the design matrix the X column and the W column are identical. Similarly, the Z column and the XZW column are identical. We summarize this situation by saying that:

xw is the defining contrast

$$\left.\begin{array}{l} (x \text{ and } w) \\ (z \text{ and } xzw) \\ (xz \text{ and } zw) \end{array}\right\} \text{ are alias pairs}$$

The two effects which constitute any alias pair are inseparable from each other. The calculated estimate of 36, for the (X and W) alias pair tells us

Table 13.7 Design matrix for a third half replicate

X	Z	W	XZ	XW	ZW	XZW		y
−	+	−	−	+	−	+		236
+	+	+	+	+	+	+		284
−	−	+	+	−	−	+		232
+	−	−	−	−	+	+		280

Table 13.8 Three-way table for a third half replicate

		Z		
	Low		High	
	W		W	
X	Low	High	Low	High
Low		232	236	
High	280			284

that a change in tensile strength of 36 kg/cm^2 can be attributed to the change made to variable x (carbon black content) or to the change made to variable w (% natural rubber) or to both. As you can see in Table 13.5, the X column and the W column are identical. An experiment in which a change in one variable is always accompanied by a change in a second variable will never allow us to separate the effect of the two variables.

Since the least important of all the seven estimates is interaction XZW we will choose a half replicate which has interaction XZW as the defining contrast. To do this we refer to Table 13.1 and select the four trials which have a plus sign in the XZW column. This third example of a half replicate is set out in Tables 13.7 and 13.8.

Without calculating effect estimates we can see which effects will be aliased with each other by spotting pairs of identical columns in the design matrix. In Table 13.7 we see that:

Interaction xzw is the defining contrast
Main effect x is aliased with interaction zw
Main effect z is aliased with interaction xw
Main effect w is aliased with interaction xz.

With this half replicate, then, we could estimate main effect x if we were able to assume that interaction zw did not exist, and we could estimate

Table 13.9 Design matrices for 2^2, 2^3 and 2^4 factorial experiments

X	Z	XZ	W	XW	ZW	XZW	V	XV	ZV	XZV	WV	XWV	ZWV	XZWV
−	−	+	−	+	+	−	−	+	+	−	+	−	−	+
+	−	−	−	−	+	+	−	−	+	+	+	+	−	−
−	+	−	−	+	−	+	−	+	−	+	+	−	+	−
+	+	+	−	−	−	−	−	−	−	−	+	+	+	+
−	−	+	+	−	−	+	−	+	+	−	−	+	+	−
+	−	−	+	+	−	−	−	−	+	+	−	−	+	+
−	+	−	+	−	+	−	−	+	−	+	−	+	−	+
+	+	+	+	+	+	+	−	−	−	−	−	−	−	−
−	−	+	−	+	+	−	+	−	−	+	−	+	+	−
+	−	−	−	−	+	+	+	+	−	−	−	−	+	+
−	+	−	−	+	−	+	+	−	+	−	−	+	−	+
+	+	+	−	−	−	−	+	+	+	+	−	−	−	−
−	−	+	+	−	−	+	+	−	−	+	+	−	−	+
+	−	−	+	+	−	−	+	+	−	−	+	+	−	−
−	+	−	+	−	+	−	+	−	+	−	+	−	+	−
+	+	+	+	+	+	+	+	+	+	+	+	+	+	+

main effect z if we were able to assume that there was no interaction between variable x and variable w. These assumptions would not be made on statistical grounds, of course. One wonders if there are any situations in which such assumptions could reasonably be made and for this reason the half replicate of a 2^3 experiment is not a widely used design. Though we have used a 2^3 factorial experiment to illustrate the ideas underlying fractional replication, the practical benefits of the technique will be enjoyed by the researcher who wishes to investigate the effect of four or more factors.

The whole of Table 13.9 constitutes a design matrix for a 2^4 factorial experiment. The sections which are boxed off constitute design matrices for a 2^3 experiment and a 2^2 experiment. To obtain the design matrix for a half replicate of a 2^4 factorial experiment we select eight rows of Table 13.9, using the full width of the table. If we want a half replicate in which interaction $xzwv$ is the defining contrast we select those eight rows which have a plus sign in the last column. Careful examination of these eight rows would indicate that:

Main effect x was aliased with interaction zwv
Main effect z was aliased with interaction xwv
Main effect w was aliased with interaction xzv
Main effect v was aliased with interaction xzw

Interaction xz was aliased with interaction wv
Interaction xw was aliased with interaction zv
Interaction xv was aliased with interaction zw.

With this particular half replicate of a 2^4 factorial design we can estimate each main effect if we can assume that the aliased three-factor interaction is negligible. We can also estimate a two-factor interaction if we are able to assume that the other two-factor interaction, with which it is aliased, does not exist.

When we have five or more factors a half replicate of the 2^n factorial experiment can yield estimates of all the main effects and two-factor interactions which are not aliased with each other. In a half replicate of a 2^5 design, with the five-variable interaction chosen as the defining contrast, we find that:

(a) each of the five main effects is aliased with a four-factor interaction;
(b) each of the ten two-factor interactions is aliased with a three-factor interaction.

13.4 QUARTER REPLICATES OF A 2^n FACTORIAL DESIGN

If we wish to investigate many independent variables, even a half replicate of the full factorial experiment may be too large. In this case we could consider carrying out only a quarter of the 2^n trials. Clearly a quarter replicate will yield much less information than would a half replicate but we may, nonetheless, be able to estimate the main effects and the two variable interactions.

To illustrate certain points let us consider the first two trials of the 2^3 factorial experiment set out in Table 13.1. These two trials constitute a quarter replicate of the full 2^3 experiment and are summarized in Table 13.10.

You will notice that in the design matrix of Table 13.10 three of the seven columns contain two plus signs. It is not possible therefore to estimate the three effects which correspond to these columns and we find that the quarter replicate of the 2^3 experiment has three defining contrasts, z, xw and xzw. The other four columns of the design matrix are identical and we can therefore calculate only one effect estimate:

Table 13.10 Design matrix for a quarter replicate of the 2^3 experiment

$$
\begin{array}{ccccccc}
X & Z & W & XZ & XW & ZW & XZW \\
\end{array}
\qquad y
$$

$$
\begin{bmatrix} -1 & +1 & -1 & -1 & +1 & -1 & +1 \\ +1 & +1 & +1 & +1 & +1 & +1 & +1 \end{bmatrix}
\qquad
\begin{bmatrix} 236 \\ 284 \end{bmatrix}
$$

$$-236 + 284 = 48.0$$

Thus the four effects (x, w, xz and zw) form an alias group and we are unable to say which one of the four has given rise to the increase in tensile strength of 48 units.

Clearly a quarter replicate of a 2^3 factorial experiment is quite useless but this particular example has served to illustrate two important points that are true of all quarter replicates:

(a) There are three defining contrasts.
(b) Each effect will fall into an alias group containing three other effects.

A quarter replicate of a 2^4 factorial experiment would consist of four trials compared with the 16 trials of a full 2^4 experiment. It would have three defining contrasts and three alias groups each containing four effects. No matter which four trials (of the possible 16) were carried out it would not be possible to separate even the four main effects.

A quarter replicate of a 2^5 experiment would consist of eight trials compared with the 32 trials in a full 2^5 experiment. The quarter replicate would have three defining contrasts and there would be seven alias groups, each containing four effects. It could be arranged that each of the five main effects was in a separate alias group but it would not be possible to ensure that each of the ten two-variable interactions was separated in this way.

If we wished to investigate the effect of seven independent variables then a quarter replicate of the full 2^7 experiment would require 32 trials. From the results of such an experiment we would be able to estimate the seven main effects and the 21 two-variable interactions independently. The $\frac{1}{4} \times 2^7$ design might therefore be more attractive than a full 2^7 design, to a researcher who wished to estimate only the main effects and the two-variable interactions.

13.5 A USEFUL METHOD FOR SELECTING A FRACTION OF A 2^n FACTORIAL EXPERIMENT

We have seen that inspection of the design matrix is a very powerful tool for evaluating the effectiveness of any experimental design before the experiment is carried out. Intercorrelation of the columns of the matrix indicate deficiencies which it may be possible to eradicate before it is too late. When the experiment is a fraction of a 2^n factorial we may get correlations of $+1.0$ or -1.0 and the effects then fall into alias groups. Inspecting the design matrix enables us to identify the alias groups and the defining contrasts.

The design matrix is not so useful to someone trying to design an experiment as to someone evaluating a design already written down. This point would become very clear if you pored over the 127 columns of a 2^7 design matrix trying to extract a suitable quarter replicate. At the design stage it is easier to use the following method which is carried out in two steps:

(a) choose defining contrast(s);
(b) generate alias groups by 'multiplying' each effect by each defining contrast.

Suppose, for example, that we wish to design a half replicate of a 2^3 factorial experiment in which the three independent variables are represented by X, W and Z. A half replicate has only one defining contrast so we will choose the three-factor interaction XWZ. Now we must 'multiply' each of the main effects (X, Z, W) and each of the interaction effects (XZ, XW, ZW, XZW) by the defining contrast. In performing these multiplications we use the normal rules of algebra plus the additional rule that $X^2 = 1$, $W^2 = 1$ and $Z^2 = 1$. The multiplication proceeds as follows:

$$X \times XZW = X^2ZW \quad = ZW \quad \text{(as } X^2 = 1)$$
$$Z \times XZW = XZ^2W \quad = XW \quad \text{(as } Z^2 = 1)$$
$$W \times XZW = XZW^2 \quad = XZ \quad \text{(as } W^2 = 1)$$
$$XZ \times XZW = X^2Z^2W \quad = W$$
$$XW \times XZW = X^2ZW^2 \quad = Z$$
$$ZW \times XZW = XZ^2W^2 \quad = X$$
$$XZW \times XZW = X^2Z^2W^2 = \, ?$$

The result of the first multiplication ($X = ZW$) tells us that main effect X and interaction ZW constitute an alias pair. This same indication is given by the sixth multiplication so we have clearly done more work than was necessary. The other two alias pairs are seen to be Z with XW, and W with XZ.

Having listed the alias pairs it is now easy to write out the design matrix for the four trials which constitute the half replicate. We know that each column except XZW must contain two plus signs and two minus signs. We also know that the X, W and Z columns must have zero correlation with each other. If we write $+ + - -$ for X, $+ - + -$ for Z and $- + + -$ for W the other columns can be obtained by multiplication.

As a second example let us design a quarter replicate of a 2^4 factorial experiment using the letters A, B, C and D to represent the independent variables. For this exercise we will, of course, require three defining contrasts. We choose two of the higher interactions as our first two defining contrasts and then we obtain the third by multiplication. If we choose the three-factor interaction ABC as the first defining contrast and the three-factor interaction ACD as the second then by multiplication we obtain the third:

$$ABC \times ACD = A^2BC^2D = BD$$

We must now multiply each main effect and interaction by all three defining contrasts. Starting with main effect A we get:

$$A \times ABC = A^2BC = BC$$
$$A \times ACD = A^2CD = CD$$
$$A \times BD \quad = ABD$$

Thus A, BC, CD and ABD form an alias group. Multiplying main effect B and then main effect C in the same way gives us the other two alias groups:

$$B, AC, D, ABCD \quad \text{and} \quad C, AB, BCD, AD$$

13.6 FINDING OPTIMUM CONDITIONS

I hope I have not given the impression that all industrial problems can be solved by carrying out one experiment. This would be wrong on two counts. Firstly, there are problems for which experimentation would be irrelevant. Secondly, there are situations in which, though experimentation is desirable, it might be better to do a series of small experiments rather than one large one.

In agriculture the researcher is under great pressure to 'get it right first time'. Because crops have an annual cycle, any deficiency in the planned experiment may result in 12 months' delay in completing the project. In industry, however, we might be able to complete an experiment in a week, or even a day. Why then should we put all our eggs in one basket? Surely, it will often be safer to allocate say 30% of our budget to the first experiment, then plan the second when we have analysed the results, then carry out a third after the second, etc.

This strategy is particularly useful when we seek optimum conditions for a production process. As each experiment is completed we learn more about the effects of the independent variables and we are able to choose better operating conditions. Better in the sense that they give higher yield, or lower impurity, or reduced cost, perhaps. The procedure has been described as 'hill climbing' since it is not unlike trying to climb a hill which is shrouded in heavy mists. We cannot see the summit, but we aim to reach it as quickly as possible by following the line of steepest ascent. Initially, the slope is steep and we have a clear sense of direction. However, as we approach the summit there is little slope and it is difficult to know which way to walk. Indeed we might pass close to the summit and start to descend without realizing we were doing so.

When seeking optimum conditions for a production process our first experiment would be a two-level factorial experiment. If the number of independent variables were five or more we would probably use a fraction of the full factorial design. From the results of this first experiment we might well eliminate one or more of the independent variables and get a clear indication of how we should change the levels of the others. At this point we could carry out two or three trials to confirm our findings or we could plunge directly into the second experiment. Thus we would carry out a series of two-level factorial experiments until the latest experiment failed to give us a clear indication of what further changes should be made. Now we would carry out a different type of experiment to locate the optimum conditions. This would require the use of three or more levels for the independent variables.

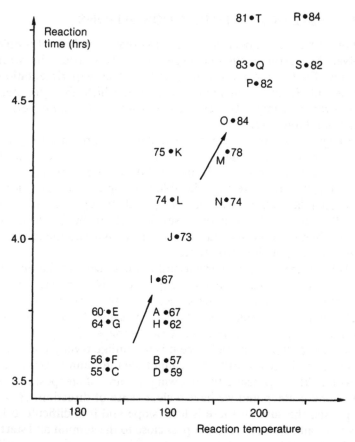

Figure 13.1 The hill-climbing approach

The hill-climbing strategy can be illustrated graphically if we have only two independent variables. Figure 13.1 displays the results of a series of experiments. These were carried out in order to find operating conditions which would give maximum yield of a product made by a batch process. Three independent variables were included in the first experiment. These were reaction temperature, catalyst concentration and reaction time. The results of the first experiment indicated that the catalyst concentration had no effect on the yield, so it was not included in the later experiments and is not shown in Figure 13.1. The eight points labelled A to H represent the eight trials of a 2^3 factorial experiment. The number written next to each point indicates the yield obtained from that batch.

Analysis of the results from this first experiment leads us to the conclusion that we should travel in the direction indicated by the arrow, which points towards I and J. Batches I and J were produced in order to confirm the findings from the first experiment. They clearly do so. The second experiment is a 2^2 factorial experiment, represented by points K, L, M and N. The results of this experiment send us in the direction indicated by the second arrow which points towards O and P. Our progress is confirmed by these two batches and we carry out the third experiment which is represented by Q, R, S, and T. The yields of these four batches differ very little and do not give a clear indication of the line of steepest ascent. Perhaps we have reached the summit.

We have made good progress. If we adopt the operating conditions at the centre of the third experiment (i.e. temperature = 202 °C, time = 4.7 hours) we can expect a mean yield of approximately 85, which is much greater than the average of 60 we got from the first experiment. However, we might wish to continue the series of experiments in order to locate more precisely the conditions that will give maximum yield. To make further progress we will need to use at least three levels of each independent variable.

There are many experimental designs we could use to locate precisely the optimum conditions. One obvious contender is the 3^2 factorial design. This would require the production of nine batches using three levels for each of the two independent variables. A plan of this experiment is shown in Figure 13.2.

Also shown in Figure 13.2 are the four trials of the final 2^2 experiment. Perhaps it would be possible to combine the results of the two experiments and use the yield of all 13 batches to locate the optimum conditions. It certainly would be possible. However, before we carry out the 3^2 experiment we should note the following points:

(a) It is easy to analyse and interpret the results of a 2^n factorial experiment, as we have seen in Chapters 12 and 13.

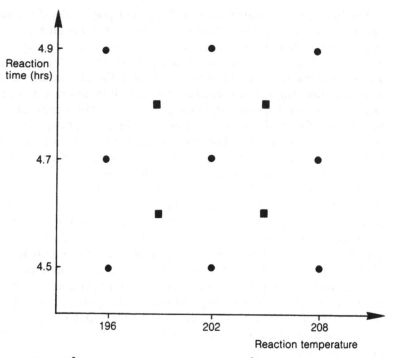

Figure 13.2   A 3^2 factorial experiment (●) and a 2^2 experiment (■).

(b) It is not quite so easy to interpret the results of a 3^n factorial experiment, but there are standard routines for analysis given in Davies (1978) and in Box, Hunter and Hunter (1978).

(c) There is no simple routine for analysing the results of all 13 trials, but we could use multiple regression analysis with safety.

(d) In any analysis of the combined results we would need to bear in mind that the yield of the process might have increased or decreased during the time interval between the two factorial experiments. The multiple regression analysis can be adapted to embrace this possibility, as we shall see in Chapter 14.

Whilst you reflected upon the benefits to be gained from the results of the two experiments in Figure 13.2, it may have occurred to you that the 3^2 factorial design is not the best supplement to a 2^2 experiment. Do not forget that the two-level design has already been executed. Its results have led us to suspect that we are close to the optimum. The important question to ask now is 'What should we do next?' Clearly we must introduce a third level for each of the two independent variables, but this could be achieved with one additional batch. Do we need to produce a further nine?

Table 13.11 Design matrix of 2^2 experiment

x	z	xz	x^2	z^2
−1	−1	1	1	1
−1	1	−1	1	1
1	−1	−1	1	1
1	1	1	1	1

13.7 CENTRAL COMPOSITE DESIGNS

Table 13.11 contains the design matrix of a 2^2 factorial design. It clearly shows what we can, and cannot, estimate from the results of the experiment. We can estimate the two-factor interaction and the two main effects. We cannot estimate the quadratic effects. If, however, we supplement the 2^2 design by adding one or more points at the centre of the square we get the design matrix in Table 13.12.

Clearly the addition of one or more central points will enable us to estimate quadratic effects. However, it is also clear in Table 13.12 that the x^2 and z^2 columns are correlated. Thus the two quadratic effects are inseparable and any evidence of curvature in the results could equally well be attributed to either x^2 or z^2. This difficulty is overcome if we supplement the 2^2 design with four 'axial points' in addition to the central points. This gives the design displayed in Figure 13.3.

By adding the four axial points and one or more central points to the 2^2 factorial design we obtain what is known as a central composite design. This is a very useful and very versatile type of experiment. By changing the number of central points and/or the distance of the axial points from the centre, we can achieve various objectives. For example, if we include only one central point and we place the axial points a distance of one unit from the centre, we obtain an orthogonal design matrix. This is given in Table 13.13.

Table 13.12 A 2^2 factorial plus two centre points

x	z	xz	x^2	z^2
−1	−1	1	1	1
−1	1	−1	1	1
1	−1	−1	1	1
1	1	1	1	1
0	0	0	0	0
0	0	0	0	0

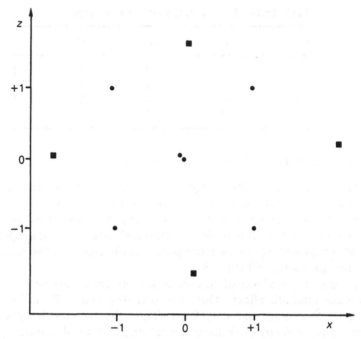

Figure 13.3   A central composite design (■ are axial points).

Table 13.13   Four axial points at ±1 and one centre point

x	z	xz	x^2	z^2
−1	−1	1	1	1
−1	1	−1	1	1
1	−1	−1	1	1
1	1	1	1	1
−1	0	0	1	0
1	0	0	1	0
0	−1	0	0	1
0	1	0	0	1
0	0	0	0	0

Because the design matrix in Table 13.13 is orthogonal the results of the experiment could be analysed to obtain unambiguous conclusions. In fact, the experiment in Table 13.13 is a 3^2 factorial. Clearly it is a good design, but we must not assume that it is the best design for all purposes. For

Table 13.14 Four axial points at ±1.414 and one centre point

x	z	xz	x^2	z^2
−1	−1	1	1	1
−1	1	−1	1	1
1	−1	−1	1	1
1	1	1	1	1
−1.414	0	0	2	0
1.414	0	0	2	0
0	−1.414	0	0	2
0	1.414	0	0	2
0	0	0	0	0

example, it does not give equally wide confidence intervals for the true response at all operating conditions which are equidistant from the centre. This would be achieved if we placed the four axial points the same distance from the centre as the four factorial points. Doing so would give the design matrix in Table 13.14.

By moving the axial points so that they are the same distance from the centre as the factorial points we have produced what is known as a rotatable design. Thus Table 13.14 represents a rotatable central composite design. It will give confidence intervals for the true response which are equally wide for all operating conditions that are the same distance from the centre of the experiment. Unfortunately it is not orthogonal. However, we could make this rotatable design orthogonal if we increased the number of centre points to eight. Alternatively we could achieve another desirable property by using five central points rather than eight. If we add four more central points to the design in Table 13.14 we would have a rotatable central composite design which gave a confidence interval at the centre equal in width to the confidence intervals at the other points. This would be a uniform precision central composite design.

Our discussion of central composite designs has been based on only two independent variables. However, the concepts can be extended to include any number of variables. Indeed, as I have said earlier, well-designed experiments become more efficient as we embrace more factors. Unfortunately, it becomes more difficult to visualize the 2^n factorial and the central composite as n increases, but their usefulness is not in doubt. Table 13.15 summarizes the rotatable central composite designs which can be achieved with two to eight independent variables.

Some of the experiments in Table 13.15 are rather large because many

Table 13.15 Rotatable central composite designs

Factorial experiment	Axial points		Number of central points	
	Number	Distance	Orthogonal	Uniform precision
2^2	4	1.414	8	5
2^3	6	1.682	9	6
2^4	8	2.000	12	7
2^5	10	2.378	17	10
$\frac{1}{2}$ of 2^5	10	2.000	10	6
2^6	12	2.828	24	15
$\frac{1}{2}$ of 2^6	12	2.378	15	9
$\frac{1}{2}$ of 2^7	14	2.828	22	14
$\frac{1}{2}$ of 2^8	16	3.364	33	20

Table 13.16 Orthogonal central composite designs (not rotatable)

Factorial experiment	Axial points		Number of central points
	Number	Distance	
2^2	4	1.000	1
2^3	6	1.215	1
2^4	8	1.414	1
$\frac{1}{2}$ of 2^5	10	1.547	1

central points are included. This is the price we must pay to achieve rotatability with orthogonality. If you feel that orthogonality is essential but rotatability is not, then you can choose a smaller design from those in Table 13.16, which leave only one central point.

13.8 SCREENING EXPERIMENTS

If you are attempting to improve production processes in the chemical and allied industries you will have been interested in the hill-climbing approach set out in section 13.6. You will recall that this involved a series of small experiments which led us towards the optimum conditions. The first experiment was a 2^3 factorial design and this was followed by a 2^2 factorial, after one of the independent variables had been eliminated.

This example of hill climbing could be criticized on the grounds that the series of experiments did not start at square one. Perhaps your lack of understanding of your process will not allow you to reduce the multitude of possible variables down to three. In fact you might wish to include such a

Table 13.17 Design vectors for Plackett–Burman experiments

$N =$ 4	+ − +	
$N =$ 8	+ + + − + − −	
$N =$ 12	+ + − + + + − − − + −	
$N =$ 16	+ + + + − + − + + − − + − − −	
$N =$ 20	+ + − − + + − + − + − + − − − − + + −	
$N =$ 24	+ + + + + − + − + + − − + + − − + − + − − − −	

large number of factors that a full factorial, a half replicate, or even a quarter replicate would be too costly. If this is so, you will be interested in screening experiments, which can be used to learn a little about a large number of variables. Screening designs help us to decide which of several independent variables are worthy of further study.

Probably the most widely used screening designs are the Plackett–Burman designs which are used to study $(N - 1)$ independent variables in N runs or trials. N must be a multiple of 4. Thus in 4 trials we could investigate 3 factors, in 8 trials we could investigate 7 factors, in 12 trials 11 factors, etc. To obtain the design matrix for a Plackett–Burman design we first select the appropriate design vector from Table 13.17.

Suppose we wish to assess the effects of 7 independent variables in 8 trials. We first take, from Table 13.17, the vector for $N = 8$, and write this as a column. See the left-hand column of Table 13.18. To generate the second column of the design matrix we copy the first column but move each entry down one row and put the bottom sign up to the top. To generate the third column we copy the second column, again shifting down one row. This process is repeated until we have seven columns, then the final step is to add a row of minus signs.

Table 13.18 Design matrix for a Plackett–Burman

A	B	C	D	E	F	G
+	−	−	+	−	+	+
+	+	−	−	+	−	+
+	+	+	−	−	+	−
−	+	+	+	−	−	+
+	−	+	+	+	−	−
−	+	−	+	+	+	−
−	−	+	−	+	+	+
−	−	−	−	−	−	−

You can see that each of the seven columns in Table 13.18 contains four + and four −. Furthermore, each column has zero correlation with any other column. It follows that this Plackett–Burman matrix must be, in essence, identical to the design matrix for a 2^3 factorial experiment in Table 13.1. This is indeed the case. If you changed all the signs and interchanged some rows you would see the resemblance.

It can be shown that, if N is a power of 2 then the Plackett–Burman will be a 2^n factorial. This will be the case if $N = 4, 8, 16, 32$, etc. Clearly, the Plackett–Burman will not be a factorial design if $N = 12, 20, 24, 28, 36$, etc., nor will the columns be orthogonal.

For a more detailed discussion of Plackett–Burman designs see Montgomery (1976).

In Table 13.18 seven independent variables A, B, C, D, E, F and G, have been assigned to the seven columns. As the columns are orthogonal we can predict that none of the main effects will be aliased with other main effects. However, it is clear that each main effect will be aliased with many interactions. We will return to this point in Chapter 16 when we examine the type of experiments recommended by Genichi Taguchi.

Plackett–Burman designs are very useful if, in the early stages of a series of experiments, we wish to assess the effects of a large number of independent variables. The sequential use of Plackett–Burman designs is illustrated in Wheeler (1987). Clearly we may be misled when we analyse the results of a Plackett–Burman experiment if either:

1. there is an outlier in the results;
2. there is a large interaction between two variables.

13.9 BLOCKING AND CONFOUNDING

When we come to consider the implementation of a planned experiment we may realize that the whole experiment cannot be carried out under homogeneous conditions. Perhaps we have designed a 2^4 factorial experiment but we find that a consignment of raw material is only sufficient for eight trials. Alternatively we might wish to carry out a 2^3 experiment but heat treatment of the eight test pieces cannot be carried out in one run because the oven will only accommodate six pieces. Because of such constraints a planned experiment may have to be carried out in two or more blocks. The subdivision of the whole experiment into suitable blocks must be done with care if the results are not to be ruined by block to block variation.

To illustrate the problems involved in blocking we will return to the 2^3 factorial experiment discussed in Chapter 12. The results of this experiment were put into a three-way table which is reproduced as Table 13.19.

You may recall that the results in the three-way table were averaged

Table 13.19 Results of a 2^3 experiment (from Table 12.18)

	z			
	Low		High	
	w		w	
x	Low	High	Low	High
Low	48	32	36	56
High	80	72	64	84

Table 13.20 (from Table 12.19)

	z			w			w	
x	Low	High	x	Low	High	z	Low	High
Low	40	46	Low	42	44	Low	64	52
High	76	74	High	72	78	High	50	70

in pairs to obtain the entries in three two-way tables. These are also reproduced here, as Table 13.20.

To complete the analysis we calculated estimates of the three main effects and the four interactions as follows:

$$\text{Main effect } x = \tfrac{1}{2}(74 + 76) - \tfrac{1}{2}(46 + 40) = 32$$

$$\text{Main effect } z = \tfrac{1}{2}(74 + 46) - \tfrac{1}{2}(76 + 40) = 2$$

$$\text{Main effect } w = \tfrac{1}{2}(78 + 44) - \tfrac{1}{2}(72 + 42) = 4$$

$$\text{Interaction } xz = \tfrac{1}{2}(74 + 40) - \tfrac{1}{2}(76 + 46) = -4$$

$$\text{Interaction } xw = \tfrac{1}{2}(78 + 42) - \tfrac{1}{2}(72 + 44) = 2$$

$$\text{Interaction } zw = \tfrac{1}{2}(70 + 64) - \tfrac{1}{2}(50 + 52) = 16$$

$$\text{Interaction } xzw = \tfrac{1}{2}\{[\tfrac{1}{2}(36 + 84) - \tfrac{1}{2}(64 + 56)]$$
$$-[\tfrac{1}{2}(48 + 72) - \tfrac{1}{2}(80 + 32)]\}$$
$$= -2$$

Suppose that the raw material used in the manufacture of the eight samples of rubber is purchased in 50 kg bags and that 10 kg are required in the manufacture of each sample. Obviously we will need to use two bags of raw material, but our past experience suggests that there may be considerable

Table 13.21 2^3 factorial experiment using two bags of raw material

		z		
	Low		High	
	w		w	
x	Low	High	Low	High
Low	Bag 1	Bag 2	Bag 2	Bag 2
High	Bag 2	Bag 1	Bag 1	Bag 1

variation from bag to bag. If we were to manufacture four samples from one bag and the other four samples from a second bag, what effect would this have on our conclusions? The answer to this question depends upon which of the eight trials are allocated to bag 1 and which to bag 2. Suppose that the experiment is implemented as shown in Table 13.21.

The reader may see immediately that the allocation of trials to bags in Table 13.21 has been done very badly. Three of the trials in the top row (i.e. at the low level of x) have been allocated to bag 2 whereas only one of the trials in the bottom row was allocated to this bag. If the response, the tensile strength of the rubber, is affected by the difference between the two bags then we would surely expect our estimate of main effect x to be in error. This is easily demonstrated if we assume that the four samples made from bag 1 have tensile strength reduced by X units from what it would have been if they had been made from bag 2. The results of the experiment would then be as in Table 13.22.

Table 13.22 The results we would have obtained if all eight samples had been made from bag 2

		z		
	Low		High	
	w		w	
x	Low	High	Low	High
Low	$48 + X$	32	36	56
High	80	$72 + X$	$64 + X$	$84 + X$

Table 13.23   The effect of using two bags of raw material

z			x	w		z	w	
x	Low	High		Low	High		Low	High
Low	$40 + \dfrac{X}{2}$	46	Low	$42 + \dfrac{X}{2}$	44	Low	$64 + \dfrac{X}{2}$	$52 + \dfrac{X}{2}$
High	$76 + \dfrac{X}{2}$	$74 + X$	High	$72 + \dfrac{X}{2}$	$78 + X$	High	$50 + \dfrac{X}{2}$	$70 + \dfrac{X}{2}$

The entries in Table 13.22 are exactly the same as those in Table 13.19 except for the addition of X, the unknown change in tensile strength. When we average the results in Table 13.22 to produce three two-way tables we find that X appears there also (Table 13.23).

We can see by the lack of balance in the two-way tables that several of the effect estimates will be affected by the difference between bags. Completing the calculations we find that four of the estimates are affected whilst the other three are in agreement with the estimates calculated in the previous chapter.

Main effect x $= \frac{1}{2}\left(74 + X + 76 + \dfrac{X}{2}\right) - \frac{1}{2}\left(46 + 40 + \dfrac{X}{2}\right) = 32 + \dfrac{X}{2}$

Main effect z $= \frac{1}{2}(46 + 74 + X) - \frac{1}{2}\left(40 + \dfrac{X}{2} + 76 + \dfrac{X}{2}\right) = 2$

Main effect w $= \frac{1}{2}(78 + X + 44) - \frac{1}{2}\left(72 + \dfrac{X}{2} + 42 + \dfrac{X}{2}\right) = 4$

Interaction xz $= \frac{1}{2}\left(74 + X + 40 + \dfrac{X}{2}\right) - \frac{1}{2}\left(76 + \dfrac{X}{2} + 46\right) = -4 + \dfrac{X}{2}$

Interaction xw $= \frac{1}{2}\left(42 + \dfrac{X}{2} + 78 + X\right) - \frac{1}{2}\left(72 + \dfrac{X}{2} + 44\right) = 2 + \dfrac{X}{2}$

Interaction zw $= \frac{1}{2}\left(64 + \dfrac{X}{2} + 70 + \dfrac{X}{2}\right) - \frac{1}{2}\left(50 + \dfrac{X}{2} + 52 + \dfrac{X}{2}\right) = 16$

Interaction $xzw = \frac{1}{2}\{[\frac{1}{2}(36 + 84 + X) - \frac{1}{2}(64 + X + 56)]$
$$- [\frac{1}{2}(48 + X + 72 + X) - \frac{1}{2}(80 + 32)]\} = -2 - \dfrac{X}{2}$$

Obviously the arrangement suggested by Table 13.21 constitutes a bad design, as we anticipated. Is it possible to allocate the eight trials to the two bags in such a way that our estimates of all three main effects will be unaffected by the difference between bags? It certainly is, provided that we

Table 13.24   A better allocation of trials to bags

	z			
	Low		High	
	w		w	
x	Low	High	Low	High
Low	Bag 2	Bag 1	Bag 1	Bag 2
High	Bag 1	Bag 2	Bag 2	Bag 1

follow the line of thought which helped us to obtain useful half replicates in the first half of this chapter. The allocation specified in Table 13.24 is much better balanced than the one we have just considered and the reader will not be surprised to learn that this new arrangement produces much better estimates.

If we again assume that the effect of using two bags of raw material is simply to increase the tensile strengths of the four bag 1 samples by X then we can calculate the effect estimates. By assuming that the level of tensile strength is changed but the effect of the independent variables is not changed we are in effect postulating that there is no interaction between the three independent variables (x, z, w) and the raw material. Working through the calculations we would find that only one estimate was affected by having carried out the experiment in two blocks. Furthermore it is the least important estimate, interaction xzw, that is in error, whilst the three main effects and the three two-variable interactions are completely unaffected by the blocking. We summarize this happy situation by saying that the difference between blocks is confounded with interaction xzw.

In the first blocked design that we considered (Table 13.21) the block effect was partially confounded with several effects. When splitting a 2^n factorial design into two blocks we can always arrange that the block effect will be confounded with one specified effect and we would usually choose the highest order interaction. If it were necessary to split a 2^n factorial experiment into four blocks then there would be three block effects to confound. Clearly there are strong similarities between confounded effects and the defining contrasts that arose in our discussion of fractional replicates.

13.10 SUMMARY

In this chapter we have discussed fractions of 2^n factorial experiments. These can be particularly useful when we wish to investigate several independent variables and we are prepared to use only two values for each variable but we are not willing to carry out the full 2^n factorial. Clearly a half replicate (or a quarter replicate) cannot offer all the benefits of a full factorial experiment, but it may give us what we need most. This is particularly true if our immediate need is an indication of the values we should use in our next experiment.

Many researchers believe that in some situations it is more efficient to proceed by means of a series of small experiments rather than by hoping to solve all of one's problems in one grand design. If this sequential approach is adopted the earlier experiments could well be fractions of 2^n factorials. When the optimum region has been located it may be necessary to use three levels for the independent variables in the final design and the use of 3^n factorials, or central composite designs, is recommended.

We have also discussed, rather briefly, the effect of splitting a 2^n factorial experiment into two or more blocks. This course of action is often forced upon us by the constraints within which the experiment must be carried out. We will encounter blocking problems again in the next chapter and will examine how blocks can be taken into account when using multiple regression analysis.

Having spent two chapters studying some of the principles underlying experimental designs we can now return to the problem which we set aside at the end of Chapter 11. We will devote the next chapter to advising the plant manager on where he went wrong, what he should have done, and what he might now do if he wishes to reduce the impurity in future batches of pigment.

PROBLEMS

13.1 It is believed that the viscosity of batches of polymer are affected by four variables: pressure (A), temperature (B), catalyst concentration (C) and reaction time (D). An experiment is suggested with the following design matrix, each variable having two levels represented by + and −.

Trial number	A	B	C	D
1	−	−	−	+
2	+	−	+	−
3	−	+	−	−
4	+	+	+	+

(a) What fraction of a full factorial design is this experiment?
(b) By multiplying the appropriate columns, generate the signs for all

two-, three- and four-variable interactions to give the full design matrix.

(c) Examine the full design matrix and decide which are the three defining contrast.

(d) Form alias groups of main effects and interactions.

13.2 The chemist who carried out the 2^3 factorial experiment in Problem 1 of Chapter 12 wishes to design a further experiment. The purpose of this new investigation is twofold:

(i) To confirm the significance of the speed × pH interaction which was suggested by the first experiment.

(ii) To assess the effect of changing the concentration of the stabilizer.

Stabilizer concentration was not included in the first experiment, but the chemist suspects that it could be significant as a main effect and he or she has good reason to believe that stabilizer concentration will interact with temperature. It is also possible that stabilizer concentration will interact with pH and with speed of agitation.

The chemist decides that he or she will represent the four variables by the symbols A, B, C and D as follows:

A is agitation speed
B is pH
C is temperature
D is stabilizer concentration.

From the results of the new experiment it is essential that he or she should be able to estimate:

Main effects A, B and D;
Two-variable interactions AB, AD, BD and CD.

The chemist is prepared to assume that main effect C and interactions AC and BC are negligible provided the same levels for variables A, B and C are used that were used in the first experiment.

(a) Can the needs of the chemist be satisfied by a half replicate of a 2^4 factorial experiment?

(b) If the answer to (a) is 'yes', list the eight treatment combinations that you would advise the chemist to use.

13.3 It was suggested in Chapter 11 that your ability to make use of multiple regression analysis would be increased if you were able to understand the terminology used by computer packages and by statisticians. You will now realize that it is wise to consult a statistician or a good text before an investigation is carried out rather than after the data have been collected. Obviously, any such consultation will be enriched if the scientist has an appreciation of the terminology used in Chapters 12 and 13. Attempting to insert the missing words in the following passage will enable you to evaluate your ability to communicate in this medium.

In Chapter 12 we discussed 2^n factorial experiments in which we have (1) _____ independent variables (or factors), each having (2) _____ values or levels. First we compared the 2^2 factorial experiment, which consists of four trials, with a classical, 'one factor at a time' experiment based on three trials. The advantages of the factorial experiment were:

(a) It gave more precise estimates because each effect estimate was based on the results of (3) _____ trials whereas each estimate in the classical experiment was based on only (4) _____ results;
(b) From the results of the 2^2 factorial it was possible to estimate not only the two main effects but also the (5) _____ effect.

The statistical significance of each effect estimate was checked by means of a (6) _____ . In order to calculate the test statistic in such a test we need to know the effect estimate, of course, and we also need an estimate of the (7) _____ . Unfortunately a 2^n factorial experiment does not yield such an estimate. By carrying out two replicates of a 2^2 factorial experiment, which requires (8) _____ trials in all, we can calculate an estimate of the residual standard deviation with (9) _____ degrees of freedom, whilst three replicates would give an estimate of the residual standard deviation with (10) _____ degrees of freedom.

The results of a 2^2 factorial experiment can be analysed by fitting multiple regression equations. If we call the response y and the two independent variables x and z, we can account for the two main effects by fitting the equation (11) _____ but to account for the interaction well we must fit (12) _____ . The extra variable in the equation is generated by multiplying each x value by the corresponding z value and is known as a (13) _____ variable. Unfortunately this generated variable will probably be highly (14) _____ with either or both of the measured variables, x and z, unless we (15) _____ these measured variables before multiplying. Provided this precaution has been taken we will have a perfect correlation matrix in which the correlation between any two independent variables is equal to (16) _____ .

When we have a large number of independent variables, even if we restrict each to two values, the number of trials required in a factorial experiment may be greater than we are prepared to carry out. Alternatively it may not be possible to carry out such a large number of trials under homogeneous conditions. We could decide, of course, to carry out only four of the eight trials which constitute a 2^3 factorial experiment. This would be known as a half (17)   and the results of the four trials could be used to calculate three effect estimates. If we had chosen the four trials such that the X column of the design matrix contained four + signs then main effect X would be known as the (18)   . The other two main effects and the four interaction effects would fall into three (19)   . Main effect Z would be aliased with interaction XZ, main effect W would be aliased with (20)   whilst interaction XZW would be aliased with (21)   . Obviously this would be a bad half replicate for it would be impossible to estimate (22)   and the estimate of main effect Z would only be useful if we were very confident that (23)   did not exist.

Similar problems arise if we decide to carry out a full 2^n factorial experiment but, because of constraints, we have to split it into two blocks. If the effect of blocking is simply to increase (or decrease) the response in one block by a fixed amount but not to change the effects of the independent variables we would say that there was no (24)   between blocks and independent variables. In the absence of such an interaction we could estimate all the effects except one and we would say that this particular effect was (25)   with the difference between blocks.

14

Improving a bad experiment

14.1 INTRODUCTION

In the last two chapters we have examined several experimental designs and highlighted some of the principles which must be respected if valid conclusions are to be drawn from experimental data. We have also discussed the possibility of carrying out only a fraction of a factorial experiment in order to reduce the number of trials whilst still being able to estimate the important effects.

Perhaps we are now in a position to return to the problems of the plant manager which we set aside at the end of Chapter 11. We have established that the experiment he carried out was less than perfect. Two of his independent variables, weight of special ingredient (x) and feedrate (z), were so very highly correlated $(+0.97)$ that the main effects, x and z, almost constituted an alias pair and it was quite impossible to decide whether the observed reduction in impurity was caused by one or the other.

The plant manager was, of course, operating under severe constraints. There were strict limits to how far he could deviate from normal operating conditions and to how many batches he could include in any experimental series. Clearly these constraints cannot be ignored when we offer advice to the plant manager.

In this chapter we will specify operating conditions that could be used in a series of eight batches which will extend the plant manager's original experiment. The results from all 18 batches will then be analysed by means of multiple regression analysis and a strategy will be formulated for future production of this pigment.

Before we attempt this salvage operation, however, we will consider what the plant manager should have done in the first place as an alternative to the ineffective experiment that he actually did carry out.

14.2 AN ALTERNATIVE FIRST EXPERIMENT

A useful starting point in any activity is a clear statement of objectives. When the activity under consideration involves the production of ten batches with operating conditions changing from batch to batch then the neglect of objectives is extremely unwise. On the other hand a clear statement of grandiose objectives will not in itself produce results. If, therefore, the plant manager hopes to investigate five independent variables in such a way that estimates of all main effects, quadratic effects and interactions can be obtained, within a strict limit of ten trials, then he is doomed to disappointment.

Let us start by considering the smallest possible experiment. With five independent variables, six trials are needed in order to obtain unambiguous estimates of all main effects. Such a small experiment would not give an estimate of residual variation and would not, therefore, enable the statistical significance of the estimates to be tested. With ten trials, however, we should be able to estimate five main effects and obtain an estimate of the residual SD with four degrees of freedom. The plant manager's main objective, then, should be to estimate the five main effects.

It is possible that one or more of the five main effects will be shown to be not significant and this would release extra degrees of freedom for the estimate of residual variance. If the plant manager uses regression analysis to analyse the results and he ends up with only two significant main effects plus a residual SD with seven degrees of freedom then he might wish to introduce a quadratic term or a cross-product term into the equation. As a secondary objective, then, it is desirable to be able to estimate interaction effects and/or curvature of the main relationships. By pursuing this secondary objective we must be very careful not to jeopardize the primary objective.

The constraints within which we must operate are concerned with the number of trials and the range of variation of the independent variables. When choosing values for those variables we will stay within the limits used in the original experiment. These are given in Table 14.1.

The proliferation of values displayed in Table 14.1 is a luxury we cannot afford if our experiment is limited to only ten batches. To estimate linear main effects and interactions we need only two values for each variable. If

Table 14.1 The original experiment

Independent variable	Values used
Weight of special ingredient (x)	0, 1, 1, 2, 3, 3, 4, 5, 5, 6
Catalyst age (w)	1, 2, 3, 4, 5, 6, 7, 8, 9, 10
Feedrate (z)	1, 2, 2, 3, 3, 5, 5, 6, 6, 7
Temperature (t)	1, 1, 1, 2, 2, 2, 2, 3, 3, 3
Agitation speed (s)	2, 2, 2, 2, 3, 3, 3, 4, 4, 4

we wish to consider quadratic effects the number of values will need to be increased to three, but any further increase is neither necessary nor desirable. Catalyst age (w) is an exception as this variable will increase in value in steps of one from batch to batch.

With these thoughts in mind let us make a start. Perhaps the first step should be to choose a 'standard' design of a suitable size and then modify this to suit our purpose. Within the upper limit of 10 trials we have:

(a) a $\frac{1}{27}$ replicate of a 3^5 factorial experiment, requiring nine trials;
(b) a $\frac{1}{4}$ replicate of a 2^5 experiment, requiring eight trials;
(c) a $\frac{1}{9}$ replicate of a 3^4 experiment, requiring nine trials;
(d) a $\frac{1}{2}$ replicate of a 2^4 experiment, requiring eight trials.

To make use of the last two experiments in the list we would set aside catalyst age (w) and consider only the four independent variables. The fifth variable (w) would re-enter the experiment when the trials were carried out in random order. Using this approach we could, purely by chance, find a high correlation between catalyst age and one of the other four variables. This could, of course, be checked before the experiment was carried out.

Whilst all four alternatives listed are worthy of consideration we will concentrate on the $\frac{1}{4}$ replicate of a 2^5 factorial experiment. To obtain a suitable quarter we could write out the design matrix for a full 2^5 experiment and then, using three defining contrasts, select 8 of the 32 rows. As an alternative we will set out the design matrix of a 2^3 factorial experiment then we will add two more columns generated by multiplication. The result is shown in Table 14.2.

In Table 14.2 the entries in the t column have been obtained by multiplying corresponding entries in the x and w columns. Similarly the s column

Table 14.2 A $\frac{1}{4}$ replicate of a 2^5 factorial experiment

Trial	Variable				
	x	w	z	$t(=xw)$	$s(=xz)$
1	-1	-1	-1	$+1$	$+1$
2	$+1$	-1	-1	-1	-1
3	-1	$+1$	-1	-1	$+1$
4	$+1$	$+1$	-1	$+1$	-1
5	-1	-1	$+1$	$+1$	-1
6	$+1$	-1	$+1$	-1	$+1$
7	-1	$+1$	$+1$	-1	-1
8	$+1$	$+1$	$+1$	$+1$	$+1$

Table 14.3　Additional trials

	Variable				
Trial	x	w	z	t	s
9	0	0	0	0	0
10	0	0	0	0	0

Table 14.4　Allocation of real values to the independent variables

Artificial value	Weight of special ingredient	Feedrate	Inlet temperature	Agitation speed	Catalyst age
	x	z	t	s	w
−1	0	1	1	2	1, 2, 4, 5
0	3	4	2	3	3, 8
+1	6	7	3	4	6, 7, 9, 10

is the cross-product of the x and z columns. The consequence of proceeding in this way will be discussed in the next section when the design is completed.

It would, of course, be absolutely impossible to estimate quadratic effects from the results of the experiment described by Table 14.2, but we still have two trials at our disposal and these will be given an intermediate level for each of the five independent variables (Table 14.3).

Having completed the design matrix we can translate it into a working design if we replace −1, 0 and +1 by realistic values of the five variables. Staying within the limits used by the plant manager we will translate as in Table 14.4.

Catalyst age (w) must be allocated values 1 to 10 and it might seem logical to replace −1 with 1, 2, 3, 4, and 0 with 5, 6, etc. On the other hand perhaps it is desirable to separate the two identical trials which have the intermediate value, '0', so these are allocated 3 and 8. Inserting the revised values of the independent variables in the design matrix gives the design of Table 14.5.

Because of the peculiar nature of one of the independent variables a further change is needed before the experiment can be carried out. In practice, catalyst age (w) must increase in steps of 1 throughout the experiment if we are using consecutive batches. The trials in Table 14.5 will,

Table 14.5

Trial	Weight of special ingredient	Catalyst age	Feedrate	Inlet temperature	Agitation speed
	x	w	z	t	s
1	0	1	1	3	4
2	6	2	1	1	2
3	0	6	1	1	4
4	6	7	1	3	2
5	0	4	7	3	2
6	6	5	7	1	4
7	0	9	7	1	2
8	6	10	7	3	4
9	3	3	4	2	3
10	3	8	4	2	3

Table 14.6 The final design

Trial	Weight of special ingredient	Catalyst age	Feedrate	Inlet temperature	Agitation speed
	x	w	z	t	s
1	0	1	1	3	4
2	6	2	1	1	2
3	3	3	4	2	3
4	0	4	7	3	2
5	6	5	7	1	4
6	0	6	1	1	4
7	6	7	1	3	2
8	3	8	4	2	3
9	0	9	7	1	2
10	6	10	7	3	4

therefore, need to be reordered to conform to this constraint. Reordering gives the final design in Table 14.6.

This design has been produced without any consideration of the difficulties of implementation. You will note, for example, that each batch requires a different weight of special ingredient from the preceding batch and that only occasionally do any of the other independent variables keep the same value in two consecutive batches. This constitutes a radical departure from routine production in which the plant operators are encouraged to repeat the same procedure as closely as possible. To leave the experiment in the hands of the operators might, therefore, be very unwise. Excessive intervention by management, on the other hand, may result in disturbance of the many other variables which have not been included in the experiment, thus making the ten batches non-typical and invalidating any conclusions. Clearly, the use of statistics cannot overcome these difficulties but it can help us to assess the quality of the experiment before it is carried out. This we will now attempt.

14.3   HOW GOOD IS THE ALTERNATIVE EXPERIMENT?

We noted in an earlier chapter that it is possible to detect the shortcomings of an experimental design by checking for intercorrelation of the independent variables. The correlation matrix for the design given in Table 14.6 is illustrated in Table 14.7.

Apart from the small positive correlation (0.16) between x and w this correlation matrix is perfect. We may be tempted to conclude, therefore, that this experiment would have given unambiguous estimates of the linear main effects of each of the five independent variables. And so it would provided that all interactions are non-existent. It is unfortunately true that, if we were to extend the matrix to include columns for cross-product variables (xz, xw, etc.), we would find several entries of 1.00 other than those on the diagonal.

These unwanted correlations could have been predicted when we decided to use Table 14.2 as the starting point for our design. By the way in which the t and s columns were generated we could see that:

Main effect t would be confounded with interaction xw;
Main effect s would be confounded with interaction xz.

So we would expect the correlation between t and xw to be 1.0 and the same could be said of the correlation between s and xz. There would be

Table 14.7   Correlation matrix for the design in Table 14.6

	x	w	z	t	s
x	1.00	0.16	0.00	0.00	0.00
w		1.00	0.00	0.00	0.00
z			1.00	0.00	0.00
t				1.00	0.00
s					1.00

Table 14.8 Alias groups of the design in Table 14.2

$$(x, \quad tw, \quad sz, \quad xwzts)$$
$$(z, \quad sx, \quad wts, \quad xzwt)$$
$$(w, \quad tx, \quad zts, \quad xzws)$$
$$(t, \quad xw, \quad wzs, \quad xzts)$$
$$(s, \quad xz, \quad wzt, \quad xwts)$$
$$(zw, xzt, \quad xws, \quad xts)$$
$$(zt, \quad xwz, xst, \quad xws)$$

several other correlations of $+1.0$ and these are indicated by the alias groups in Table 14.8.

An extended correlation matrix would show a correlation of $+1.0$ between any pair of variables in the same alias group. Thus main effect x would be perfectly correlated with the interaction variables tw, sz and $xwzts$. Any estimate of main effect x would only be valid, therefore, if the interactions did not exist.

In summary, then, we can see that this alternative design has certain shortcomings but it is an improvement on the experiment carried out by the plant manager. Furthermore it is fair to say that any experiment which attempts to investigate five independent variables in only ten trials will have similar imperfections. If the plant manager were to abandon the results of his first experiment and to start again he could not hope to achieve very much if he were only prepared to include ten batches. Perhaps he would achieve much more by extending his original experiment in such a way as to remove its major defects.

14.4 EXTENDING THE ORIGINAL EXPERIMENT

You will recall that we first expressed reservations about the plant manager's experiment when we examined a correlation matrix in Chapter 11. That matrix is reproduced in Table 14.9.

Table 14.9 Correlation matrix of the original experiment

	x	w	z	t	s
x	1.00	−0.26	0.97	−0.07	0.00
w		1.00	−0.20	−0.05	0.44
z			1.00	0.07	−0.06
t				1.00	0.00
s					1.00

The major defect in the experiment is indicated by the correlation of 0.97 between x and z. Any extension of the experiment would need to contain trials in which the values of x and z were carefully selected so as to reduce this correlation. How this might be achieved is indicated in Figure 14.1 which shows the original ten trials plus a further eight trials which could be used to obtain a better overall design.

By introducing the eight additional points into Figure 14.1 we reduce the correlation between x and z from +0.97 to 0.05. If we now find suitable values for t, s and w to go with these x and z values we then have an additional experiment of eight trials which will form a very useful extension to the original investigation. Note that the eight additional trials by themselves do not constitute a good experiment but the whole series of 18 trials does.

In the original experiment there was zero correlation between s and t. In fact we can see in Figure 14.2 that the values of agitation speed and inlet temperature must have been carefully planned. The eight additional points in Figure 14.2 make use of the same values and the whole grid of 18 points constitutes two replicates of a 3^2 factorial experiment.

Catalyst age (w) is again a special case. If we start the series of eight batches with a new catalyst then w will vary from 1 to 8. When allocating these values to the trials we will attempt to reduce the correlation of 0.44 which exists between w and s in the original experiment. Combining at random the values of the independent variables from Figure 14.1 and Figure 14.2, then carefully adding values for catalyst age we get the experiment listed in Table 14.10.

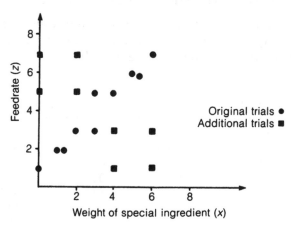

Figure 14.1   Reducing the correlation between x and z

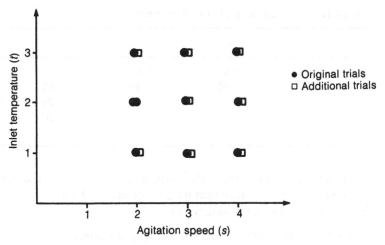

Figure 14.2   Maintaining the zero correlation between s and t

Table 14.10   An extension to the original experiment

Trial	Weight of special ingredient	Catalyst age	Feedrate	Inlet temperature	Agitation speed
	x	w	z	t	s
11	4	1	3	2	4
12	0	2	7	1	4
13	4	3	1	3	4
14	2	4	5	3	2
15	2	5	7	1	2
16	0	6	5	2	3
17	6	7	1	3	3
18	6	8	3	1	3

The effectiveness of this extension to the original experiment can be assessed by inspection of the correlation matrix in Table 14.11. (Note that this matrix is based on all 18 trials.) Clearly the large correlations of Table 14.9 have been greatly reduced by the extension and we have managed to avoid any increase in the correlations which were already satisfactory in the original experiment.

It is safe to conclude from the very small intercorrelations of the independent variables in Table 14.11 that the major defect of the original experiment has been eliminated. Can we, therefore, expect to be able to

Table 14.11 Correlation matrix of all 18 trials

	x	w	z	t	s
x	1.00	0.02	0.05	0.10	0.03
w		1.00	−0.19	−0.05	0.05
z			1.00	−0.29	−0.20
t				1.00	0.00
s					1.00

estimate all five main effects after the additional eight trials have been carried out? We can be sure that no pair of main effects will form an alias pair but there is still a possibility that:

(a) either a main effect will be aliased with an interaction
(b) or two interactions will form an alias pair.

To check these possibilities we will extend the correlation matrix by introducing cross-product variables. Before doing so we will, of course, scale the independent variables. The regression package used for this analysis subtracts the mean then divides by the standard deviation:

$$X = (x - \bar{x})/s_x \qquad W = (w - \bar{w})/s_w \qquad \text{etc.}$$

Table 14.12 The extended experiment

	X	W	Z	T	S	XW	XZ	XT	XS	WZ	WT	WS	ZT	ZS	TS	XX	WW	ZZ	TT	SS
X		02	05	10	03	30	00	08	20	27	35	01	22	11	16	00	20	00	17	40
W			19	05	05	20	21	32	01	05	28	68	07	03	09	33	12	03	03	15
Z				29	20	26	00	23	13	03	09	03	19	15	20	00	04	00	06	00
T					00	34	21	12	19	09	02	09	04	20	00	09	28	25	00	00
S						01	10	16	32	03	10	10	19	00	00	21	71	19	00	00
XW							56	10	05	18	09	03	17	47	24	19	20	00	10	04
XZ								39	05	03	16	39	40	11	27	04	09	04	19	15
XT									08	17	10	23	06	32	29	08	02	45	07	14
XS										59	28	01	35	20	00	41	00	25	53	03
WZ											15	06	10	10	24	54	32	34	34	16
WT												17	43	22	12	18	14	27	04	09
WS													22	08	11	01	31	13	09	04
ZT														28	51	16	18	63	20	33
ZS															53	29	25	15	48	13
TS																09	16	45	00	00
XX																	14	43	08	38
WW																		23	20	01
ZZ																			37	13
TT																				00

This differs from what we did in earlier chapters when we divided by half of the range.

The complete correlation matrix, containing five quadratic variables in addition to the ten interactions, is rather large and so it has been printed without decimal points and with negative signs placed under the numbers, as Table 14.12.

The largest correlation coefficients in Table 14.12 are:

0.71 between S and WW
0.68 between W and WS
−0.63 between ZT and ZZ
etc.

Clearly the design is not perfect but it is doubtful if we could improve upon it to any great extent without increasing the number of trials. Perhaps the next step forward should be to carry out the eight additional trials then make use of the regression package to analyse the whole set of results. We can refer back to Table 14.12 to check particular correlation coefficients when we see which variables are included in the regression equation.

14.5 FINAL ANALYSIS

The plant manager has carried out the eight additional trials using eight consecutive batches starting with a new catalyst. The impurity determinations made on the eight batches are given in the full set of results in Table 14.13.

The data in Table 14.13 are fed into the regression package. The five independent variables are scaled [e.g. $X = (x - 3)/2.114$] then the quadratic variables (i.e. X^2, W^2, etc.) and the interaction variables (i.e. XW, XZ, etc.) are generated. This gives us a total of 20 independent variables and all of these are contenders for inclusion in the regression equation. The equation builds up as follows:

$y = 4.56 - 1.62X$ 61.97% fit

$y = 4.56 - 1.71X + 0.919T$ 81.64% fit

$y = 4.52 - 1.70X + 0.775T + 0.596XZ$ 92.02% fit

$y = 4.56 - 1.68X + 0.835T + 0.485XZ$
$\quad - 0.291XT$ 93.68% fit

$y = 4.55 - 1.69X + 0.827T + 0.406XZ$
$\quad - 0.264XT + 0.260WS$ 95.01% fit

$y = 4.84 - 1.77X + 0.831T + 0.431XZ$
$\quad - 0.277XT + 0.254WS - 0.289SS$ 95.81% fit

etc.   etc.

Table 14.13 The full set of data for the final analysis

Batch number	Impurity	Weight of special ingredient	Catalyst age	Main ingredient		Agitation speed
				Feedrate	Inlet temperature	
	y	x	w	z	t	s
1	4	3	1	3	1	3
2	3	4	2	5	2	3
3	4	6	3	7	3	3
4	6	3	4	5	3	2
5	7	1	5	2	2	2
6	2	5	6	6	1	2
7	6	1	7	2	2	2
8	10	0	8	1	3	4
9	5	2	9	3	1	4
10	3	5	10	6	2	4
11	3	4	1	3	2	4
12	4	0	2	7	1	4
13	4	4	3	1	3	4
14	6	2	4	5	3	2
15	4	2	5	7	1	2
16	7	0	6	5	2	3
17	2	6	7	1	3	3
18	2	6	8	3	1	3
Mean	4.56	3.00	5.06	4.00	2.00	3.00
SD	2.121	2.114	2.754	2.142	0.840	0.840

To test the significance of each variable we will use the t-test, which has served us well in this capacity in previous chapters. The test statistic at each stage has been calculated and is given in Table 14.14.

Had we allowed the computer program to decide which variables were to be included in the equation and specified a 5% significance level, then the program would have stopped after X, T and XZ had been included. The 'best' regression equation would have been printed out as:

$$Y = 4.52 - 1.70X + 0.775T + 0.596XZ$$

There are reasons why we should not accept, without question, the recommendation of the regression package. These include:

(a) When we descale the variables using $x = 3.0 + 2.114X$, $t = 2.0 + 0.840T$ and $z = 4.0 + 2.142Z$ we will find that descaling the cross-product variable XZ has introduced z into the equation. For this reason many

Table 14.14 Significance of the independent variables

Variable entering the equation	% fit	Test statistic	Degrees of freedom	Critical values	
				5%	1%
X	61.97	5.11	16	2.12	2.92
T	81.64	4.01	15	2.13	2.95
XZ	92.02	4.26	14	2.15	2.98
XT	93.68	1.85	13	2.16	3.01
WS	95.01	1.79	12	2.18	3.05
SS	95.81	1.45	11	2.20	3.11
etc.	etc.	etc.	etc.	etc.	etc.

statisticians would advise that whenever a cross-product variable is entered into a regression equation, the two linear variables should also be included. Following this advice would lead us to fit an equation with X, T, XZ and also Z as independent variables.

(b) The fourth variable to enter the equation in Table 14.14 was XT. Though its entry was not statistically significant the interaction between X and T is worth pursuing because both X and T were already in the equation. Further analysis of the data might therefore include an equation with X, T and XT as independent variables. We see in Table 14.12 that the correlation between XZ and XT is -0.39, which is not negligible, so the inclusion of XT as the third variable might give a significant increase in percentage fit.

With these thoughts in mind we run the computer program again to obtain the following equations:

$$y = 4.52 - 1.68X + 0.699T + 0.610XZ - 0.241Z \quad 93.25\% \text{ fit}$$
$$y = 4.61 - 1.68X + 0.978T - 0.517XT \quad\quad\quad\quad 88.04\% \text{ fit}$$

In the first of these equations we see that the introduction of Z as the fourth independent variable increases the percentage fit from 92.02% to 93.25%. This increase is not statistically significant but the purpose of introducing Z was simply to 'balance' the equation.

The second of the two equations has a surprisingly high percentage fit and the introduction of XT as a third variable has resulted in an increase from 81.64% to 88.04%. This increase is statistically significant with a test statistic equal to 2.74.

In the light of what has been revealed by these two equations we might return to the computer for further use of the regression package. One possibility is to fit an equation which includes X, T, Z, XZ and XT as independent variables. We might even try ZT as a sixth variable. All of this analysis can be carried out very quickly on our desktop computer but we must call a halt at some point and attempt to translate our findings into language that will be understood by plant personnel. Further improvements in the equation are of doubtful value as we already have sufficient evidence on which to base confident conclusions.

For the sake of simplicity we might wish to base conclusions on the second regression equation ($y = 4.56 - 1.71X + 0.919T$) which accounts for 81.64% of the batch to batch variation in impurity. Descaling X and T gives us:

$$y = 4.81 - 0.83x + 1.13t$$

We can expect therefore a reduction in impurity of 0.83% to result from a unit increase in the weight of special ingredient (x). We could expect a further increase in impurity of 1.13% for a unit decrease in the inlet temperature (t) of the main ingredient. Note that we can expect both of these benefits, not just one or the other, since there is very little correlation between x and t.

The absence of z in the above equation resolves the dilemma that remained unresolved in Chapter 11. You will recall that we were unable to decide whether the pigment impurity was dependent on the weight of special ingredient (x) or on the feedrate (z). We can now conclude that x is important whereas z can be ignored. This statement must be qualified with the reservation that changing the value of feedrate (z) might have some effect on pigment impurity if it were set to a value outside the range used in the experiment. It is always possible that we will fail to detect the effect of an independent variable if its range of values in the investigation is too small.

Many regression packages would print out confidence intervals for the coefficients in the equation. If we had a confidence interval for the coefficient of x we would predict a decrease in impurity within a certain range rather than quoting one figure, 0.83%. Such confidence intervals can be useful but far more important to the plant manager is a graphical representation which will help him to choose values of x and t which can be expected to give a tolerable level of impurity. Figure 14.3 may be useful in this respect.

From Figure 14.3(a) or (b) we can see that a low inlet temperature and a high weight of special ingredient are required in order to obtain a low level of impurity. There are, of course, many combinations of the two independent variables which will give an expected impurity of less than 2%, say. This point may be even more clear in Figure 14.4, which is a contour diagram representing the same equation as Figure 14.3.

A contour diagram is even more useful if the regression equation con-

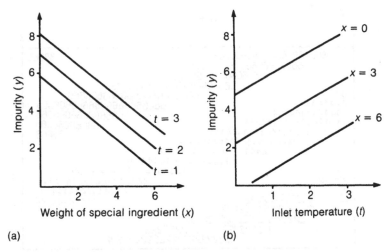

Figure 14.3   Graphical representation of $y = 4.81 - 0.83x + 1.13t$

tains quadratic or cross-product terms. The alternative equation which contained the XT interaction (i.e. $y = 4.61 - 1.68X + 0.978T - 0.517XT$) after descaling becomes:

$$y = 2.82 - 0.192x + 2.12t - 0.308xt$$

This equation is represented by the contour diagram in Figure 14.5.

A comparison of Figure 14.4 and Figure 14.5 reveals strong similarities overall. Nonetheless, the small differences between the two contour diagrams may be of practical importance. We will therefore concentrate on Figure 14.5.

The numbers (2, 4, 6, 8) in the contour diagram represent contour bands or ranges of impurity. Thus a 4 indicates that the predicted impurity level lies between 4.00% and 4.99% for that particular combination of temperature (t) and weight of special ingredient (x). For minimum impurity we need to be in the bottom left-hand corner of Figure 14.5. In practical terms we need to use a low inlet temperature and a high weight of special ingredient. This conclusion could easily have been drawn from an inspection of the regression equation; so why do we need a contour diagram? A contour diagram would have been more useful to us if:

(a) the regression equation had been more complex;
(b) we had wished to optimize two or more dependent variables.

Confronted with a contour diagram for impurity and a second for yield we might well find that the operating conditions which give minimum impurity differ somewhat from the operating conditions which give maximum yield.

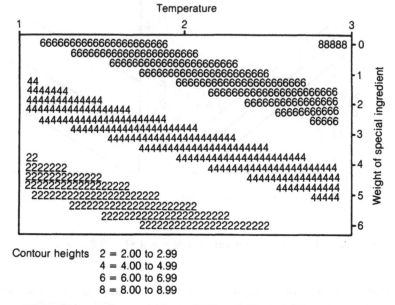

Contour heights   2 = 2.00 to 2.99
                   4 = 4.00 to 4.99
                   6 = 6.00 to 6.99
                   8 = 8.00 to 8.99

Figure 14.4   Contour diagram of $y = 44.81 - 0.83x + 1.13t$

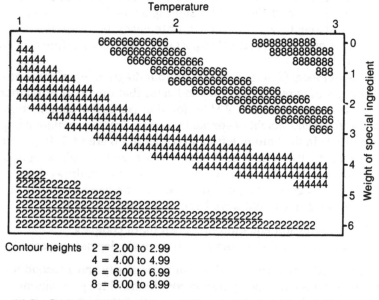

Contour heights   2 = 2.00 to 2.99
                   4 = 4.00 to 4.99
                   6 = 6.00 to 6.99
                   8 = 8.00 to 8.99

Figure 14.5   Contour diagram of $y = 2.82 - 0.192x + 2.12t - 0.308xt$

A compromise could be sought and the contour diagrams would be particularly useful in this search for operating conditions which are acceptable on both counts.

There is a third independent variable (feedrate) which may be important but which is completely ignored in Figure 14.5. We had an indication that there is an interaction between feedrate (z) and weight of special ingredient (x) but xz is not included in the equation which gave Figure 14.5. To accommodate this interaction and also the xt interaction we could fit an equation involving X, Z, T, XZ and XT. After descaling of the variables this equation could be represented by a series of contour diagrams similar to Figure 14.5 but each having a different value of feedrate (z). Alternatively we could produce a series of contour diagrams using x and z as the independent variables with each diagram having a different value of temperature (t). Such a set of diagrams would be a useful aid in selecting suitable operating conditions for future batches.

This discussion has embraced many possibilities, but has said little about the immediate needs of the plant manager. He is paid a handsome salary for making decisions concerning the safety and efficient operation of the plant, and he must now specify operating conditions for future batches. These will include values for the five independent variables in the investigation and his specification is:

Weight of special ingredient	5 units
Feedrate of main ingredient	4 units
Temperature of main ingredient	1.5 units
Agitation speed	3 units

Regulations concerning catalyst changes are to continue as before.

The plant manager is not entirely happy but he hopes that future experiments will add further to his growing understanding of the plant.

14.6 INSPECTION OF RESIDUALS

In the preceding analysis we have concentrated on reaching decisions about which of the independent variables are important and what values they need to have if the impurity of future batches is to be reduced. Multiple regression proved to be a very useful tool in this analysis.

We have seen in earlier chapters some of the dangers inherent in this very powerful technique but we cannot claim to have given thorough coverage to all the assumptions which underlie regression analysis. Restrictions on the size of this book prevent such coverage and the reader is referred to one of the many excellent texts in the Bibliography for completion of his/her regression education. It would, however, be irresponsible to close without pointing to one simple way of checking that assumptions are not being violated. This method involves examination of the residuals from the regression equation, and such examination should be carried out before using the equation as a basis for drawing conclusions. This examination is facilitated if the regression package fitting the equation also prints out a graphical representation of the residuals.

It is highly desirable that the residuals should have a distribution which approximates to the normal distribution. Whilst a normal distribution of residuals does not guarantee that all assumptions are satisfied, any clear departure from normality is a strong indication that something is amiss. For this reason some regression packages print out a probability plot of the residuals.

It is highly desirable that the residuals should not contain a serial pattern when plotted against 'time' or against any of the independent variables. To aid detection of such patterns some regression packages print out a cusum plot of the residuals. Such a graph can be very revealing.

Figure 14.6 is a cusum plot of the residuals from one of the equations examined earlier ($y = 4.61 - 1.68X + 0.978T - 0.517XT$). Clearly there is a pattern here which cannot be ignored. Furthermore, the maximum cusum is found to be statistically significant, indicating a change in the mean level of the residuals at batch number 11.

Even if the regression package did not produce a cusum plot of the residuals we would almost certainly have spotted the change simply by glancing at a list of the residuals. The pattern of + and − signs is very striking in Table 14.15 which contains the residuals represented in Figure 14.6.

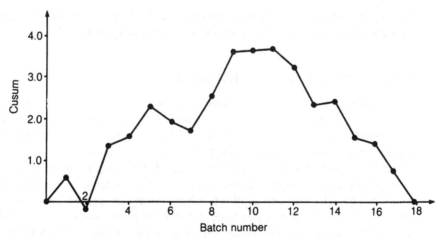

Figure 14.6 Cusum plot of residuals from regression equation

Table 14.15   Residuals from the regression equation

Batch number	Residual	Batch number	Residual	Batch number	Residual
1	0.59	7	−0.24	13	−0.93
2	−0.79	8	0.82	14	0.12
3	1.57	9	1.08	15	−0.93
4	0.20	10	0.03	16	−0.06
5	0.76	11	0.08	17	−0.79
6	−0.39	12	−0.43	18	−0.68

Most of the early batches have positive residuals whilst a majority of the later batches have negative residuals. The mean of all 18 residuals is, of course, zero and there appears to be a change around batch number 11 from positive to negative. An explanation for this change springs to mind immediately. The whole experiment of 18 trials was actually carried out in two halves or two blocks with an interval of many days between batch 10 and batch 11. The plant manager made every effort to repeat the circumstances surrounding the first ten batches when manufacturing the last eight but the cusum plot suggests that he was not entirely successful.

The change in the mean level of the residuals is statistically significant, but is it important? The mean of the residuals for batches 1 to 10 is +0.363 and the mean for batches 11 to 18 is −0.453 which indicates a decrease in impurity of 0.816% in the later batches. Certainly a decrease of almost 1% cannot be ignored especially as this is a change over and above the variation in impurity accounted for by the independent variables in the regression equation.

So the change in the residuals is indicative of a change in impurity which is both significant and important, but does this new finding invalidate our early conclusions? Probably not, because both of the prime variables, x and t, have similar mean values in both halves of the experiment. We can, however, resolve the uncertainty by introducing an extra independent variable which distinguishes between the early batches and those produced later. This variable would be equal to 0 for trials 1 to 10 and equal to 1 for trials 11 to 18. Submitting this extended data set to the regression package would probably result in X being the first independent variable and T being the second (as we found earlier); then it is possible that the new variable would enter followed by an interaction, probably XZ or XT.

Perhaps the inspection of residuals has emphasized, yet again, that we cannot expect to analyse a set of data in one run of a regression package. The researcher would be well advised to regard data analysis as an interactive process in which computer runs are interspersed with periods of contemplation in which results and graphical representations are scrutinized.

14.7 SUMMARY

In this chapter we have completed the circle by returning to the problem which arose in Chapter 9, and we have, at last, advised the plant manager. In order to draw unambiguous conclusions we had first to extend the rather bad experiment which had already been carried out. The additional trials were carefully selected in order to eliminate the deficiencies that had been identified.

A multiple regression package was used in the analysis of the experimental results and several equations were fitted before we identified inlet temperature (t) and weight of special ingredient (x) as the most important variables to control. We also found that two interactions were quite important.

The use of contour diagrams was helpful when translating the regression equation into a practical strategy for future production. We noted, however, that this final step should not be taken until the residuals from the regression equation have been examined. Such an examination can serve as a check on some of the assumptions underlying regression analysis and, by means of a cusum plot, we spotted a change in the residuals which indicated a need for further analysis. To achieve further progress we could introduce a 'blocking variable' which was equal to 0 for the first block of trials and equal to 1 for the second block of trials.

In the final chapter we will examine a statistical technique which could have been used throughout the second part of this book. We managed very well without this technique but its study will certainly enlarge the reader's understanding of both design and analysis.

PROBLEMS

14.1 Knaresborough Breweries, well known for their 'light headed' beers, have decided to launch a new 'Strong-ail' beer. Ten experiments have been carried out to determine the effect of three variables on the quality of 'Strong-ail'. Only main effects are being investigated with no consideration being given to interactions. The three variables are:

Malt strain	F1 or F4
Fermentation temperature	High or low
Yeast	Enzyme or bioenzyme

The design of the experiment was as follows:

Cell number	Malt	Temperature	Yeast
1	F1	Low	Bioenzyme
2	F1	High	Enzyme
3	F4	Low	Enzyme
4	F4	Low	Bioenzyme
5	F4	Low	Enzyme
6	F1	High	Enzyme
7	F1	High	Enzyme
8	F4	Low	Bioenzyme
9	F1	Low	Bioenzyme
10	F1	Low	Bioenzyme

(a) List the design matrix in coded form using the values -1 and $+1$. Calculate the correlation between malt and temperature and hence complete the following correlation matrix:

	Malt	Temperature	Yeast
Malt	1.0		0.00
Temperature		1.0	-0.65
Yeast	0.00	-0.65	1.0

Refer to the solution for part (a) before attempting part (b).

(b) Extend the design by choosing two cells which will greatly improve the correlation matrix.

Refer to the solution for part (b) before attempting part (c).

(c) Complete the new correlation matrix given below:

	Malt	Temperature	Yeast
Malt	1.0	-0.17	$+0.17$
Temperature	-0.17	1.0	
Yeast	$+0.17$		1.0

14.2 Dr Scratchplan has decided to investigate the effect of temperature, pressure and concentration of catalyst on the whiteness of polymer. He has chosen three levels for each factor.

Temperature (°C)	270, 275, 280
Pressure (p.s.i.)	600, 700, 800
Concentration of catalyst (%)	0.15, 0.20, 0.25

Since he is away for the next week, suffering a course on experimental design, he produces the following plan for his laboratory assistant to carry out.

Stage 1: Set temperature at 275 °C and pressure at 700 p.s.i. and investigate the three levels of concentration of catalyst.

Stage 2: Choosing the concentration of catalyst which gave the whitest polymer, investigate the two remaining levels of pressure keeping temperature at 275 °C.

Stage 3: Choosing the 'best' pressure and concentration of catalyst, investigate the two remaining levels of temperature.

While on the course he realizes that temperature may have a curved relationship with whiteness and also that the interaction (pressure × concentration of catalyst) may be important. He phones his laboratory assistant and is given the levels for the seven cells in the experiment.

Cell	Temperature	Pressure	Concentration
1	275	700	0.15
2	275	700	0.20
3	275	700	0.25
4	275	600	0.25
5	275	800	0.25
6	270	600	0.25
7	280	600	0.25

(a) List the design matrix for temperature, pressure, concentration, $(\text{temperature})^2$ and (pressure × concentration) in coded form using -1, 0 and $+1$ for the three levels of each variable.

Refer to the solution for part (a) before attempting part (b).

(b) Complete the following correlation matrix:

	Temp.	Press.	Conc.	$(\text{Temp.})^2$	(Press × conc.)
Temp.	1.0	0.0	0.0	0.0	0.0
Press.		1.0	−0.24	−0.65	
Conc.			1.0	+0.37	−0.24
$(\text{Temp.})^2$				1.0	
(Press. × conc.)					1.0

Refer to the solution for part (b) before attempting part (c).

(c) Dr Scratchplan is somewhat concerned about the quality of his experimental design and wishes to extend the experiment but is informed by the laboratory assistant that there is only enough feedstock for a further two cells. Which two should Dr Scratch-plan choose?

Refer to the solution for part (c) before attempting part (d).

(d) Evaluate the design by completing the following correlation matrix:

	Temp.	Press.	Conc.	(Temp.)^2	(Press. × conc.)
Temp.	1.0	−0.19	−0.26	0.27	0.30
Press.		1.0	−0.24	−0.70	
Conc.			1.0	0.08	−0.24
(Temp.)^2				1.0	−0.10
(Press. × conc.)					1.0

15

Analysis of variance

15.1 INTRODUCTION

It is hoped that the reader will have acquired a working knowledge of multiple regression analysis before starting to read this chapter. If such a worthy objective has been achieved then the reader will realize that multiple regression analysis can be understood without any knowledge at all of a particular statistical technique known as 'analysis of variance'. Why, then, is it thought necessary to introduce this technique in the penultimate chapter? Three reasons can be offered:

(a) Many computer packages print out multiple regression results in analysis-of-variance 'language'.
(b) The majority of texts use analysis of variance as a foundation for multiple regression analysis.
(c) Presenting analysis of variance at this time offers a different perspective on many of the problems we have considered in earlier chapters.

It is hoped, therefore, that studying this chapter will help the reader to consolidate some important concepts and to broaden his/her statistical knowledge whilst exploring a technique which has an even wider field of application than multiple regression analysis itself.

15.2 TESTING ERROR AND SAMPLING VARIATION

In the plant manager's experiment we had one impurity determination on each of ten batches. No mention was made of how these determinations were obtained. In fact the analytical procedure for determining the percentage of this particular impurity in digozo blue pigment had previously been the subject of extensive investigation and considerable controversy. It had been suggested by production personnel that variation in sampling and testing of pigment caused the introduction of additional error which in some cases may have prevented the pigment from meeting the specification.

In order to compare the variation of impurity from sample to sample

Table 15.1 Five tests on each of four samples

Sample	A	B	C	D
Impurity determinations	5.1 5.2 5.3 5.3 5.1	5.3 5.5 5.8 5.5 5.4	5.8 5.5 5.8 5.8 5.6	5.2 5.0 5.3 5.3 5.2
Mean	5.20	5.50	5.70	5.20
SD	0.1000	0.1871	0.1414	0.1225

with the variation due to testing error, an investigation was carried out as follows. One batch was selected from the many batches produced under normal operating conditions. Four samples of pigment were extracted from this batch and five tests were carried out on each sample. The 20 impurity determinations are given in Table 15.1.

The data in Table 15.1 will form the basis for comparing the variation due to testing error with the variation due to heterogeneity of material. From these data we can estimate:

(a) the testing standard deviation (σ_t), which is a measure of the variation in impurity determinations that we would get if the test method were applied to pigment that was perfectly homogeneous;

(b) the sampling standard deviation (σ_s), which is a measure of the variation in impurity determinations that we would get if there were no testing error but the pigment exhibited its normal degree of heterogeneity;

(c) a confidence interval for the true impurity (μ) of the whole batch of pigment from which the samples were taken.

In order to obtain these three estimates we will use a technique known as one-way analysis of variance and the first step in this procedure is to break down the total variation of impurity into two components. To measure the total variation we will use the total sum of squares which is calculated in columns 2, 3 and 4 of Table 15.2. From each measurement we subtract the overall mean (5.40), square the deviations and then sum the squared deviations to obtain the total sum of squares (1.22).

We have been carrying out similar calculations since Chapter 1. Obviously the total sum of squares is just an old friend with a different name. The reader will realize that dividing the total sum of squares by its degrees of freedom (19) and then taking the square root would give us the standard deviation of the 20 impurity measurements. We do not need this standard deviation, however, for standard deviations cannot be subdivided in the way that we will partition the total sum of squares.

The within sample sum of squares is calculated in columns 5, 6 and 7 of Table 15.2. Once again we square deviations from a mean but we use the appropriate sample mean rather than the overall mean. The deviations in column 6 tend to be smaller than those in column 3 and the within sample sum of squares (0.32) is certainly less than the total sum of squares (1.22).

The between samples sum of squares is calculated in columns 8 and 9 of Table 15.2. For this calculation we square the deviations of the sample means from the overall mean. The sum of the squares is equal to 0.90. The calculations which are set out in Table 15.2 are also presented in a graphical form in Figures 15.1, 15.2 and 15.3. In all three diagrams the thick vertical lines represent the deviations which are squared and summed.

A third method of calculation, and one which would be preferred in practice, involves the use of the formulae below:

Total sum of squares

= (overall SD)^2(degrees of freedom)

Within samples sum of squares

= $\Sigma\{(\text{sample SD})^2(\text{degrees of freedom})\}$

Between samples sum of squares

= (SD of sample means)^2(number of samples − 1)

× (number of observations on each sample)

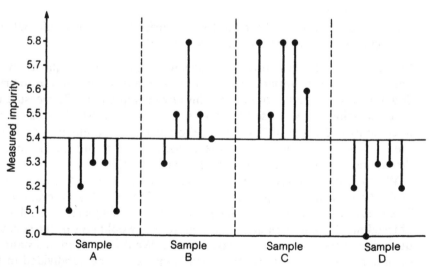

Figure 15.1   Total sum of squares

Table 15.2 Calculation of sums of squares

Sample	Measured impurity	Deviation of impurity from overall mean	Squared deviation	Sample mean	Deviation of impurity from sample mean	Squared deviation	Deviation of sample mean from overall mean	Squared deviation
1	2	3	4	5	6	7	8	9
A	5.1	-0.3	0.09		-0.1	0.01	-0.2	0.04
	5.2	-0.2	0.04	5.20	0.0	0.00	-0.2	0.04
	5.3	-0.1	0.01		0.1	0.01	-0.2	0.04
	5.3	-0.1	0.01		0.1	0.01	-0.2	0.04
	5.1	-0.3	0.09		-0.1	0.01	-0.2	0.04
B	5.3	-0.1	0.01		-0.2	0.04	0.1	0.01
	5.5	0.1	0.01	5.50	0.0	0.00	0.1	0.01
	5.8	0.4	0.16		0.3	0.09	0.1	0.01
	5.5	0.1	0.01		0.0	0.00	0.1	0.01
	5.4	0.0	0.00		-0.1	0.01	0.1	0.01

Table 15.2 Cont.

Sample	Measured impurity	Deviation of impurity from overall mean	Squared deviation	Sample mean	Deviation of impurity from sample mean	Squared deviation	Deviation of sample mean from overall mean	Squared deviation
1	2	3	4	5	6	7	8	9
C	5.8	0.4	0.16		0.1	0.01	0.3	0.09
	5.5	0.1	0.01		-0.2	0.04	0.3	0.09
	5.8	0.4	0.16	5.70	0.1	0.01	0.3	0.09
	5.8	0.4	0.16		0.1	0.01	0.3	0.09
	5.6	0.2	0.04		-0.1	0.01	0.3	0.09
D	5.2	-0.2	0.04		0.0	0.00	-0.2	0.04
	5.0	-0.4	0.16		-0.2	0.04	-0.2	0.04
	5.3	-0.1	0.01	5.20	0.1	0.01	-0.2	0.04
	5.3	-0.1	0.01		0.1	0.01	-0.2	0.04
	5.2	-0.2	0.04		0.0	0.00	-0.2	0.04
Total	108.0	0.0	1.22	—	0.0	0.32	0.0	0.90
Overall mean	5.40		↑			↑		↑
			Total sum of squares			Within sample sum of squares		Between samples sum of squares

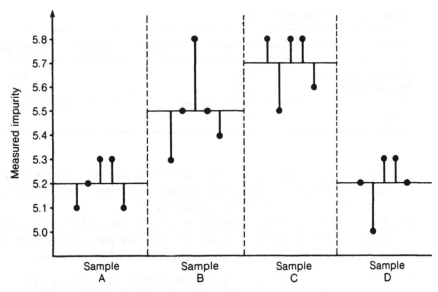

Figure 15.2 Within samples sum of squares

Use of these formulae involves the calculation of six standard deviations but this can be done very quickly with a modern calculator. The 'overall SD' is simply the standard deviation of all 20 measurements, and is equal to 0.2534. The four 'sample SDs' are given in Table 15.1 and the 'SD of sample means' is obtained by entering the four sample means into the calculator. Using these standard deviations we get the following results:

Total sum of squares

$$= (0.2534)^2(20 - 1)$$

$$= 1.22$$

Within samples sum of squares

$$= (0.1000)^2(5 - 1) + (0.1871)^2(5 - 1)$$
$$+ (0.1414)^2(5 - 1) + (0.1225)^2(5 - 1)$$

$$= 0.32$$

Between samples sum of squares

$$= (0.2449)^2(4 - 1)(5)$$

$$= 0.90$$

The reader will notice that the use of these formulae has given identical results to those in Table 15.2. Two points should be noted, however:

Table 15.3   Analysis of variance table

Source of variation	Sum of squares	Degrees of freedom	Mean square
Between samples	0.90	3	0.30
Within samples	0.32	16	0.02
Total	1.22	19	—

(a) It is important that the standard deviations are not rounded prematurely. Four significant figures have been used but it would be better to carry all the figures given by the calculator.
(b) The third of the three equations must only be used when the same number of tests has been made on each sample, but the first two equations can be used with unequal numbers of observations.

Point (b) need never be a restriction in practice for we only ever need to use two of the equations. This is because the 'within samples sum of squares' plus the 'between samples sum of squares' will always equal the total sum of squares. This is easily seen in Table 15.3 where all three sums of squares have been gathered together.

We see also in Table 15.3 that the degrees of freedom can be added in

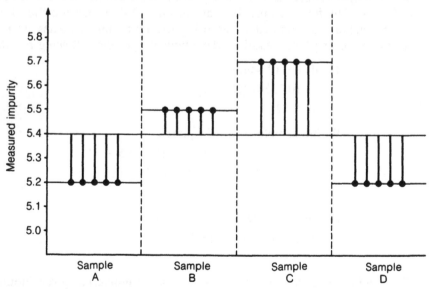

Figure 15.3   Between samples sum of squares

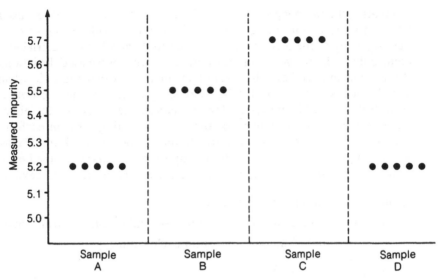

Figure 15.4   Hypothetical data with no testing error

the same manner as the sums of squares. The total sum of squares has 19 degrees of freedom, of course, because it is based on the standard deviation of 20 measurements. The within samples sum of squares has 16 degrees of freedom because it was calculated from four standard deviations, each of which had 4 degrees of freedom. The between samples sum of squares has 3 degrees of freedom because it is based on the standard deviation of four sample means.

The mean squares in an analysis of variance table are calculated by dividing each sum of squares by its degrees of freedom. It is usual to omit the total mean square from the table for two reasons – firstly because it is not needed, and secondly because the mean squares are not additive like the sums of squares and degrees of freedom. From the mean squares in Table 15.3 we can obtain estimates of the sampling standard deviation (σ_s) and the testing standard deviation (σ_t). The latter is easier to estimate so we will turn to that one first.

The within samples sum of squares is not due to heterogeneity of material but is entirely the result of testing error. If there were no testing error then the data from our experiment would resemble the hypothetical data in Figure 15.4 and the within samples sum of squares would be equal to zero. It follows, therefore, that we can use the square root of the within samples mean square as an estimate of the testing standard deviation.

$$\text{Estimate of testing standard deviation } (\sigma_t) = \sqrt{0.02}$$
$$= 0.1414$$

Estimating the sampling standard deviation is not quite so easy because the between samples sum of squares is influenced by both testing error and heterogeneity of the pigment. It is obvious that variation in material would contribute to the between samples sum of squares for we used deviations of the sample means from the overall mean in its calculation. If all four sample means were equal, the between samples sum of squares would be zero. But we would not expect identical sample means even if the pigment were perfectly homogeneous, for the random testing error in each measurement would work its way through to the sample means. For this reason we take into account both the within samples mean square and the between samples mean square when we estimate the sampling standard deviation.

Estimate of sampling standard deviation (σ_s)

$= \sqrt{[}$(between samples mean square – within samples mean square)/

(number of measurements on each sample)$]$

$$= \sqrt{\left(\frac{0.30 - 0.02}{5}\right)}$$

$= 0.237$

Before we make use of this estimate it would be wise to check the possibility that one of its two contributing influences might not exist. We have suggested that the between samples mean square may be partly due to testing error and partly due to heterogeneity of material. We would only get a mean square equal to zero if both of these were absent. Is it possible, then, that our between samples mean square (0.30) is the result of only one of these two causes? Perhaps $\sigma_t = 0$ or perhaps $\sigma_s = 0$.

We can rule out, immediately, the possibility that the testing standard deviation is equal to zero because the within samples mean square is not equal to zero. The other suggestion $(\sigma_s = 0)$ cannot be rejected so easily, of course, because the testing error alone would give a positive value to the between samples mean square.

It can be shown, however, that the two mean squares would be approximately equal if there were no sampling variation. We have found the between samples mean square (0.30) to be considerably larger than the within samples mean square (0.02). Should we conclude that:

(a) the material is heterogeneous (i.e. $\sigma_s > 0$), or
(b) an extremely unusual combination of random testing errors has resulted in very large differences between the sample means.

To help us decide whether (a) or (b) is the most reasonable conclusion we can carry out a significance test. The reader may recall from Chapter 5 that we use the F-test when we wish to compare two sources of variation.

Null hypothesis – pigment impurity does not vary from sample to sample throughout the batch (i.e. $\sigma_s = 0$).

Alternative hypothesis: $\sigma_s > 0$

$$\text{Test statistic} = \frac{\text{between samples mean square}}{\text{within samples mean square}}$$

$$= \frac{0.30}{0.02}$$

$$= 15.0$$

Critical values – from the one-sided F-table with 3 and 16 degrees of freedom:

3.24 at the 5% significance level
5.29 at the 1% significance level.

Decision – we reject the null hypothesis with great confidence.

Conclusion – we conclude that there is variation in impurity throughout the batch.

The very large test statistic implies that the heterogeneity of the material gives rise to variation in impurity measurements which is considerably greater than the variation due to testing error.

Having estimated the sampling and testing standard deviations we can now make use of these estimates. Whenever we attempt to assess the impurity of a batch of this particular pigment, we must take one or more samples from the bulk and make one or more determinations on each sample. Obviously, the larger the number of samples and the larger the number of determinations on each, the greater will be the precision of our estimate. If we take m samples and make p determinations on each then a 95% confidence interval for the true mean impurity of the batch is given by

$$\bar{x} \pm 1.96\sqrt{[\sigma_s^2/m + \sigma_t^2/mp]}$$

where $\bar{x}$ is the mean of the mp determinations, σ_s the sampling standard deviation, σ_t the testing standard deviation and 1.96 is from the normal distribution.

We cannot use this formula as it stands, because we do not know the true values of σ_s and σ_t. However, we do have estimates and these can be inserted into the formula provided we replace the 1.96 with a suitable value from the t-table. What degrees of freedom should we use? We know that the within samples sum of squares has 16 degrees of freedom and the between samples sum of squares has only 3. If we take the geometric mean of these two, i.e. $\sqrt{(16 \times 3)}$, we get a rough estimate of the appropriate

Table 15.4 Confidence limits for the true mean impurity of a batch

Number of samples taken from the batch (m)	Number of determinations on each sample (p)			
	1	2	4	8
1	0.65	0.61	0.58	0.57
2	0.46	0.43	0.41	0.40
4	0.33	0.31	0.29	0.29

degrees of freedom. The formula which would give us the correct degrees of freedom is rather involved, unfortunately. It can be found in Wilson (1979). We can see in Table 15.4 that one determination on each of four samples gives a much narrower interval than 4 determinations on one sample. Clearly it is more effective to increase the number of samples than to increase the number of determinations on each.

15.3   TESTING ERROR AND PROCESS CAPABILITY

In Chapter 3 we briefly discussed process capability. You may recall that we examined the viscosity data for 80 consecutive batches of RGX 200. Four of these batches were found to have viscosity outside the specification limits of 180 and 220 cSt. The standard deviation of the 80 viscosity determinations was 8.336 cSt which gave a capability index, Cp, of 0.80. Clearly the process is incapable of producing RGX 200 which will meet the specification every time. Some batches will require rework or blending.

There are two ways by which we could increase the capability index. We could either widen the specification interval or we could reduce the variability of the process. The first of these alternatives would be easier to achieve. We could ask the customer, 'Is it really important that the viscosity of the RGX 200 should be greater than 180 and less than 220?' The customer may agree to a change in the specification. Within the chemical and allied industries there must be many specifications which were derived quite arbitrarily or which reflect requirements that have long since changed.

If it is not possible to change the specification we must turn our attention to reducing the process variability. Of course, the quality gurus would suggest that the reduction of process variability should have been the focus of our attention before now. We should be pursuing a policy of continuous improvement and one of our objectives should be to increase the consistency of the product.

How are we to reduce the variability of the RGX process? The task would certainly be less daunting if we knew what factors caused the variability. Well we do. At least we know some of the causes, if not all of them. Obviously, part of the variation in viscosity must be due to variation in the materials we use. We can also assert with confidence that part of the variation in viscosity must be due to testing error, part due to variability in operator performance and part due to climatic changes, etc. However, what we do not know is how much of the variation in viscosity is caused by each of these factors.

If we carried out a suitable experiment and analysed the results using analysis of variance, we could estimate what percentage of the total variation was due to each cause. We could then decide what action should be taken to reduce the variation. If, for example, a substantial proportion of the variation in viscosity were due to unpredictable fluctuations in a particular raw material, then we would be wise to focus our attention on the supplier. If, on the other hand, we found that a large percentage of the variation was due to testing error, we would be better employed developing a superior analytical method for the determination of viscosity.

Let us return to the viscosity data which had a standard deviation of 8.336 cSt. Clearly, part of this variation is due to real changes in viscosity from batch to batch, and part due to testing error. Unfortunately, we cannot separate the two components of variation because we have only one determination per batch. Suppose, however, that we had 80 viscosity results which were obtained by making repeat determinations on 40 batches. Using analysis of variance we could separate the true variation in viscosity from the testing error. Let us say this gave a testing standard deviation of 5.219 cSt and a process standard deviation of 6.500 cSt. These results are compatible with the total standard deviation of 8.336 cSt that we

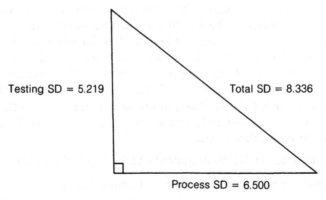

Figure 15.5 Testing error and process variability

calculated earlier. The relationship between the three is illustrated in Figure 15.5 and is encapsulated in the formula

$$\text{Total SD} = \sqrt{[(\text{testing SD})^2 + (\text{process SD})^2]}$$

The term 'process standard deviation' covers all the variation in viscosity determinations that remains when we have removed the extra variation due to testing error. If there were no variation in true viscosity from batch to batch, the process SD would be equal to zero and the total SD would reduce to 5.219 cSt. If, on the other hand, we had the perfect test method, the testing SD would be zero and the total SD would reduce to 6.500 cSt. With this analytical excellence in mind let us reassess the capability of the process. Clearly, any reduction in the total SD would increase the capability index

$$Cp = 2T/6\sigma$$
$$= (220 - 180)/(6 \times 6.500)$$
$$= 1.03$$

Thus, if we were able to eliminate the testing error completely, we would have a process capability index of 1.03 compared with the 0.8 calculated in Chapter 3. This suggests that our process could be far more capable than it is at the moment. Indeed, it could be argued that the process actually is more capable, with the Cp of 1.03 being more realistic than the 0.8. Those who put forward this point of view would maintain that the testing error does not increase the real variation in viscosity, it simply increases our estimate of the variation. However, the customer could balance this argument by asserting that the testing error does influence the process capability, for its presence reduces our ability to maintain the mean viscosity close to the target value of 200 cSt.

The total variation in the 80 viscosity determinations could be further subdivided. If the 40 batches had been made using 5 consignments of raw material we could separate the between consignment variation from that within consignments. This would allow us to estimate the consignments SD and to add a second triangle to Figure 15.6. Suppose the consignment SD were found to be 4.642 cSt. Figure 15.6 shows how the total SD can be broken down into three components, testing SD, consignments SD and residual SD. The latter includes all the process variation that remains after we have extracted the variation in viscosity due to using different consignments of the raw material. The relationships in Figure 15.6 can also be expressed in the three equations:

Process SD $= \sqrt{[(\text{consignments SD})^2 + (\text{residual SD})^2]}$

Total SD $= \sqrt{[(\text{process SD})^2 + (\text{testing SD})^2]}$

Total SD $= \sqrt{[(\text{consignments SD})^2 + (\text{residual SD})^2 + (\text{testing SD})^2]}$

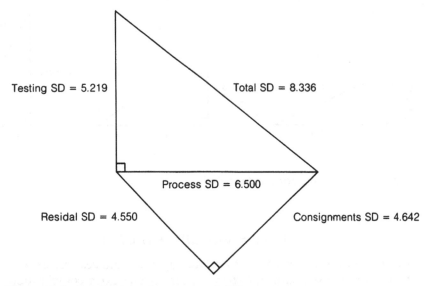

Figure 15.6   Subdivision of process variability

Clearly this subdivision of the total variation in viscosity could continue many stages further. The residual SD could be split into 'operator SD' and a new but smaller 'residual SD'. The consignments SD could be split into 'between suppliers SD' and 'within suppliers SD'. The testing SD could be split into 'between laboratories SD' and 'within laboratories SD'. Of course, the experiment that gives us the viscosity determinations would need to be planned so as to enable the subdivisions to take place. Increased understanding of the process usually has a price tag. However, it can be argued that ignorance of the process also has a price tag, which can include the cost of lost business.

   We have subdivided the total variation to obtain several standard deviations. Each of these standard deviations can be expressed as a capability index if we divide the width of the specification interval by 6 standard deviations. Thus we could define:

$Cp = 2T/(6 \times \text{total SD})$ Process capability as defined in Chapter 3

$CT = 2T/(6 \times \text{testing SD})$ Test capability

$CP = 2T/(6 \times \text{process SD})$ True process capability

Using these definitions we can express the standard deviation equation

$$\text{Total SD} = \sqrt{[(\text{process SD})^2 + (\text{testing SD})^2]}$$

in terms of capability indices

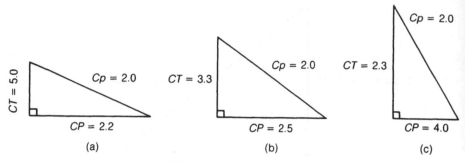

Figure 15.7   Testing error and capability

$$(1/Cp) = \sqrt{[(1/CP)^2 + (1/CT)^2]}$$

Clearly we can use triangles to relate capability indices just as we used them to relate standard deviations. There is one extra complication; we must use reciprocals of the indices.

The three triangles in Figure 15.7 represent three combinations of CT and CP which give a Cp of 2.0. Thus in all three cases the measured variability in the product is the same. However, the proportion of this variability which is due to testing error changes from triangle to triangle. With the very imprecise test method in triangle (c) we require an extremely capable process to meet the customer's demand for a Cp of 2.0. It is hard to imagine how we could achieve this if our SPC system were dependent on the same test method. In triangle (a) the much more precise test method will allow us to meet the customer's demand if the true process capability is as low as 2.2.

15.4   ANALYSIS OF VARIANCE AND SIMPLE REGRESSION

In Chapter 10 we fitted a simple regression equation ($y = a + bx$) and then we calculated a residual sum of squares. It was pointed out at the time that this calculation could be based upon the distances of the points from the regression line. These distances, known as residuals, would be squared and the squares added to obtain the residual sum of squares. The method is illustrated in Figure 15.8.

Compare Figure 15.8 with Figure 15.9. The latter illustrates the calculation of the total sum of squares which is based on deviations from the overall mean ($\bar{y} = 5.0$). Clearly the deviations of the points from the horizontal line in Figure 15.9 tend to be greater than the deviations from

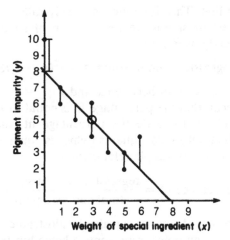

Figure 15.8   Residual sum of squares

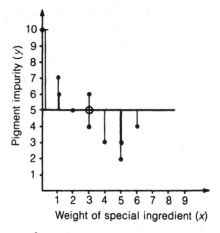

Figure 15.9   Total sum of squares

the regression line in Figure 15.8 and the total sum of squares (50.0) is therefore greater than the residual sum of squares (14.0).

The horizontal line in Figure 15.9 and the regression line in Figure 15.8 both pass through the centroid. The method of least squares can be regarded as a process which starts with the horizontal line and rotates it to a new position which minimizes the residual sum of squares. The difference between the total sum of squares and the residual sum of squares is a measure of how much of the variation in impurity has been explained by the regression line. This difference is usually called the 'due to regression sum of squares' or simply 'regression sum of squares' and it could be calculated directly as follows:

$$\text{Regression sum of squares} = r^2(n - 1)(\text{SD of } y)^2$$

where r is the correlation between x and y.

You may recall that the percentage fit is equal to 100 r^2. It would seem possible, therefore, to relate the percentage fit, the regression sum of squares and the total sum of squares [which is equal to $(n - 1)(\text{SD of } y)^2$]. Indeed, there is a very simple relationship:

$$\text{Percentage fit} = \frac{\text{regression sum of squares}}{\text{total sum of squares}} \times 100$$

The regression sum of squares can also be illustrated graphically. In Figure 15.10 each of the ten points represents a batch but the predicted impurity values were used in plotting the points, rather than the actual impurity values used in Figures 15.8 and 15.9.

It is interesting to contrast Figures 15.8, 15.9 and 15.10 on the one hand with Figures 15.1, 15.2 and 15.3 on the other. Comparing Figure 15.9 with Figure 15.1 we see that the total sum of squares is calculated in the same way for both sets of data, as it would be calculated for any set of data. Comparing Figure 15.8 with Figure 15.2 we see a similarity between the residual sum of squares and the within samples sum of squares. In both

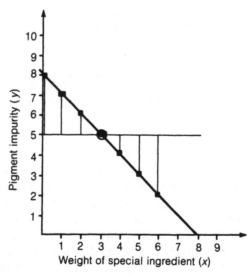

Figure 15.10   Due to regression sum of squares

Table 15.5 Analysis of variance table (simple regression)

Source of variation	Sum of squares	Degrees of freedom	Mean square
Due to regression on x	36.0	1	36.0
Residual	14.0	8	1.75
Total	50.0	9	—

cases we are measuring the deviations of the points from a fitted line; in Figure 15.8 it is a regression line, whilst in Figure 15.2 it is a stepped horizontal line which passes through the sample means. To complete the comparison we see a similarity between the regression sum of squares in Figure 15.10 and the between samples sum of squares in Figure 15.3. In both of these diagrams we are measuring the deviation of the fitted line from the horizontal line.

With both sets of data we are using analysis of variance to break down the total variation into two components. In both cases we could refer to one component as 'assignable variation' because it is introduced intentionally, by changing weight of special ingredient in one case and by taking different samples in the other. The second component of variation could be described as 'unassignable variation' or 'residual variation' since it has not been accounted for.

To complete our application of analysis of variance to the simple regression example we can draw up an analysis of variance table and follow this with an F-test. In Table 15.5 we see again the additive property of the sums of squares and the degrees of freedom. The residual sum of squares has $(n - 2)$ degrees of freedom (as it did in Chapter 10) because we calculated two constants, the slope and the intercept, when fitting the equation. The due to regression sum of squares has 1 degree of freedom because we have one independent variable in the equation.

Null hypothesis – there is no relationship between weight of special ingredient (x) and pigment impurity (y) (i.e. the true slope, β, is equal to zero).

Alternative hypothesis – there is a relationship between x and y (i.e. $\beta \neq 0$).

$$\text{Test statistic} = \frac{\text{due to regression on } x \text{ mean square}}{\text{residual mean square}}$$

$$= \frac{36.0}{1.75}$$

$$= 20.57$$

Critical values – from the one-sided F-table with 1 and 8 degrees of freedom:

5.32 at the 5% significance level
11.26 at the 1% significance level.

Decision – we reject the null hypothesis at the 1% level of significance.

Conclusion – we conclude that, within the population of batches, pigment impurity (y) is related to the weight to special ingredient (x).

We have, of course, reached this same conclusion twice already. The first occasion was when we tested the significance of the correlation between x and y, in Chapter 10. The second occasion was also in Chapter 10, when we tested the significance of the slope b (or the significance of the percentage fit, if you prefer that viewpoint). The one-sided F-test above will always lead us to the same conclusion as the two-sided t-test or the two-sided correlation test.

15.5 ANALYSIS OF VARIANCE WITH MULTIPLE REGRESSION

The reader may recall that, in Chapter 11, a second independent variable was introduced into the regression equation and the percentage fit increased from 72% to 93.4%. It was demonstrated by means of a t-test that this increase in percentage fit was statistically significant and we concluded that the second independent variable (t) was related to the pigment impurity. We concluded that it was important to control both the weight of special ingredient (x) and the inlet temperature (t) in the production of future batches.

Let us now apply analysis of variance to this second step of the multiple regression analysis. After fitting the second equation we calculated a new set of residuals in Table 11.4. By squaring these residuals and summing the squares we obtain the new residual sum of squares which is equal to 3.30. This figure is, of course, less than the residual sum of squares from the first equation (14.00). Since the total sum of squares remains unchanged by the introduction of another independent variable, any reduction in the residual sum of squares must be matched by an increase in the regression sum of squares as we can see in Table 15.6.

Note that the regression sum of squares now has 2 degrees of freedom because we have two independent variables in the equation. The residual degrees of freedom become ($n - 3$) because we have calculated three constants from the data (i.e. one intercept and two slopes). We can, as always, follow the analysis of variance table with a one-sided F-test to check the significance of the two independent variables.

Table 15.6 Analysis of variance with two independent variables

Source of variation	Sum of squares	Degrees of freedom	Mean square
Due to regression on x and t	46.70	2	23.35
Residual	3.30	7	0.471
Total	50.00	9	—

Null hypothesis – pigment impurity is not related to weight of special ingredient (x) and/or inlet temperature (t).

Alternative hypothesis – pigment impurity is related to weight of special ingredient (x) or inlet temperature (t).

$$\text{Test statistic} = \frac{\text{due to regression on } x \text{ and } t \text{ mean square}}{\text{residual mean square}}$$

$$= \frac{23.35}{0.471}$$

$$= 49.57$$

Critical values – from the one-sided F-table with 2 and 7 degrees of freedom:

4.74 at the 5% significance level
9.55 at the 1% significance level.

Decision – we reject the null hypothesis at the 1% significance level.

Conclusion – we conclude that pigment impurity is related to weight of special ingredient and/or inlet temperature.

This significance test is perfectly valid but it has served no useful purpose. To reach a conclusion that 'y is dependent on x and/or t' does not constitute a step forward as we have already concluded that 'y is dependent on x'. The increase in percentage fit when t is introduced may or may not be significant, but Table 15.6 and the subsequent F-test do not help us to decide. To progress further we must take the regression sum of squares from Table 15.6 and subdivide it into two components. These are the two entries above the broken line in Table 15.7.

The sums of squares above the broken line add up to the sum of squares immediately below the line. The same is true of the degrees of freedom. The entires below the broken line are taken from Table 15.6 whilst the 'due to regression on x' row is taken from Table 15.5. Obviously, we have

Table 15.7 Separating the effects of the two variables

Source of variation	Sum of squares	Degrees of freedom	Mean square
Due to regression on x	36.00	1	—
Due to introduction of t	10.70	1	10.70
Due to regression on x and t	46.70	2	—
Residual	3.30	7	0.471
Total	50.00	9	—

broken down the regression sum of squares (46.70) into two components, the first being the 36.00 which was tested earlier and the second being 10.70 which has resulted from the introduction of the second independent variable and which will now be subjected to an F-test.

Null hypothesis – pigment impurity is not related to inlet temperature (t).

Alternative hypothesis – pigment impurity is related to inlet temperature (t).

$$\text{Test statistic} = \frac{\text{due to inclusion of } t \text{ mean square}}{\text{residual mean square}}$$

$$= \frac{10.70}{0.471}$$

$$= 22.72$$

Critical values – from the one-sided F-table with 1 and 7 degrees of freedom:

5.59 at the 5% significance level
12.25 at the 1% significance level.

Decision – we reject the null hypothesis at the 1% significance level.

Conclusion – we conclude that the pigment impurity (y) is dependent on the inlet temperature (t) as well as the weight of special ingredient (x).

Use of analysis of variance and the one-sided F-test has led us to exactly the same conclusion that we reached in Chapter 11 when using the two-sided t-test. Many multiple regression packages print out analyses of variance tables and carry out F-tests. This may be quite acceptable to the experienced user but the beginner would surely find the t-test easier to follow. In expressing this opinion the author does not wish to denigrate analysis of variance. There can be no better foundation on which to build an understanding of many interrelated multivariate techniques but analysis of variance can prove a difficult obstacle for the novice to overcome.

15.6 ANALYSIS OF VARIANCE AND FACTORIAL EXPERIMENTS

This very versatile technique can be used to analyse the results of a factorial experiment. By means of analysis of variance we can calculate a mean square for each main effect and for each interaction. The calculations are similar to those in Table 15.2 when we carried out a one-way analysis of variance to separate the within samples and the between samples sums of squares. For a 2^2 factorial experiment we would use two-way analysis of variance as follows.

We will use the results of the 2^2 factorial experiment in Table 15.8 to calculate:

(a) between feedrates sum of squares;
(b) between temperatures sum of squares;
(c) interaction sum of squares.

These sums of squares could equally well be called 'between rows sum of squares', 'between columns sum of squares' and 'between diagonals sum of squares' because of the way in which they are calculated. Each of the sums of squares is based upon the deviation of group means from the overall mean. There are three possible groupings, however, and in Table 15.9 the yields of the four batches are split into two groups depending upon the temperature that was used in the manufacture.

Clearly the calculations in Table 15.9 are based on exactly the same procedure as that followed in Table 15.2. We have carried out a one-way analysis of variance using one of the independent variables, temperature, as a criterion for splitting the yield values into two groups. If we had used

Table 15.8 Results of a 2^2 factorial experiment (from Table 12.4)

Feedrate	Temperature (°C)	
	Low (60)	High (70)
Low (40)	76	72
High (50)	70	78

Table 15.9

Temperature	Yield	Deviation of yield from overall mean	Squared deviation	Group mean yield	Deviation of yield from group mean	Squared deviation	Deviation of group mean from overall mean	Squared deviation
1	2	3	4	5	6	7	8	9
Low	76	2	4	73	3	9	−1	1
	70	−4	16		−3	9	−1	1
High	72	−2	4	75	−3	9	1	1
	78	4	16		3	9	1	1
Total	296	0	40	—	0	36	0	4
Mean	74		← Total sum of squares			← Within temperatures mean square		← Between temperatures mean square

the other independent variable, feedrate, as a grouping criterion then the group means would have been the two row means of Table 15.9 and we would have obtained:

total sum of squares = 40

within feedrate sum of squares = 40

between feedrates sum of squares = 0

A third method of classification using the diagonal means would have given:

total sum of squares = 40

within diagonals sum of squares = 4

between diagonals (i.e. interaction) sum of squares = 36

As a result of carrying out these three one-way analyses of variance we can extract the sum of squares due to each independent variable and their interaction. These are given in Table 15.10.

Comparing the sums of squares in Table 15.10 with those in Table 15.9 we can see that the 'within temperatures sum of squares' (36) cannot be classed as unassignable variation (due to sampling or testing error perhaps) but must be credited to the other independent variable (feedrate) and the interaction. Similarly the 'within feedrates sum of squares' (40) must be attributed to the effect of temperature and to the temperature × feedrate interaction. In fact all of the variation in yield must be credited to the three sources listed in Table 15.10 with no residual variation remaining. It is impossible, therefore, to test the mean squares in Table 15.10 by means of an F-test.

We had the same problem in Chapter 12, of course, when we attempted to analyse this data using the t-test. On that occasion we resorted to carrying out a second replicate of the 2^2 factorial experiment and the results of both replicates are given in Table 15.11.

Now that we have more than one observation in each cell of Table 15.11

Table 15.10 Two-way analysis of variance table

Source of variation	Sum of squares	Degrees of freedom	Mean square
Between temperatures	4	1	4.0
Between feedrates	0	1	0.0
Interaction	36	1	36.0
Total	40	3	—

Table 15.11 Yield from 2×2^2 factorial experiment (from Table 12.8)

	Temperature (°C)	
Feedrate	Low (60)	High (70)
Low (40)	77 75	73 71
High (50)	67 73	80 76

we could estimate the residual standard deviation to use in a t-test. In Chapter 12 we did just that. Alternatively we can use analysis of variance to subdivide the total variation into two components, one which can be assigned to the independent variables and one which we will call residual. This is done by using one-way analysis of variance with the eight yield values split into four groups such that all members of a group were produced under identical conditions. Each group corresponds to one of the four cells in Table 15.11. We will calculate the sums of squares by means of the formulae given earlier in this chapter.

Total sum of squares
= (overall SD)^2(total number of observations − 1)
= $(3.9641)^2(8 − 1)$
= 110.0

Within groups sum of squares
= Σ(group SD)^2(number of observations in group − 1)
= $(1.4142)^2(2 − 1) + (1.4142)^2(2 − 1)$
  $+ (4.2426)^2(2 − 1) + (2.8284)^2(2 − 1)$
= 30.0

Between groups sum of squares
= (SD of group means)^2(number of groups − 1)
  × (number of observations in each group)
= $(3.6515)^2(4 − 1)(2)$
= 80.0

Carrying out an F-test on the mean squares in Table 15.12 would show that the between groups mean square was not significantly greater than the within groups mean square. The within groups mean square is a measure of the variability in yield that we would expect to find between batches produced with the same values of temperature and feedrate. The between

Table 15.12   One-way analysis of variance on 2×2^2 factorial experiment

Source of variation	Sum of squares	Degrees of freedom	Mean square
Between groups	80.0	3	26.67
Within groups	30.0	4	7.5
Total	110.0	7	—

Table 15.13   Two-way analysis of variance on 2×2^2 factorial experiment

Source of variation	Sum of squares	Degrees of freedom	Mean square
Between temperatures	8.0	1	8.0
Between feedrates	0.0	1	0.0
Interaction	72.0	1	72.0
Between groups	80.0	3	—
Within groups (residual)	30.0	4	7.5
Total	110.0	7	—

groups mean square is also subject to this residual variation but may also be inflated by the effects of temperature and feedrate changes. Are we unable to conclude, then, that the temperature and feedrate effects exist? Despite the inconclusive F-test it would be unwise to abandon the analysis at this point for we can test the two main effects and the interaction separately. To do so we must break down the between groups sum of squares into three components (Table 15.13).

The sums of squares above the broken line in Table 15.13 are calculated by considering the variation in the row means, column means, and the diagonal means of Table 15.11. F-tests can now be carried out to compare the mean squares above the line with the residual mean square and such tests would reveal that only the interaction between temperature and feedrate was significant. This is precisely the same conclusion that we reached in Chapter 12 when using the t-test.

15.7 SUMMARY

In this chapter we have travelled a familiar route riding in a new vehicle, analysis of variance. It is hoped that you have recognized many landmarks first encountered in earlier chapters and that you have viewed the whole terrain from a new perspective. Perhaps any concepts which appeared disjoint when first encountered will now have assumed their rightful place in the whole ethos of design and analysis.

PROBLEMS

15.1 The production manager of Indochem knows that the quality of the product is related to the concentration of a particular impurity in a liquid feedstock. Furthermore it is suspected that this impurity gradually settles during storage of the feedstock. To investigate this possibility an R & D chemist is asked to take three samples of feedstock from a storage tank that has been undisturbed for several days. The samples are taken at depths of 0.5 m, 2.5 m and 4.5 m, then five determinations of impurity content are made on each sample.

Sample	Top			Middle			Bottom		
Determinations	3.2	3.3	3.1	3.3	3.4	3.3	3.6	3.8	3.7
of impurity	3.3	3.1		3.5	3.0		3.4	4.0	

(a) Calculate the total sum of squares.
(b) Calculate the within samples sum of squares.
(c) Calculate the between samples sum of squares.
(d) Draw up a one-way analysis of variance table.
(e) Carry out an F-test to compare the between sample variation with the within sample variation.
(f) Calculate an estimate of the testing standard deviation.
(g) What conclusions can you draw concerning the settlement of the impurity within the tank?
(h) What other method of analysis would be appropriate in this situation?

15.2 We could carry out a simple regression analysis on the data in Problem 1, using depth (x) as the independent variable and impurity (y) as the response. Feeding the 15 pairs of numbers into a regression analysis program results in the following equation being printed out:

$$y = 3.0875 + 0.1250x \qquad 57.87\% \text{ fit}$$

(a) Calculate the total sum of squares.
(b) Calculate the residual sum of squares.
(c) Calculate the regression sum of squares.
(d) Draw up an analysis of variance table.
(e) Carry out an F-test to check the statistical significance of the relationship between depth and impurity.

15.3 We could carry out a multiple regression analysis on the data in Problem 1 using depth (x) and $(depth)^2$ as independent variables.

Feeding the 15 pairs of numbers into a regression package, and asking for scaling of x into $X = (x - 2.5)/2.0$ before the generation of X^2, we get the following equations printed out:

$$y = 3.4000 + 0.2500X \qquad\qquad \text{57.87\% fit}$$
$$y = 3.3000 + 0.2500X + 0.1500X^2 \qquad \text{64.815\% fit}$$

The analysis of variance table for the first equation will be identical with the one produced in problem 2(d). Scaling the independent variable has not changed either the standard deviation of y or the correlation between x and y, so the sums of squares are unchanged.

(a) Complete the analysis of variance table below:

Source of variation	Sum of squares	Degrees of freedom	Mean square
Due to regression on X		—	
Due to introduction of X^2			
Due to regression on X and X^2		—	
Residual			
Total		—	

(b) Carry out an F-test on the mean squares in the above table.
(c) Compare the analysis of variance table from Problem 1 with the one produced in this problem and list any similarities.

16
An introduction to Taguchi techniques

16.1 INTRODUCTION

Almost certainly you have heard of 'Taguchi methods'. Very likely you have read that they are in widespread use throughout Japanese industry. They are certainly attracting a great deal of attention from engineers in the UK and the USA. In this chapter I will introduce the various techniques and concepts that are currently packaged as 'Taguchi methods'. The scope of these techniques is enormous. I cannot, in one chapter, equip you to make full use of them, but I can offer you an overview which will help you to get the various techniques into perspective.

Genichi Taguchi is an engineer and speaks the language of engineering, in Japanese. Several of his texts have been translated into English, but I, and many others, find them very difficult to understand in parts. I am tempted to blame the translators. There can be no doubt that Taguchi's publications contain some very important and novel ideas for anyone who wishes to improve the quality of manufactured products. He has been awarded the annual Deming prize on several occasions in recognition of his achievements. I ask you at the outset to read this chapter with an open mind. Please do not judge Taguchi techniques until you have embraced their full scope. You may find some of his ideas very exciting, as I do, but you would be very unwise to accept the whole package, without considering the criticism that has been directed at certain elements of 'Taguchi methods'.

16.2 ORTHOGONAL ARRAYS

Taguchi strongly recommends the use of orthogonal arrays in the planning of experiments. Indeed, it is no overstatement to say that orthogonal arrays are the prime focus of his two-volume text, *System of Experimental Design*. Clearly, therefore, it is important to grasp the essentials of these arrays if you are to evaluate the usefulness of Taguchi methods.

Table 16.1 Orthogonal array L8

No.	Column						
	1	2	3	4	5	6	7
1	1	1	1	1	1	1	1
2	1	1	1	2	2	2	2
3	1	2	2	1	1	2	2
4	1	2	2	2	2	1	1
5	2	1	2	1	2	1	2
6	2	1	2	2	1	2	1
7	2	2	1	1	2	2	1
8	2	2	1	2	1	1	2

Table 16.2 Design matrix for a 2^3 factorial experiment

	X	Z	XZ	W	XW	ZW	XZW
1	−	−	+	−	+	+	−
2	+	−	−	−	−	+	+
3	−	+	−	−	+	−	+
4	+	+	+	−	−	−	−
5	−	−	+	+	−	−	+
6	+	−	−	+	+	−	−
7	−	+	−	+	−	+	−
8	+	+	+	+	+	+	+

Table 16.1 contains an orthogonal array. It is presented exactly as in Taguchi's publications. You can see that it has eight rows and seven columns, and that each column contains four 1's and four 2's. If you selected any two columns from Table 16.1 and calculated their correlation coefficient you would find that it was equal to zero. If you calculated the mean for each column you would find that all the column means were equal. Perhaps you recall similar tables in earlier chapters of this book.

Table 16.2 is a copy of part of Table 13.9 and is very similar to Tables 13.1 and 12.19. In earlier chapters we have referred to such a table as a design matrix. Clearly the design matrix for a 2^3 factorial experiment bears a striking resemblance to Taguchi's L8 orthogonal array. Both contain eight rows and both contain seven uncorrelated columns. (Orthogonal simply means uncorrelated.) In fact the L8 array would be identical to the matrix in Table 16.2, if we replaced '2' by '−' and replaced '1' by '+', then turned it upside down. We would then find that Taguchi's column 1 was the same as our *W* column, 2 was identical to *Z*, 3 was identical to *ZW*, etc.

As Taguchi is recommending the use of the L8 array, and as this is identical to the design matrix of a 2^3 experiment, you might conclude that he favours the use of 2^3 factorial experiments. However, this is certainly not the case. To see precisely what Taguchi does advocate let us examine Table 16.3, which is taken from volume 1 of Taguchi (1987).

Table 16.3 contains the design of an experiment which is described in several of Taguchi's publications. The purpose of this experiment was to find the best conditions at the compounding stage in the production of tiles. This objective was to be achieved by manufacturing 100 tiles at each set of conditions. The response, or dependent variable, was the percentage of tiles that were found to be defective on inspection. Each of the seven independent variables or factors had two levels.

If you compare Table 16.3 with Table 16.1 you will see that the experiment was designed by assigning one independent variable to each column of the L8 orthogonal array. Thus the scientist wished to assess the effects of 7 factors using only the 8 sets of conditions specified by the 8 rows of Table 16.3. Clearly this experiment offers a considerable saving in effort and cost compared with a full 2^7 factorial experiment which requires 128 sets of conditions. However, as we have seen in Chapter 13, there is always a price to pay for reducing the size of an experiment. To appreciate the consequences of using the L8 array to assess the effects of seven factors we need to give further consideration to fractional factorial experiments. Let me

Table 16.3 Taguchi's tile experiment

			Prescribed content of experiment			
A	B	C	D	E	F	G
5	Coarse	43	Current	1300	0	0
5	Coarse	43	New	1200	4	5
5	Fine	53	Current	1300	4	5
5	Fine	53	New	1200	0	0
1	Coarse	53	Current	1200	0	5
1	Coarse	53	New	1300	4	0
1	Fine	43	Current	1200	4	0
1	Fine	43	New	1300	0	5

Key: A = quantity of additive; B = granularity; C = agalmatolite quantity; D = agalmatolite type; E = quantity charged; F = chamotte quantity; G = feldspar quantity.

Table 16.4 Design matrix of $\frac{1}{2}$ of 2^4

	X ZWY	Z XWY	XZ WY	W XZY	XW ZY	ZW XY	Y XZW
1	−	−	+	−	+	+	−
2	+	−	−	−	−	+	+
3	−	+	−	−	+	−	+
4	+	+	+	−	−	−	−
5	−	−	+	+	−	−	+
6	+	−	−	+	+	−	−
7	−	+	−	+	−	+	−
8	+	+	+	+	+	+	+

Table 16.5 The 2^3 design matrix can represent various experiments

Experiment	No. of alias groups	No. of effects in each alias group	No. of defining contrasts
Full 2^3	7	1	0
Half of 2^4	7	2	1
Quarter of 2^5	7	4	3
Eighth of 2^6	7	8	7
Sixteenth of 2^7	7	16	15

remind you that the design matrix of the 2^3 experiment in Table 16.2 can also be regarded as the design matrix of a half replicate of a 2^4 factorial experiment. However, as you may recall, the 7 estimates that we could calculate from the results of the 8 trials would not allow us to separate the 15 main effects and interactions. These would group into seven alias pairs, plus a defining contrast, as we see in Table 16.4.

There is no reason why we cannot take this process a step further by using the same design matrix to represent a quarter-replicate of a 2^5 design. The 31 main effects and interactions would then fall into 7 groups of 4, with 3 defining contrasts. Proceeding yet further we could use the 2^3 design matrix to represent a one-eighth replicate of a 2^6 experiment and then we could take the ultimate step to a sixteenth of a 2^7 design. This progression is summarized in Table 16.5.

The tile experiment in Table 16.3 is a sixteenth of a 2^7 factorial experiment. Thus each of the 7 main effects will be aliased with 15 interactions. Some of these will be higher interactions, of course, but three of them will be two-factor interactions. For example, main effect A is aliased with BC, DE and FG. (You could easily check this assertion by converting Table 16.3 to the $+$ and $-$ notation, then multiplying the pairs of columns.) Main effect B is aliased with interactions AC, DF and EG, in addition to 12 higher-order interactions. Thus, if the results of the tile experiment showed that factor B had a significant effect, we could not be sure that this really was due to the change in granularity. It might be due to an interaction between the agalmatolite type (D) and the chamotte quantity (F), or to an interaction between the quantity charged (E) and the feldspar quantity (G), or it might be due to an interaction between quantity of additive (A) and agalmatolite quantity (C).

The scientist conducting the experiment might have considerable understanding of the process. Thus he might be able to say with confidence that these interactions do not exist and that the change in response is, therefore, due to the change in granularity. Perhaps Taguchi has been privileged to work with such very knowledgeable scientists, or perhaps he has confined his attention to relatively simple processes. My personal experience is in the chemical industry, where many production processes are very complex. I cannot conceive of a process chemist, who needed to assess the effects of 7 factors, being able to dismiss the existence of 21 two-factor interactions. A very wise plant manager once said to me, 'Even the process operators have a feel for the main effects. The big savings result from scientists discovering interactions.'

In fairness to Professor Taguchi I should point out that, so far, I have described only a small part of his experimental philosophy. When the full story is revealed you will realize that the Taguchi approach is far more ambitious than anything I have included in earlier chapters. Indeed you may conclude that the inability to assess interactions is a small price to pay for the other benefits that we will discuss in subsequent sections. I will return to his point later but I must quote one of the clearest recommendations in Taguchi (1987): 'There is no guarantee that it is possible to express the total difference in y, by the main effects alone. Nowhere is there a guarantee that it is possible to omit entirely the terms of interactions . . . Nevertheless, the author advises experiments on main effects only.' The L8 orthogonal array, which we have examined in some detail, is just one of several arrays recommended by Taguchi. His publications also include L4, L16, L32, L9 and L27. The L4 orthogonal array is set out in Table 16.6. You can see, I am sure, that it is essentially the same as the design matrix of a 2^2 factorial experiment.

If you followed Taguchi's recommendation and assigned three factors to the three columns of Table 16.6 you would be carrying out a half replicate of a 2^3 factorial experiment. Each main effect would be aliased with a two-factor interaction. The L16 orthogonal array has 15 columns each con-

Table 16.6 The L4 orthogonal array

No.	Column		
	1	2	3
1	1	1	1
2	1	2	2
3	2	1	2
4	2	2	1

Table 16.7 The L9 orthogonal array

No.	Column			
	1	2	3	4
1	1	1	1	1
2	1	2	2	2
3	1	3	3	3
4	2	1	2	3
5	2	2	3	1
6	2	3	1	2
7	3	1	3	2
8	3	2	1	3
9	3	3	2	1

taining eight 1's and eight 2's. The L32 orthogonal array has 31 columns. I am sure you can envisage their appearance, from your knowledge of 2^4 and 2^5 factorial designs.

The L4, L8, L16 and L32 orthogonal arrays are directly applicable in situations where the scientist wishes to explore the effect of several factors, with each factor at only two levels. The L9 and L27 arrays use three levels for each factor. The L9 orthogonal array is set out in Table 16.7.

The four columns of Table 16.7 are orthogonal. Thus, if you selected any two columns and calculated their correlation coefficient you would find it was zero. We will not be able to discuss the L9 orthogonal array in such detail as we did the L8 array, since I have not laid the foundation for such discussion by presenting a thorough treatment of three-level factorial designs. However, I should point out that you would have a 3^2 factorial experiment if you assigned two factors to any two columns of the L9 array.

From the results of this experiment you could estimate the two main effects, each with 2 degrees of freedom, and the interaction with 4 degrees of freedom. Alternatively, if you wished to investigate three factors, you could assign them to any three columns of Table 16.7 and you would have a one-third replicate of a 3^3 factorial experiment. To gain Taguchi's full approval you could assign four factors to the four columns of the L9 array, and thus obtain a one-ninth replicate of a 3^4 factorial experiment. I am sure you will realize that the person who assigns three or four factors to the L9 array will have main effects aliased with two-factor interactions.

16.3 LINEAR GRAPHS

In practice you may wish to investigate the effects of several factors, with some at two levels, some at three levels, and perhaps one or two factors with even more levels.

Clearly, none of the orthogonal arrays directly matches your requirements. However, there are ways in which the arrays can be modified. Some of these techniques are quite ingenious and Taguchi devotes several chapters to them in his 1987 text. Furthermore, he provides a series of diagrams, known as linear graphs, to facilitate the modification of orthogonal arrays and the assignment of factors to columns. An understanding of linear graphs is important if you intend to make extensive use of orthogonal arrays, but a detailed treatment would not be appropriate in this book. None the less, I can find space for an illustration of the use of linear graphs.

John Roberts is a plant manager in charge of the AXF 200 process at R G Chemical's Barnborough site. He is concerned about low viscosity in recent batches of AXF and wishes to assess the effect of three factors on the viscosity of the final product. He decides to use only two levels for each of these factors. Thus the mol ratio (A) will be at 2.1 and 2.4, the feedrate (B) will be at 26.0 and 29.0, the reaction time (C) will be 4.5 and 5.0. With three factors, each at two levels, John could make use of the L4 orthogonal array, which would involve the production of only four batches. However, he also wishes to check the possibility that there may be an interaction between mol ratio and feedrate (AB). Furthermore he suspects that there could be an interaction between mol ratio and reaction time (AC). Three factors and two interactions cannot be fitted into L4 so he shifts his attention to the L8 orthogonal array. Producing a plan for the experiment simply involves assigning the letters A, B and C to three columns of Table 16.1. However, John does not have complete freedom to assign these three letters to any three columns because he wishes to assess interactions. This is where the linear graphs prove useful.

Taguchi offers us two linear graphs for the L8 orthogonal array. These

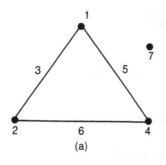
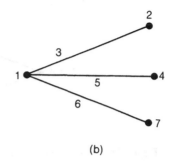

(a)   (b)

Figure 16.1   Linear graphs for the L8 array

are shown in Figure 16.1. They illustrate the relationships between the seven columns of the array. We see that the points labelled 1 and 2 are joined by a line labelled 3. This is telling us that, if we assign factor X to column 1 and factor Z to column 2, then interaction XZ should be assigned to column 3, or we will not be able to assess XZ. Of course, we could assign some other factor, W, to column 3 in which case it would be aliased with interaction XZ. We also see that there is a line labelled 5 joining the points labelled 1 and 4. Thus we must use column 5 of the array if we wish to assess the interaction between the two factors assigned to columns 1 and 4.

Figure 16.2 contains a linear graph depicting what John Roberts wishes to assess from the results of his experiment. To help us assign the letters A, B and C to three columns of the L8 orthogonal array we must attempt to match up one of the linear graphs in Figure 16.1 with Figure 16.2. This is quite easily achieved as we see in Figure 16.3.

We can see in Figure 16.3 that factors A, B and C are to be assigned to columns 1, 2 and 4 of the array, columns 3 and 5 will be used to assess the interactions, leaving columns 6 and 7 available for estimation of the residual variation or error. Thus the L8 orthogonal array becomes the experimental plan of Table 16.8.

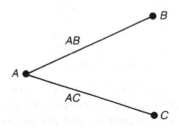

Figure 16.2   A linear graph of our requirements

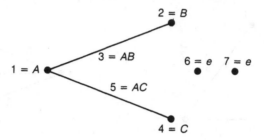

Figure 16.3   A linear graph of the planned experiment

Table 16.8   Independent variables assigned to appropriate columns

1 A	2 B	3 AB	4 C	5 AC	6 e	7 e
1	1	1	1	1	1	1
1	1	1	2	2	2	2
1	2	2	1	1	2	2
1	2	2	2	2	1	1
2	1	2	1	2	1	2
2	1	2	2	1	2	1
2	2	1	1	2	2	1
2	2	1	2	1	1	2

Table 16.9   A practical plan for the experiment

Batch no.	Mol ratio (A)	Feedrate (B)	Reaction time (C)	Viscosity
633	2.1	26.0	4.5	*
629	2.1	26.0	5.0	*
632	2.1	29.0	4.5	*
635	2.1	29.0	5.0	*
628	2.4	26.0	4.5	*
634	2.4	26.0	5.0	*
630	2.4	29.0	4.5	*
631	2.4	29.0	5.0	*

To carry out the experiment we need to communicate the information in columns 1, 2 and 4 to the process operators. This must be translated into real world values of course, and the series of eight trials should be randomized, to give the more practical plan in Table 16.9.

Table 16.10   Analysis of variance

Source of variation	DF	SS	MS
Mol ratio (A)	1		
Feedrate (B)	1		
Reaction time (C)	1		
Interaction AB	1		
Interaction AC	1		
Residual	2		
Total	7		

The viscosity determinations could be analysed by the methods we used in Chapter 12, to obtain effect estimates and to check their statistical significance. However, as Genichi Taguchi makes great use of analysis of variance in his publications, we will follow his lead. Each of the three main effects and the two interactions will have 1 degree of freedom, leaving 2 degrees of freedom for the residual (Table 16.10).

The mean squares for each of the five effects could be tested against the residual mean square using the F-test. We realize that a residual mean square with only 2 degrees of freedom is not a good estimate of error, but Taguchi advises that the non-significant effects should be merged with the residual to give greater degrees of freedom. His cavalier attitude to the merging of sums of squares has been criticized by many statisticians.

This illustration of the use of linear graphs may have struck you as very simple, even trivial. However, I should point out that we were using a relatively simple array, L8. Furthermore, if you have studied earlier chapters of this book you have some knowledge of the 2^3 factorial experiment. Should you attempt to use the larger arrays you would meet greater complexity. With the L16 array, for example, you would first need to choose one from the six linear graphs offered by Taguchi. With the L32 orthogonal array you could choose any one from 13 linear graphs. Furthermore, the graphs become more complex as the size of the array increases. One of the L32 linear graphs is presented in Figure 16.4.

16.4   DESIGN, QUALITY AND NOISE

Genichi Taguchi has written a great deal about the design of experiments. However, he has also written at length about the design of products and the design of production processes. Indeed, it could be argued that his greatest achievement lies in bringing these together. There can be no doubt that he has had considerable influence on the engineering professions as a

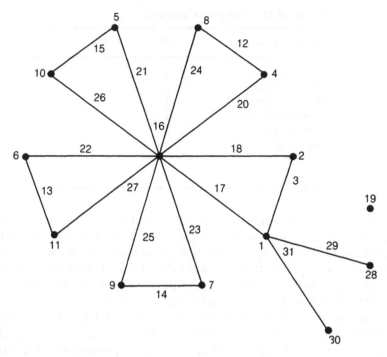

Figure 16.4   A linear graph for the L32 array

result of introducing the design of experiments to engineers responsible for the design of products and processes. 'Design for designers' would be an appropriate subtitle for many of his publications.

Why, you may wonder, has Taguchi aroused such interest amongst engineers, when statisticians have had so little influence over the past 60 years? No doubt there are many reasons, but one important point to note is that Taguchi goes well beyond the conventional statistics text. He demonstrates how the design of experiments can be used in the design of products and processes. If you are to appreciate the power of Taguchi techniques, therefore, you must study his philosophy of design.

He suggests that the design of products and/or processes should proceed in three stages: systems design, parameter design and tolerance design. In the first stage, systems design, the engineer draws upon his/her knowledge of materials, mechanics, electronics, fluid flow, chemistry, etc. to produce an initial design of a product or process. The use of experimentation may well be irrelevant during this phase, but will become an essential element at the next stage, parameter design, when the objective is to choose suitable values for the parameters of the product or process. If, for example, we were designing an electronic circuit we would carry out experiments to determine optimum levels for resistors, capacitors, types of transistor, etc. Taguchi is very concerned with cost, as we will see later, and he would advocate the use of low-cost components during parameter design. In the third stage, tolerance design, costs may rise as inexpensive components are replaced by better ones to achieve quality within the desired tolerance.

For experimentation during parameter design, Taguchi recommends a rather novel approach. This differs somewhat from what you would find in other texts on experimental design. In earlier chapters of this book we have discussed various experiments which were carried out in order to achieve lower impurity, greater tensile strength, increased yield, etc. In all these experiments we were primarily concerned with means. We estimated the effect of each factor by calculating the difference between two means. When the dependent variable was impurity, for example, we sought the conditions which gave lowest mean impurity. It is true that we also estimated standard deviations, but these were used simply to calculate confidence limits for means or to check the statistical significance of means. Taguchi advocates that we assess the effect of each factor on both the mean response and the variation in response. When we have identified those factors which affect the variability we can adjust these factors to reduce variability in the product. Of course, this may leave us with an undesirable mean level of response, so we must then adjust those factors which affect the mean but not the variability. For example, suppose that the viscosity of our product is known to be dependent on feedrate and reaction time. Let us suppose further that an experiment has revealed the nature of these relationships, which are displayed in Figures 16.5 and 16.6. We see in Figure 16.5 that a change in the level of feedrate will affect both mean viscosity and variation in viscosity. On the other hand, Figure 16.6 reveals that a change in reaction time will cause a change in mean viscosity but will not affect variation in viscosity. Thus we would first increase the feedrate to reduce the variation in viscosity. Then we would decrease reaction time to achieve a suitable level of viscosity.

The variation in viscosity within the normal distribution curves in Figures 16.5 and 16.6 is due to random fluctuations in feedrate and reaction time. If, in practice, we have difficulty controlling reaction time and feedrate then these diagrams are telling us something very important. On the other hand, we may be able to control feedrate and reaction time very accurately. In that case, what at first appeared to be a very important message from these diagrams, may be irrelevant. I am inclined to the view that with many processes in the chemical industry we need to probe a little deeper than Figures 16.5 and 16.6 imply.

Suppose we can control reaction time and feedrate very accurately. We will find, even when these factors are tightly controlled, that there is still

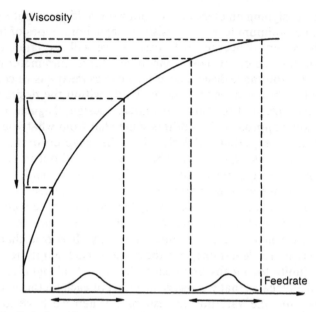

Figure 16.5 Relationship between viscosity and feedrate

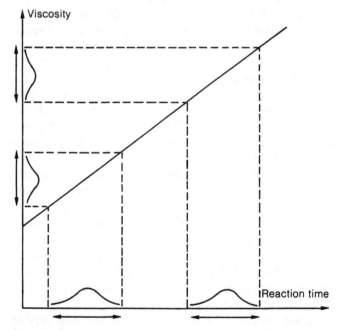

Figure 16.6 Relationship between viscosity and reaction time

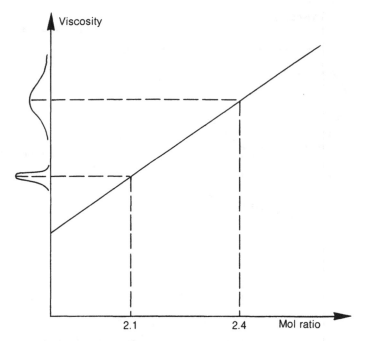

Figure 16.7   Relationship between viscosity and mol ratio

variation in viscosity. This will be due to factors which have not been controlled, such as humidity of the atmosphere, conductivity of the water supply, impurity in the raw materials, ambient temperature, human error, etc. These are independent variables we cannot control and are classed by Taguchi as noise factors, to distinguish them from feedrate, temperature and mol ratio which were controlled during the experiment and which he classes as control factors. Suppose the influence of these noise factors is such as to give the variation in viscosity shown in Figure 16.7.

Figure 16.7 is telling us that the variation in viscosity will be greater if we set the mol ratio at 2.4 than it would be if we set the mol ratio at the lower level of 2.1. Thus the variation in viscosity, which is caused by the noise factors, depends upon the level of mol ratio. In other words the control factor, mol ratio, is interacting with one or more of the noise factors. This interaction is illustrated in Figure 16.8.

The importance of Figures 16.7 and 16.8 cannot be stressed too highly. They are telling us that our process will be more robust if we operate with a mol ratio of 2.1 rather than 2.4. One of Taguchi's stated objectives is to help engineers and scientists to achieve more robust design of products and processes. The purpose of the parameter design stage is to choose levels for the control factors to render the response less dependent on the noise

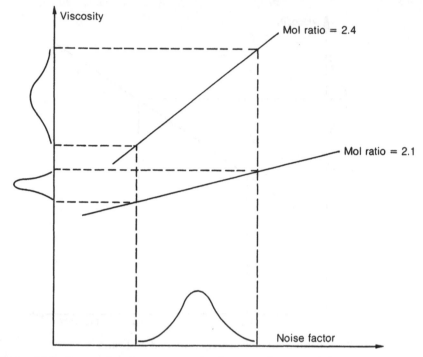

Figure 16.8   Interaction between a noise factor and a control factor

factors. This is achieved by seeking interactions between control factors and noise factors. This last sentence may appear to contradict an earlier comment I made on Taguchi's attitude towards interactions. Let me make the point very clear:

(a) Taguchi is not in favour of pursuing all possible two-factor interactions between pairs of control factors.
(b) However, he is not opposed to the investigation of a small number of pre-specified interactions between control factors.
(c) But, most importantly, he is strongly in favour of assessing interactions between control factors and noise factors.
(d) Furthermore, he advocates a novel type of experimental design to help us find such interactions during parameter design.

16.5 EXPERIMENTS FOR ROBUST DESIGN

Let us now examine the type of experiment that Taguchi would recommend for parameter design. We will build upon the experiment in Table 16.8. Clearly, this experiment, as it stands, will not help us to assess the effect of

Table 16.11 Four replicates of the L8 design

Mol ratio (A)	Feedrate (B)	Reaction time (C)	Viscosity	Mean viscosity	SD of viscosity
2.1	26.0	4.5	****	—	—
2.1	26.0	5.0	****	—	—
2.1	29.0	4.5	****	—	—
2.1	29.0	5.0	****	—	—
2.4	26.0	4.5	****	—	—
2.4	26.0	5.0	****	—	—
2.4	29.0	4.5	****	—	—
2.4	29.0	5.0	****	—	—

the three control factors upon the variation in viscosity. To achieve that we would need two or more response values for each of the eight sets of operating conditions. Suppose we were to carry out four replicates of the experiment. The results of the 32 trials could be tabulated as in Table 16.11.

The results of the experiment in Table 16.11 could be analysed as follows. First we would calculate the mean and the standard deviation of the four viscosity results in each of the eight rows. Then we would use analysis of variance, or some other technique, to assess the effect of the control factors on the mean viscosity. The analysis of variance would then be repeated using SD of viscosity as the response variable. After these two analyses were complete we could decide which of the three control factors and which of the two interactions, AB and AC,

(a) influenced the mean viscosity;
(b) influenced the variation in viscosity;
(c) influenced both;
(d) influenced neither.

You may have rather mixed feelings about this experiment and about the data analysis. Obviously we can assess the effect of the control factors on the mean viscosity with greater precision than we could from the results of the smaller experiment in Table 16.8. We should get better estimates from four replicates than we would get from one. The price we pay for this increased precision is having to carry out 32 trials rather than 8. However, we could expect little sympathy from Taguchi concerning the size of the experiment. Had we taken his advice and included only the three main effects, without interactions AB and AC, we could have accommodated the three control factors in the L4 orthogonal array. Four replicates of L4 would have needed only 16 trials. Better still, he would argue, to carry out four replicates of the L8 array but to include seven control factors, thus getting much more useful information from the 32 trials than we could get from Table 16.9.

Perhaps your strongest reservation concerning this experiment is in relation to the variation in viscosity. How good is our assessment of the effect of the control factors on this variation? We know that a standard deviation based on four observations can be a poor estimate of the population standard deviation. Furthermore, you have been advised in earlier chapters that it is not valid to combine standard deviations by simply averaging them, but that is exactly what we would be doing when we calculate the effect estimates from the eight standard deviations in Table 16.11. Finally, you may be concerned that those trials which give a higher mean viscosity might reasonably be expected to give a higher standard deviation. Pehaps we should be using the coefficient of variation as the response variable (you may recall that the coefficient of variation is obtained by expressing the standard deviation as a percentage of the mean).

Taguchi may well have had similar thoughts whilst considering the use of the standard deviation of replicates as a response variable. He recommends that we use the signal to noise ratio, or SNT, as an alternative:

$$SNT = -10 \log [(SD/mean)^2]$$

The minus sign is included to ensure that the SNT increases as the relative variability decreases. Alternatively, we could eliminate the minus sign if we invert the ratio in brackets:

$$SNT = 10 \log [(mean/SD)^2]$$

Signal to noise ratio is an important concept in electrical engineering, so it is natural that Taguchi, himself an engineer, should wish to use it. However, regarding the SNT equation from a statistical viewpoint, it clearly involves the relative standard deviation (i.e. SD/mean) and it also includes a logarithmic transformation. We have seen in earlier chapters that there are situations in which the logarithmic transformation is useful. There are also situations in which it is preferable to use the relative standard deviation or coefficient of variation rather than the simple standard deviation. However, it would seem very unwise to use either or both of these without first examining data to assess the appropriateness of such refinements. Taguchi has been criticized for the emphasis he places on the SNT.

For the sake of completeness let me offer you two additional formulae for calculating SNTs. These are:

$$SNH = -10 \log [(\Sigma y^2)/r]$$
$$SNL = -10 \log [(\Sigma^1/y^2)/r]$$

in which y represents the response values and r is the number of response values at each set of conditions. SNH is used when we wish to maximize the mean response (e.g. yield), whilst SNL is used when we wish to minimize the response (e.g. impurity). The SNT formula is appropriate when we wish to direct the mean response to a target value as we are attempting to do with the viscosity of AXF 200 in John Roberts' experiment.

You may recall that this digression into SNTs was triggered by our discussion of how we might analyse the standard deviation in Table 16.11. A further point to note about these standard deviations is that they can be attributed to chance variation. Though we deliberately changed the levels of the three control factors we did not attempt to vary the noise factors. Thus the variation in viscosity, which gives the eight standard deviations in Table 16.2, is simply due to random fluctuations in the noise factors. Naturally, these random fluctuations will differ from row to row within the table. Surely the experiment would be more efficient if we took control of this variation in the noise factors, so as to exert the same influence on each row. This line of argument leads us to one of Taguchi's most important recommendations.

Taguchi advocates that we use two orthogonal arrays in a parameter design experiment, one for the control factors and the other for the noise factors, These are combined to form one experiment in which the total number of trials will be equal to the number of rows in one array multiplied by the number of rows in the other array. Thus if we use an L27 array for the control factors and an L8 for the noise factors we will require 216 trials. The control factor array is known as the 'inner array' whilst the noise factors are in the 'outer array'. Care should be taken when using this terminology, however, for Taguchi also classes noise factors into inner noise, outer noise and variational noise.

Let us return to John Roberts' problem and design an experiment using inner and outer arrays. For the control factors, mol ratio (A), feedrate (B) and reaction time (C) we will use the L8 orthogonal array. We will assign the control factors to the columns as we did earlier so that we will again be able to assess interactions AB and AC, leaving two columns for the assessment of residual variation. For the outer array, or noise array, we will use L4 to which we can assign three noise factors. Let me remind you that these noise factors will be independent variables which we would not normally be able to control, but which we will attempt to control during this experiment.

John Roberts, the plant manager, has not given much consideration to such variables in the past, but he produces a very long list of noise factors when he puts his mind to the task. After balancing what is desirable against that which is practicable he suggests that the three noise factors should be density of the additive (W), shift (X) and wind direction (Z). The additive is supplied by one of our own subsidiaries and is always within specification. Using the certificates of conformance we select a quantity of additive which has density close to the lower specification limit and a similar quantity at the upper limit. Sixteen batches will be made using low-density additive and 16 using high density. The process is manned on a two-shift system. Thus 16 batches will be made during the night shift and the other 16 during the day shift. Many plant personnel subscribe to the view that low quality and low yield are more likely to occur when the wind is in the east. Fortunately the prevailing wind is westerly. It is planned that 16 batches will be made on days when there is a strong westerly wind with the other 16 being manufactured during periods of east wind, which always blows strongly.

I am sure I do not need to point out that this experiment will be difficult to schedule. As easterly winds occur only one day in five on average and as the westerly is not always strong it may take some time to manufacture the 32 batches. Randomization of the 32 trials is obviously not possible. 'Is it worth it?' you may ask. Let me remind you of what we would hope to achieve by carrying out such an experiment. The purpose of experiments during the parameter design stage is to find the 'best' levels or values for the control factors. Taguchi suggests that our definition of 'best' should include robustness, which is an insensitivity to environmental variation. Thus we hope to find values for the control factors which give a quality product regardless of the fluctuations in the noise factors. In John Roberts' experiment we wish to establish what mol ratio, feedrate and reaction time will give us the required viscosity, despite variation in density of additive, despite changes in wind direction and regardless of whether the AXF 200 is

Table 16.12 An experiment with inner and outer arrays

L4 outer array

A	B	AB	C	AC	e	e	2	1	1	2	W
							1	2	1	2	X
							1	1	2	2	Z
1	1	1	1	1	1	1	*	*	*	*	
1	1	1	2	2	2	2	*	*	*	*	
1	2	2	1	1	2	2	*	*	*	*	
1	2	2	2	2	1	1	*	*	*	*	
2	1	2	1	2	1	2	*	*	*	*	
2	1	2	2	1	2	1	*	*	*	*	
2	2	1	1	2	2	1	*	*	*	*	
2	2	1	2	1	1	2	*	*	*	*	

L8 inner array

made during the day or during the night. Surely, this is a very noble objective, but John Roberts must decide whether or not the potential gains are worth the trouble and cost.

Table 16.12 contains a plan of the experiment in Taguchi notation. The inner array is exactly like Table 16.8. The outer array is an L4 array turned on its side. The 32 asterisks represent the viscosity results from the 32 experimental batches. When the experiment has been carried out the asterisks can be replaced by the viscosity determinations and data analysis can commence. From the 32 results we can calculate 31 effect estimates each of which has 1 degree of freedom. These estimates can be put into three groups:

(a) the seven control factors A, B, AB, C, AC, e and e;
(b) the three noise factors W, X and Z;
(c) interactions between the control factors and the noise factors. There will be 21 of these, AW, BW, ABW, CW, ACW, eW, eW, AX, BX, ABX, CX, ACX, eX, eX, AZ, BZ, ABZ, CZ, ACZ, eZ and eZ.

We could combine the sums of squares associated with some of these estimates to obtain a residual. One obvious possibility would be to combine all the effects containing a letter 'e'. This would give a residual standard deviation with 8 degrees of freedom against which statistical significance of the other 23 effect estimates could be assessed. Taguchi suggests other ways of combining insignificant effects to obtain a residual mean square, but it should be cautioned that these arbitrary methods can lead to false conclusions. I would recommend that the 31 effect estimates be plotted on normal probability paper and conclusions drawn from that.

Having established the statistical significance of the largest effect estimates we must interpret our findings into language that John Roberts and his colleagues would understand. This understanding would surely be facilitated by diagrams based on the results of the experiment. Let us suppose that we produce a normal plot of the 31 effect estimates and find that 26 of the points lie roughly on a straight line whilst the other 5 do not. The five divergent points represent A, B, AC, BX and CZ.

To illustrate the nature of these significant effects we could calculate appropriate means and confidence limits from the 32 viscosity results in Table 16.12. These are displayed in Figures 16.9 to 16.13.

The two points plotted in Figure 16.9 are each based on 16 of the 32 viscosity results. The same can be said for Figure 16.10. Each point in Figures 16.11 to 16.13 is based on only eight viscosity results. Naturally the points based on fewer observations have wider confidence intervals. The standard deviation used to calculate the confidence intervals could be obtained from the 26 effect estimates that were not classed as significant.

First, we will examine Figures 16.12 and 16.13, as each of these diagrams

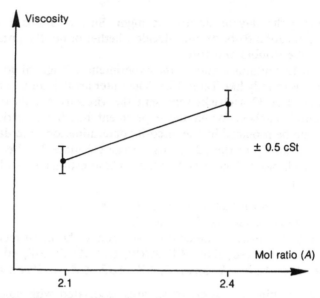

Figure 16.9   The effect of mol ratio on viscosity

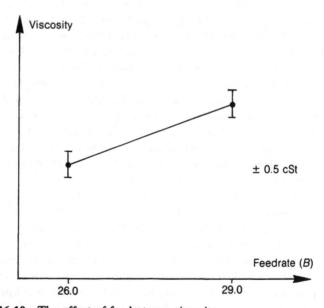

Figure 16.10   The effect of feedrate on viscosity

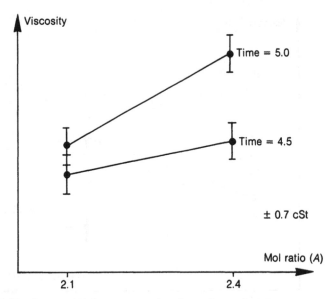

Figure 16.11 Interaction between mol ratio and reaction time

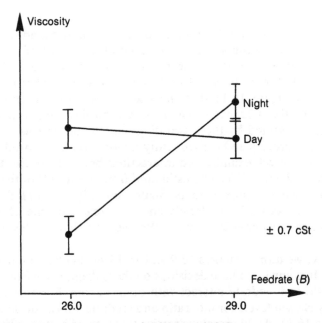

Figure 16.12 Interaction between feedrate and shift

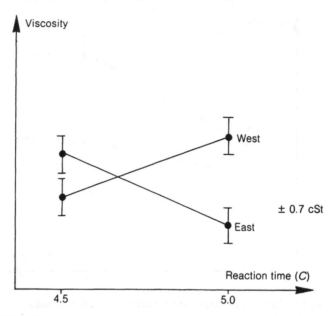

Figure 16.13   Interaction between reaction time and wind direction

portrays an interaction between a control factor and a noise factor. They will help us to choose levels for the control factors which will reduce the effect on viscosity of the noise factors. We see in Figure 16.12 that the difference between shifts will be less if we use the higher feedrate, 29.0. We see in Figure 16.13 that the use of the lower reaction time, 4.5, will reduce the influence of changes in wind direction. Acting upon the evidence presented by these two diagrams should give us a more robust production process and consequently a more consistent product.

Second, we will examine the interactions between control factors. Two of these, AB and AC, can be estimated from the results of the experiment but only one has proved to be significant. Figure 16.11 illustrates this interaction. As we have already chosen a reaction time of 4.5 only the lower of the two lines is relevant. We use it to choose a level for the mol ratio.

Lastly, we turn to Figures 16.9 and 16.10 which illustrate main effects A and B. It is unwise to base decisions on these diagrams in isolation, as both mol ratio and feedrate interact with other factors. Furthermore, we have already chosen levels for mol ratio and feedrate whilst considering Figures 16.11 to 16.13. At this point we realize that we had better go back to Figure 16.10 and reconsider our choice of mol ratio in order to achieve the desired level of viscosity.

16.6   QUALITY, VARIABILITY AND COST

I hope that I have managed to convey to you the importance Taguchi places on the reduction of variability in the product. The process of never-ending improvement, which is advocated by all of the quality gurus, should include the never-ending reduction in variation. Consistency is an objective, in its own right.

How does this philosophy fit your conception of quality? Perhaps you define quality in terms of 'meeting customer requirements' or 'fitness for purpose' or 'conformance to specification'. It is true that, in Europe and the USA, most definitions of quality do include one or more of these phrases. It is also true that customer requirements, however they are assessed, need to be communicated to production personnel in very specific terms. This is often done by quoting specification limits.

Perhaps this emphasis on specifications gives rise to our 'goal-post obsession'. In football, games are won by scoring goals. Near misses do not count. Furthermore, a goal is a goal regardless of whether the ball passed just inside the post, just beneath the crossbar or is right in the centre of the goal-mouth. It is natural, perhaps, to think along similar lines when we are assessing the quality of a product. If we find that the quality measurement is between the specification limits we class the product as good. If the measurement is outside the specified band we class the product as bad. Thus two items which differ by very little might be classed as good on the one hand and bad on the other.

The goal-post philosophy is represented diagrammatically in Figure 16.14. The abrupt change from good to bad as we deviate further from the

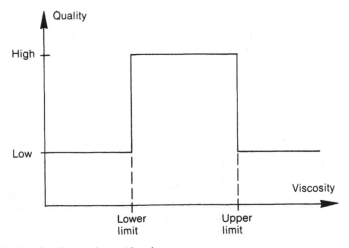

Figure 16.14   Quality and specification

centre of the specification interval does not accord with the customer's view of quality, Taguchi suggests. He maintains that the value of a product to the customer will be highest if the measurement is equal to some target value, which might well be midway between the specification limits. Furthermore, the value of the product declines smoothly, not abruptly, as the measurement deviates from the target. Actually, Taguchi does not use the phrase 'value to the customer', preferring to speak of 'loss to the customer' or 'loss to society from the time the product is shipped'. This loss is least when the viscosity is on target, as we see in Figure 16.15.

The relationship between loss to society and viscosity can be approximated by a quadratic function of the form

$$\text{Loss} = k(y - T)^2$$

in which T is the target value and k is a constant. If we could estimate k we could calculate the loss for any value of viscosity. If we manufacture a large number of batches which have a mean viscosity $\bar{y}$ and a standard deviation s, then the average loss per batch can be written as

$$\text{Average loss} = k[s^2 + (\bar{y} - T)^2]$$

To reduce the average loss, which by definition would increase the average quality, we can take two kinds of corrective action. Firstly, we can attempt

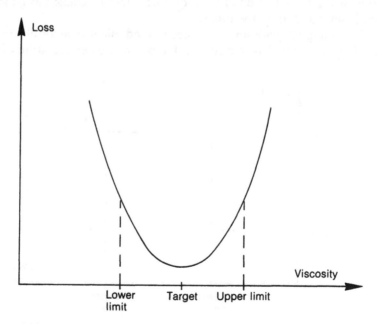

Figure 16.15   A quadratic loss function

to keep the mean viscosity, $\bar{y}$, on target. This is the purpose of statistical process control (SPC). If SPC is completely successful $(\bar{y} - T)^2$ will be equal to zero and the average loss will reduce to ks^2. Secondly, we can attempt to reduce the variability of the product. This may be achieved by reducing the inherent variability of the production process. Clearly, Taguchi's design philosophy can be said to spring directly from his definition of quality as 'the loss to society, from the time the product is shipped'.

Some people react against this very novel definition of quality. Others have difficulty accepting the quadratic loss function which is clearly only an approximation to the true loss function, if such exists. Perhaps, every customer has a unique loss function. On the other hand, it has been suggested that the loss function may prove to be one of the greatest of Taguchi's many contributions to quality improvement. It may be the means by which quality can be brought into the business plan, and thus achieve the status it deserves.

16.7 SUMMARY

Let me remind you that this chapter is an introduction to Taguchi techniques. It was never intended to equip you to make full use of all that Taguchi has to offer. That would require many, very detailed, chapters.

I hope that I have touched upon all of the topics you would expect to find in a summary of Taguchi's work. These include orthogonal arrays, linear graphs, parameter design, inner and outer arrays, signal to noise ratios and the loss function. This whole bundle of techniques is presented as an indivisible package by those who claim to hold the copyright of the expression 'Taguchi methods'. Clearly, there is much of value in such a package. On the other hand, each of the elements I have listed has been criticized to a greater or lesser extent by eminent statisticians. Might I suggest that it would be extremely foolish to dismiss Taguchi techniques out of hand, but it would be very unwise to accept the whole package without considering the strengths and weaknesses of the individual elements. Let me offer you some brief comments on these elements, in the hope that they will help you to select those which will be useful to you.

Orthogonal arrays are excellent, but they were not invented by Genichi Taguchi. Some of them are the fractional factorial design matrices that we considered in earlier chapters. All are to be found in Diamond (1981).

If you assign a control factor to each column of an orthogonal array, you will not be able to assess the effect of interactions between these factors. However, Taguchi suggests that interactions between control factors are less important than interactions between control factors and noise factors. By discovering such interactions we can produce a more robust design in which the product is less influenced by noise factors and thus less variable.

The reduction of process and product variability is a means of reducing 'the loss to society from the time the product is shipped'. Thus the reduction in variability gives an increase in quality, by definition.

I will leave you with one recommendation. Give serious thought to carrying out a parameter design experiment using an inner and an outer array. Consider the long-term benefits you might gain from such an experiment which could well result in a more robust process or product. You may decide that the size of the experiment is prohibitively large. There are many engineers, scientists and managers in British companies who have tried Taguchi methods and found that the benefits were so great that the cost of the experiment proved to be a good investment.

PROBLEMS

16.1 In Chapter 12 we analysed the results of a 2^3 factorial experiment carried out by an R & D chemist in the hope of increasing the tensile strength of rubber. The experiment was successful. However, despite the increased mean strength being achieved in routine production, variation in strength from batch to batch is just as great as before.

The chemist had assumed that he, and the customer, would have to tolerate this variation. However, his recent study of Taguchi methods has opened up the possibility of choosing operating conditions with a view to reducing variability whilst still achieving a suitable mean level. He suspects that much of the variation in tensile strength is due to variation in the raw materials. The carbon black, for example, is obtained from two suppliers and stored in different locations. Thus it is possible that differences between suppliers and deterioration in storage might magnify the variation between deliveries of this material. The chemist decides to carry out a further experiment in which he will include three 'noise factors':

- Supplier of carbon black (A)
    A low = Johnson Graphite
    A high = PX Powders
- Delivery of carbon black (B)
    B low = routine delivery
    B high = special delivery
- Type of storage of carbon black (C)
    C low = storage in damp warehouse
    C high = storage in dry store.

In all the experimental batches 45 parts per hundred of carbon black will be included in the mix. Three control factors will be used:

- Mill speed (X)
     X low = 50 rpm
     X high = 60 rpm
- Percentage natural rubber (W)
     W low = 70%
     W high = 80%
- Type of accelerator (Z)
     Z low = MD accelerator
     Z high = HLX accelerator

The control factors X, W and Z are assigned to the three columns of an L4 orthogonal array, which is described as the 'inner array'. The noise factors A, B and C are assigned to the three columns of a second L4 orthogonal array, known as the 'outer array'. The plan of the experiment and the tensile strength results for the 16 batches are given in Table 16.13.

(a) In order to analyse the data in Table 16.13 we need to re-organise it into a response vector and a design matrix (see section 13.2)

  (i) Put the 16 tensile strength results into a single column response vector.
  (ii) Draw up a design matrix with six columns labelled X, Z, W, A, B and C. In each of the 16 rows of the design matrix put + and − signs to correspond with those in Table 16.13

(b) Use the design matrix and the response vector to calculate estimates of the six main effects X, Z, W, A, B and C.

(c) In order to estimate the two-factor interaction effects we need to extend the design matrix by multiplying pairs of columns. From the existing 6 columns we could generate 15 cross product

Table 16.13   Tensile strength of the 16 samples

			+	−	+	−	A
			+	+	−	−	B
			+	−	−	+	C
−	−	+	267	227	227	246	
+	−	−	240	239	253	255	
−	+	−	223	226	212	218	
+	+	+	241	208	226	256	
X	Z	W					

columns. However, you would find that 6 of the 15 were perfectly correlated with the 6 main effect columns. Extend your design matrix by generating 9 cross product columns which are orthogonal to the 6 columns you produced in part (a).

(d) Use the extended design matrix and the response vector to calculate estimates of the nine interactions which can be assessed from this data.

(e) Plot the 15 effect estimates on half normal probability paper.

(f) Use your half normal plot to decide which of the main effects and interactions are statistically significant.

(g) Report your findings in simple language paying particular attention to the choice of mill speed (X), percentage natural rubber (W) and type of accelerator (Z) which will make the tensile strength less dependent on the variation in carbon black.

(h) List the interactions which could not be assessed from this data. Do you consider these interactions less important than those which could be estimated?

(i) Clearly your extended design matrix closely resembles the design matrix of a 2^4 factorial experiment. However, we have 6 independent variables in this experiment. How would you describe the experiment?

(j) (More difficult) List the defining contrasts and the aliass groups for this design.

16.2   In question 16.1 the data analysis you carried out was admirable. However, the procedure you followed differed from that which would be used by true Taguchi disciples. They would make use of signal-to-noise ratios.

The data from question one is reproduced in Table 16.14 but the outer array has been removed, as it will play no part in this analysis.

(a) Calculate the mean, the standard deviation and the signal to noise ratio for each of the four rows in Table 16.14

$$\text{S/N ratio} = 10 \log [(\text{Mean/SD})^2]$$

Table 16.14   Calculation of the S/N ratio

Control Factors			Response (Tensile strength)				Mean	SD	S/N
X	Z	W							
−	−	+	267	227	227	246			
+	−	−	240	239	253	255			
−	+	−	223	226	212	218			
+	+	+	241	208	226	256			

(b) Using the mean column in Table 16.14 as the response variable, calculate estimates of the effect of the control factors on the mean tensile strength.

(c) Using the SD column of Table 16.14 as the response variable calculate estimates of the effect of the control factors on the variation in tensile strength.

(d) Using the S/N column of Table 16.14 as the response variable, calculate estimates of the effect of the control factors on the signal to noise ratio.

(e) What conclusions do you draw from the three sets of estimates obtained in (b), (c) and (d). How do your conclusions from this analysis compare with your conclusions in question 16.1?

Appendix A
The sigma (Σ) notation

Throughout this book the calculation of certain statistics involves adding sets of numbers. The simplest case is where we add a set of observations. For example the first six batches of digozo blue gave the number of overloads as 0, 4, 2, 2, 1, 4 giving a total of 13. We can either refer to this total as 'the sum of the observations' or we can use a statistical shorthand which is far more concise. Thus Σx is the shorthand for 'the sum of the observations' where x is the symbol for an observation and Σ is the symbol for add. Therefore Σx is an instruction telling us to 'add the observations'.

$$\Sigma x = 0 + 4 + 2 + 2 + 1 + 4 = 13$$

We can now use this shorthand to represent other statistics. For example 'the sum of the squared observations' is denoted by Σx^2. Using the data given above:

$$\Sigma x^2 = 0^2 + 4^2 + 2^2 + 2^2 + 1^2 + 4^2 = 41$$

For another example let us calculate $\Sigma(x - 2)^2$. This gives:

$$\Sigma(x - 2)^2 = (0 - 2)^2 + (4 - 2)^2 + (2 - 2)^2$$
$$+ (2 - 2)^2 + (1 - 2)^2 + (4 - 2)^2 = 13$$

The gain in simplicity and unambiguity can clearly be seen if we compare the mathematical expression with the written expression which is 'the sum of squares of the observations after two has been subtracted from each observation'.

Appendix B
Notation and formulae

NOTATION

n	sample size, number of observations or number of points
$\bar{x}$	sample mean
s	sample standard deviation
s^2	sample variance
p	sample percentage
μ	population mean
σ	population standard deviation
σ^2	population variance
π	population percentage
z	standardized value
t	critical value from the t-table
c	a change (or difference) which we wish to detect
Σ	'the sum of' (see Appendix A)
r	$\begin{cases} \text{number of occurrences (in Chapter 6)} \\ \text{correlation coefficient (in later chapters)} \end{cases}$
r_{xy}	correlation coefficient, between x and y
a	intercept $\Big\}$ of least squares regression line, $y = a + bx$
b	slope
RSD	Residual standard deviation
k	number of independent variables in a regression equation
df	degrees of freedom

FORMULAE

Sample mean: $\bar{x} = \Sigma x / n$

Sample standard deviation:
$$s = \sqrt{[\Sigma(x - \bar{x})^2/(n - 1)]} = \sqrt{[(\Sigma x^2 - n\bar{x}^2)/(n - 1)]}$$

Combined standard deviation
$$= \sqrt{\{\Sigma[(df)(SD^2)]/[\Sigma\,df]\}}$$

Poisson distribution: $\mu^r e^{-\mu}/r!$

Binomial distribution:

$$\frac{n!}{r!(n-r)!}\,(\pi/100)^r(1-\pi/100)^{n-r}$$

Normal distribution: $z = (x - \mu)/\sigma$

Confidence interval for μ: $\bar{x} \pm ts/\sqrt{n}$

Confidence interval for π: $p \pm \left\{k\,\sqrt{\left[\dfrac{p(100-p)}{n}\right] + \left(\dfrac{50}{n}\right)}\right\}$

Confidence interval for the difference between two population means:

$$|\bar{x}_1 - \bar{x}_2| \pm ts\,\sqrt{\left(\frac{1}{n_1} + \frac{1}{n_2}\right)}$$

Sample size needed to estimate the population mean within $\pm c$:

$$n = (ts/c)^2$$

Sample sizes needed to estimate the difference between two population means within $\pm c$:

$$n_1 = n_2 = 2\left(\frac{ts}{c}\right)^2$$

Sample size needed to estimate a population percentage within $\pm c$:

$$n = (100/c)^2$$

Localized standard deviation (for use in cusum test):

$$\sqrt{\left[\frac{\Sigma(x_c - x_{c+1})^2}{2(n-1)}\right]}$$

Correlation coefficient of x and y:

$$\frac{\Sigma[(x-\bar{x})(y-\bar{y})]/(n-1)}{(\text{SD of } x)(\text{SD of } y)} \quad \text{or} \quad \frac{[\Sigma xy - n\bar{x}\bar{y}]/(n-1)}{(\text{SD of } x)(\text{SD of } y)}$$

Least squares regression line $y = a + bx$

$$\text{Slope } b = \frac{[\Sigma xy - n\bar{x}\bar{y}]/(n-1)}{(\text{SD of } x)^2}$$

$$\text{Intercept } a = \bar{y} - b\bar{x}$$

Significance test	Test statistic
One-sample t-test	$\dfrac{\|\bar{x}-\mu\|}{s/\sqrt{n}}$
t-test for two independent samples	$\dfrac{\|\bar{x}_1-\bar{x}_2\|}{s\sqrt{\left(\dfrac{1}{n_1}+\dfrac{1}{n_2}\right)}}$ $s=\sqrt{\{[(n_1-1)s_1^2+(n_2-1)s_2^2]/[(n_1-1)+(n_2-1)]\}}$
t-test for two matched samples (paired comparison test)	$\dfrac{\|\bar{d}-\mu_d\|}{s_d/\sqrt{n}}$ ($\bar{d}$ and s_d are calculated from the differences)
F-test	Larger variance/smaller variance
One-sample percentage test	$\dfrac{\|p-\pi\|-(50/n)}{\sqrt{\left[\dfrac{\pi(100-\pi)}{n}\right]}}$
Chi-square test	$\sum\dfrac{(O-E)^2}{E}$
Dixon's test ($2<n<8$)	$\dfrac{x_2-x_1}{x_n-x_1}$ or $\dfrac{x_n-x_{n-1}}{x_n-x_1}$
Dixon's test ($7<n<13$)	$\dfrac{x_2-x_1}{x_{n-1}-x_1}$ or $\dfrac{x_n-x_{n-1}}{x_n-x_2}$
Dixon's test ($12<n$)	$\dfrac{x_3-x_1}{x_{n-2}-x_1}$ or $\dfrac{x_n-x_{n-2}}{x_n-x_3}$
Cusum test	$\dfrac{\|\text{maximum cusum}\|}{\text{localized standard deviation}}$
Regression t-test	$\sqrt{\{[(\text{new \% fit}-\text{old \% fit})(n-k-1)]/[(100-\text{new \% fit})]\}}$
Effect estimate from p replicates of a 2^n factorial experiment	$\dfrac{\text{effect estimate}}{\text{RSD}/\sqrt{(p2^{n-2})}}$

Residual = actual value − predicted value

Residual sum of squares =

$\Sigma(\text{residual})^2$

or $(1-r^2)(n-1)(\text{SD of }y)^2$
with only one independent variable

Residual standard deviation (RSD) =

$\sqrt{[(\text{residual sum of squares})/(\text{residual degrees of freedom})]}$

Residual degrees of freedom $= n - k - 1$

Confidence interval for true intercept (a) is:

$$a \pm t(\text{RSD}) \sqrt{\left[\frac{1}{n} + \frac{\bar{x}^2}{(n-1)(\text{SD of } x)^2}\right]}$$

Confidence interval for true slope (β) is:

$$b \pm t(\text{RSD}) \sqrt{\left[\frac{1}{(n-1)(\text{SD of } x)^2}\right]}$$

Confidence interval for the true value of y corresponding to a particular value of x (say X) is:

$$a + bX \pm t(\text{RSD}) \sqrt{\left[\frac{1}{n} + \frac{(X - \bar{x})^2}{(n-1)(\text{SD of } x)^2}\right]}$$

Confidence interval for an observed value of y corresponding to a particular value of x (say X) is:

$$a + bX \pm t(\text{RSD}) \sqrt{\left[1 + \frac{1}{n} + \frac{(X - \bar{x})^2}{(n-1)(\text{SD of } x)^2}\right]}$$

Percentage fit $= 100\, r_{xy}^2$ (for $y = a + bx$)

Percentage fit $= 100(r_{xy}^2 + r_{zy}^2 - 2r_{xy}r_{zy}r_{xz})/(1 - r_{xz}^2)$
   (for $y = a + bx + cz$)

Percentage fit $= 100\, (\text{regression sum of squares})/(\text{total sum of squares})$

Appendix C
Sampling distributions

It is suggested throughout this book that you can carry out significance tests or calculate confidence limits without being able to derive the formulae which are used in these activities. If you have already worked through the later chapters you will probably agree with this assertion. Nonetheless you may feel very uneasy about making use of formulae when you have little understanding of how they are derived and you might therefore wish to explore the theoretical basis of the methods advocated in this book.

Unfortunately a deeper understanding of statistical inference must necessarily be based on a knowledge of 'sampling distributions'. These are rather special forms of probability distributions and any discussion of them has been carefully avoided throughout this book. The reason why no mention has been made of sampling distributions is because they demand that you view the sampling process from a very different standpoint to that normally adopted by the scientist or technologist.

The scientist takes one sample of n items from a population and uses the information in the sample to infer certain characteristics of the population. The whole purpose of the scientist's investigation is to learn more about the population.

The mathematical statistician, on the other hand, starts with a population about which he knows everything and from this he takes many samples, each containing n items. The purpose of this investigation is to learn about how the mean (or some other statistic) varies from sample to sample.

The contrast between the two approaches cannot be emphasized too strongly. The scientist/technologist takes one sample whereas the mathematical statistician takes many (perhaps even an infinite number). The scientist/technologist is operating in the real world whereas the mathematical statistician is operating in the abstract world of mathematics where sampling is effortless. To bridge the gap between these two worlds we will consider a hypothetical example in which the peculiar approach of the mathematical statistician is applied to a concrete situation. Please note that what follows is not a recommended strategy for the practical scientist.

We will start with a population which consists of a very large consignment of bottle tops. Imagine that the diameter of each bottle top is measured, then the measurement is written on a red ticket and the ticket placed in a large drum. Clearly we can regard the numbers on the red tickets as constituting a second population and we will refer to this as the parent population. Furthermore we will refer to the probability distribution of the numbers on the red tickets as the parent distribution. The situation is illustrated in Figure C.1.

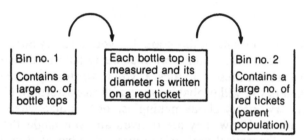

Figure C.1   A population of individual measurements

From the parent population we now take a random sample of n red tickets, calculate the mean diameter for the sample ($\bar{x}$), write this sample mean on a blue ticket, place the blue ticket in a third bin and finally return the sample of red tickets to bin number 2 (Figure C.2). This sampling procedure is repeated. As we are sampling with replacement we could continue indefinitely.

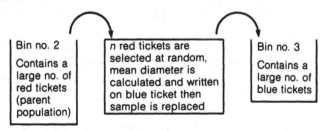

Figure C.2   A population of sample means

The distribution of the numbers on the blue tickets is known as the sampling distribution of the mean. Clearly this sampling distribution tells us how the sample mean varies from sample to sample. The sample means in bin number 3 will have an average value which is known as the 'mean of the sampling distribution'. The sample means will be scattered around this average value and we can speak of the standard deviation of the sample means. It is the accepted convention to refer to this standard deviation as the 'standard error of sample means'.

Many interesting questions can be asked concerning the relationship between the sampling distribution and the parent distribution. For example:

(a) How is the mean of the sampling distribution related to the mean of the parent distribution?
(b) How is the standard error of the sampling distribution related to the standard deviation of the parent distribution?
(c) How is the shape of the sampling distribution related to the shape of the parent distribution?

The mathematical statistician has given us answers to these questions. (His answers are based upon mathematics, not upon bottle tops or coloured tickets.) He tells us that, whatever the parent distribution:

(a) the mean of the 'sampling distribution of the mean' is equal to the mean of the parent distribution;
(b) the standard error of the 'sampling distribution of the mean' is equal to the standard deviation of the parent distribution (σ) divided by the square root of the sample size, i.e. $\sigma/\sqrt{n}$.

The standard error ($\sigma/\sqrt{n}$) will always be less than the standard deviation (σ) and will be very much less if n is large. This confirms what common sense would suggest, that sample means are not as widely scattered as individual observations, i.e. the mean diameters written on the blue tickets will have a smaller spread than the individual diameters written on the red tickets.

Concerning the shape of the sampling distribution of the mean it follows that it will be narrower than the parent distribution. The mathematical statistician further informs us that:

(a) the sampling distribution of the mean will be more like a normal distribution than is the parent distribution;
(b) the sampling distribution of the mean is closer to a normal distribution if n is large.

The relationship between the parent distribution and the sampling distribution is illustrated in Figure C.3.

If the parent distribution is actually normal then the sampling distribution of the mean will also be normal. If the parent population does not have a normal distribution we can nonetheless be confident that the sampling distribution of the mean will be approximately normal provided n

is 'large'. In a nutshell, if individual observations come from a population with mean μ and standard deviation σ then sample means can be considered to come from a population with mean μ and standard deviation (or standard error) $\sigma/\sqrt{n}$. From our knowledge of the normal distribution we can say that 95% of sample means will lie in the range $\mu \pm 1.96\sigma/\sqrt{n}$.

From the point of view of the scientist who intends to take one sample we can say that there is a 95% chance that his sample mean will lie in the interval $\mu \pm 1.96\sigma/\sqrt{n}$. Much more important, we can look at the problem from the point of view of the scientist who has already taken one sample of n observations, and say that there is a 95% chance that the population mean (μ) lies in the interval $\bar{x} \pm 1.96\sigma/\sqrt{n}$. In Chapter 4 we referred to such intervals as 95% confidence intervals but we used a slightly different formula. This formula ($\bar{x} \pm ts/\sqrt{n}$) is more useful because we would not normally know the value of the population standard deviation (σ) and would need to use the sample standard deviation (s). The extra

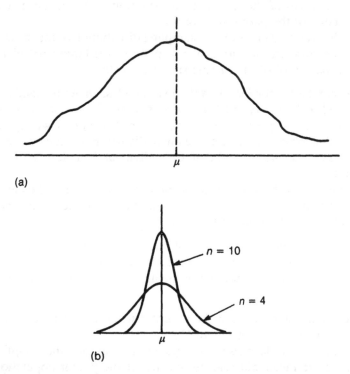

(a)

(b)

Figure C.3 (a) Parent distribution (mean μ, standard deviation σ) (b) Sampling distribution of the mean (mean μ, standard error $\sigma/\sqrt{n}$)

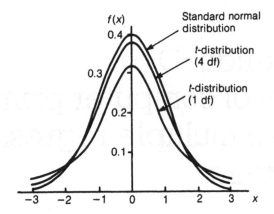

Figure C.4 *t*-distributions and the normal distribution

uncertainty of using *s* instead of σ requires that the 1.96 be replaced by a critical value from the *t*-table. This *t* value will always exceed 1.96.

The critical values in the *t*-table are actually obtained by finding areas under the *t*-distribution which is the sampling distribution of the statistic *t* [where $t = (\bar{x} - \mu)/(s/\sqrt{n})$]. Just as the sample mean ($\bar{x}$) varies from sample to sample so does *t*.

Mathematical statisticians have studied the sampling distribution of *t* in order to obtain the *t*-table. The *t*-distribution is rather similar to the standard normal distribution especially if *n* is large. This point is illustrated in Figure C.4.

Appendix D
Copy of computer print-out from a multiple regression program

DATA INPUT

Number of cases   10
Number of variables   6
Variable names   Imp., Wt, Age, Feed, Temp., Agit.

Imp.	Wt	Age	Feed	Temp.	Agit.
4	3	1	3	1	3
3	4	2	5	2	3
4	6	3	7	3	3
6	3	4	5	3	2
7	1	5	2	2	2
2	5	6	6	1	2
6	1	7	2	2	2
10	0	8	1	3	4
5	2	9	3	1	4
3	5	10	6	2	4

DATA ANALYSIS

Variable	Low	High	Mean	SD
Imp.	2	10	5.00	2.3570
Wt	0	6	3.00	2.0000
Age	1	10	5.50	3.0277
Feed	1	7	4.00	2.0548
Temp.	1	3	2.00	0.81650
Agit.	2	4	2.90	0.87560

CORRELATION MATRIX

	Imp.	Wt	Age	Feed	Temp.	Agit.
Imp.	1.00	−0.849	0.234	−0.780	0.520	0.108
Wt	−0.849	1.00	−0.257	0.973	−0.068	0.000
Age	0.234	−0.257	1.00	−0.196	−0.045	0.440
Feed	−0.780	0.973	−0.196	1.00	0.066	−0.062
Temp.	0.520	−0.068	−0.045	0.066	1.00	0.00
Agit.	0.108	0.000	0.440	−0.062	0.00	1.00

REGRESSION ANALYSIS

Dependent variable Imp.

First equation

Variable	Coeff.
Wt	−1.00
Const.	8.00

Percentage fit = 72.0%

Analysis of variance:

Source	SS	d.f.	MS
Regression on Wt	36.0	1	36.0
Residual	14.0	8	1.75
Total	50.0	9	—

Partial correlation with dependent variable

Wt	0.000
Age	0.030
Feed	0.378
Temp.	0.875
Agit.	0.203

Second equation

Variable	Coeff.
Wt	−0.9628
Temp.	1.3395
Const.	5.2093

Percentage fit = 93.4%

Analysis of variance:

Source	SS	d.f.	MS
Regression on Wt	36.00	1	—
Inclusion of Temp.	10.70	1	10.70
Regression on Wt and Temp.	46.70	2	—
Residual	3.30	7	0.47
Total	50.0	9	—

No other independent variables are statistically significant.

Appendix E
Partial correlation

The correlation coefficient was introduced in Chapter 10 and is used extensively in later chapters of this book. It is a simple and very useful measure of the strength (and direction) of relationship between two variables. It is equally useful whether we are discussing the relationship between an independent variable and a dependent variable, or the relationship between two independent variables. It is not, however, quite so useful when we want to consider the relationships between three variables.

Suppose we have measured three variables; a dependent variable (y) and two independent variables (x and z). We can calculate three simple correlation coefficients, r_{xy}, r_{zy} and r_{xz}. If r_{xy} is greater than r_{zy} (ignoring any negative signs) then we would choose x as the first independent variable and fit the equation $y = a + bx$. If the magnitude of r_{xy} were greater than the critical values from Table H it would be reasonable to conclude that the variation in the independent variable, x, was responsible for part of the variation in the dependent variable, y. Having made this decision we can now turn our attention to the second independent variable (z) and its relationship with the dependent variable. In doing so we must not ignore the possibility that y may appear to depend on z simply because both y and z are dependent on x; and with this possibility in mind it is reasonable to ask 'what would be the correlation between z and y if x did not vary?'

To answer this question we calculate what is known as the 'partial correlation between z and y with x constant' using the formula:

$$\frac{r_{yz} - r_{xz}r_{xy}}{\sqrt{[(1 - r_{xz}^2)(1 - r_{xy}^2)]}}$$

To illustrate the use of this formula we will use the data from Chapter 10 and concentrate on the three variables, weight of special ingredient (x), feedrate (z) and impurity (y). The relevant correlation coefficients taken from Appendix D are:

$$r_{xy} = -0.849 \qquad r_{zy} = -0.780 \qquad r_{xz} = 0.973$$

Calculation of the 'partial correlation between z and y with x fixed' proceeds as follows:

$$\frac{r_{yz} - r_{xz}r_{xy}}{\sqrt{[(1 - r_{xz}^2)(1 - r_{xy}^2)]}} = \frac{-0.780 - (0.973)(-0.849)}{\sqrt{[1 - 0.973^2)(1 - 0.849^2)]}}$$

$$= \frac{-0.780 + 0.826}{\sqrt{[(0.053)(0.279)]}}$$

$$= \frac{0.046}{0.122}$$

$$= 0.377$$

This result is in agreement with the partial correlation coefficient in the computer print-out of Appendix D. Note that the partial correlation between z and y (with x fixed) is positive whereas the simple correlation between z and y is negative. This is telling us that, when the weight of special ingredient (x) is taken into account, an increase in feedrate (z) can be expected to give an increase in impurity (y) and not a decrease in impurity as indicated by the simple correlation ($r_{zy} = -0.780$).

Though a partial correlation coefficient is much more complex than a simple correlation coefficient it is nonetheless just as easy to test the statistical significance of the former as the latter. We use the modulus of the partial correlation as the test statistic and we obtain the critical values from Table H, but we must reduce the sample size by one. Thus to test the partial correlation coefficient just calculated we would use a sample size of nine and the critical values would be 0.666 at the 5% level and 0.798 at the 1% level of significance. As the test statistic is equal to 0.377 we cannot reject the null hypothesis and we are unable to conclude that the percentage impurity (y) of a batch is dependent upon the feedrate (z). This conclusion is in agreement with that drawn in Chapter 11 after fitting the multiple regression equation $y = a + bx + cz$.

Clearly it is meaningless to speak of partial correlation when referring to a situation in which there are only two variables. We need a third, 'fixed' variable if we are to calculate a partial correlation coefficient. We are not, however, restricted to this level for the concept is useful when we have four or more variables. Suppose that we add a third independent variable, w, to the two already considered. After fitting the equation $y = a + bx + cz$ we would be interested in 'the partial correlation between y and w with x and z fixed'. This would be known as a second order partial correlation coefficient to distinguish it from the first order partial correlation in which we have only one fixed variable. When testing the significance of a second order coefficient we would again use critical values from Table H but we would reduce the sample size by two. When testing partial correlation coefficients of higher order we would reduce the sample size by subtracting the number of fixed variables.

A part correlation coefficient is rather similar to a partial correlation coefficient. With part correlation, however, the effect of the fixed variable is removed from only one of the other two variables. Thus we can calculate 'the part correlations of y and z with the effect of x removed from z' using:

$$\frac{r_{yz} - r_{xz}r_{xy}}{\sqrt{(1 - r_{xz}^2)}} = 0.200$$

Alternatively we can calculate 'the part correlation of y and z with the effect of x removed from y' using:

$$\frac{r_{yz} - r_{xz}r_{xy}}{\sqrt{(1 - r_{xy}^2)}} = 0.087$$

Obviously there is a distinction between these two part correlation coefficients and the partial correlation coefficient calculated earlier. If you were restricted to using correlation rather than regression analysis the distinction would be very important. Hopefully you will have access to a multiple regression package and need only consider correlation for the light it sheds on the regression analysis. Perhaps the relation between the two will be made clear by the block diagrams, Figures E.1 and E.2, which

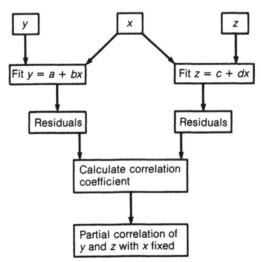

Figure E.1   Calculation of partial correlation coefficient

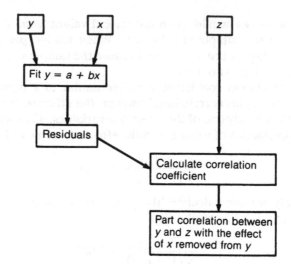

Figure E.2   Calculation of part correlation coefficient

illustrate an alternative method of calculation using residuals from regression equations. The important point to note is that 'removing the effect of x from y' can be achieved by fitting $y = a + bx$, then replacing the y values by the residuals from the equation.

Appendix F
Significance tests on effect estimates from a $p2^n$ factorial experiment

It was stated in Chapter 12, in a discussion of factorial experiments, that we can distinguish between real and chance effects by comparing:

$$\text{Test statistic} = \frac{\text{effect estimate}}{\text{RSD}/\sqrt{(p2^{n-2})}} \qquad (F.1)$$

with a critical value from the t-table. The foundation on which this test is based can be readily understood if we consider the way in which the effect estimates are calculated. For simplicity we will start with the 2^2 factorial design, which gives the four response values y_1, y_2, y_3 and y_4 in Table F.1.

Table F.1 Results of a 2^2 factorial experiment

Feedrate	Temperature		Mean
	Low	High	
Low	y_1	y_2	$\frac{1}{2}(y_1 + y_2)$
High	y_3	y_4	$\frac{1}{2}(y_3 + y_4)$
Mean	$\frac{1}{2}(y_1 + y_3)$	$\frac{1}{2}(y_2 + y_4)$	

To calculate an estimate of the feedrate main effect we subtract the mean response at the low level of feedrate from the mean response at the high level of feedrate:

$$\text{Feedrate main effect} = \tfrac{1}{2}(y_3 + y_4) - \tfrac{1}{2}(y_1 + y_2)$$
$$= \tfrac{1}{2}(y_3 + y_4 - y_1 - y_2)$$

The other estimates can be expressed in a similar form:

$$\text{Temperature main effect} = \tfrac{1}{2}(y_2 + y_4 - y_1 - y_3)$$
$$\text{Interaction effect} = \tfrac{1}{2}(y_1 + y_4 - y_2 - y_3)$$

If we wish to test the statistical significance of the effect estimates we must first answer the question 'How would the effect estimates vary from experiment to experiment if the 2^2 factorial experiment were repeated many, many times?' It is clear that each response value would vary from experiment to experiment because each response value is subject to random error. This is made explicit in the model that was introduced in Chapter 12:

$$y_1 = \mu - \frac{T}{2} - \frac{F}{2} + \frac{I}{2} + e_1$$

$$y_2 = \mu - \frac{T}{2} - \frac{F}{2} + \frac{I}{2} + e_2$$

etc.

Within the model T, F and I are constants (although their values are unknown), whilst e_1, e_2, e_3 and e_4 are random errors. It is assumed that these random errors come from a normal distribution that has a mean equal to zero and a standard deviation equal to σ. (σ is also unknown and cannot be estimated from a simple factorial experiment, as we saw in Chapter 12.) It follows that:

$$\text{SD}(y_1) = \sigma \qquad \text{SD}(y_2) = \sigma \qquad \text{SD}(y_3) = \sigma \qquad \text{SD}(y_4) = \sigma$$

Note that, in speaking of the standard deviation of y_1, we are not referring to the variation amongst the response values within the experiment that was actually carried out. We are referring to the hypothetical variation, from experiment to experiment, of the response in the top left-hand corner of Table F.1 (i.e. the yield of the batch made with low feedrate and low temperature). So, the variation of y_1 from experiment to experiment will have a standard deviation equal to σ and the same can be said of y_2, y_3 and y_4. We want to know how the effect estimates will vary from experiment to experiment, and each estimate is calculated from four response values. To help us with this step in the argument we will make use of the following well-known results in mathematical statistics, in which 'Var' is an abbreviation for variance:

Theorem 1

$\text{Var}(x + z) = \text{Var}(x) + \text{Var}(z)$ if x and z are independent

Theorem 2

$\text{Var}(x - z) = \text{Var}(x) + \text{Var}(z)$ if x and z are independent

Theorem 3

$\text{Var}(kx) = k^2 \text{Var}(x)$ where k is a constant

Using these three theorems we can express the variance of each effect estimate in terms of σ^2. For example:

Feedrate estimate (FE) $= \frac{1}{2}(y_3 + y_4 - y_1 - y_2)$

$$\begin{aligned}\text{Var(FE)} &= \tfrac{1}{4}\text{Var}(y_3 + y_4 - y_1 - y_2) \\ &= \tfrac{1}{4}[\text{Var}(y_3) + \text{Var}(y_4) + \text{Var}(y_1) + \text{Var}(y_2)] \\ &= \tfrac{1}{4}(\sigma^2 + \sigma^2 + \sigma^2 + \sigma^2) \\ &= \sigma^2\end{aligned}$$

Thus the standard deviation of the feedrate estimate is equal to the standard deviation of individual errors, i.e.

$$\text{SD(FE)} = \sigma$$

Similarly we could show that the standard deviation of the temperature estimate and of the interaction estimate were also equal to σ. To carry out a *t*-test on any of these effect estimates we calculate:

$$\text{Test statistic} = \frac{\text{effect estimate} - \text{true effect}}{s}$$

where s is an estimate of σ.

As our null hypothesis says that the true effect is equal to zero, and σ is estimated by a residual standard deviation (RSD), this becomes:

$$\text{Test statistic} = \frac{\text{effect estimate}}{\text{RSD}}$$

which is in keeping with equation (F.1) if we let $n = 2$ and $p = 1$.

Table F.2 Two replicates of a 2^2 factorial experiment

Feedrate	Temperature		
	Low	High	Mean
Low	y_1 y_5	y_2 y_6	$\frac{1}{4}(y_1 + y_5 + y_2 + y_6)$
High	y_3 y_7	y_4 y_8	$\frac{1}{4}(y_3 + y_7 + y_4 + y_8)$
Mean	$\frac{1}{4}(y_1 + y_5 + y_3 + y_7)$	$\frac{1}{4}(y_2 + y_6 + y_4 + y_8)$	

If we had carried out two replicates of a 2^2 factorial experiment to obtain the results in Table F.2 then the estimate of the feedrate main effect would be calculated from:

$$\text{Feedrate main effect} = \tfrac{1}{4}(y_3 + y_7 + y_4 + y_8) - \tfrac{1}{4}(y_1 + y_5 + y_2 + y_6)$$
$$= \tfrac{1}{4}(y_3 + y_7 + y_4 + y_8 - y_1 - y_5 - y_2 - y_6)$$
$$\text{Var(FE)} = \tfrac{1}{16}(\sigma^2 + \sigma^2 + \sigma^2 + \sigma^2 + \sigma^2 + \sigma^2 + \sigma^2 + \sigma^2)$$
$$= \tfrac{1}{2}\sigma^2$$
$$\text{SD(FE)} = \sigma/\sqrt{2}$$

Comparing this result with the corresponding result for a single replicate of the 2^2 factorial experiment we see that effect estimates from the larger experiment are more precise. For the 2×2^2 experiment it follows that the test statistic will be given by:

$$\frac{\text{Effect estimate}}{\text{RSD}/\sqrt{2}}$$

which is in agreement with equation (F.1) if we let $p = 2$ and $n = 2$.

Following similar arguments we could show that equation (F.1) gives the correct test statistic for any number of replicates of any 2^n factorial experiment.

Solutions to problems

CHAPTER 2

Solution to 2.1

(a) Mean   4 (sum of ten values is 40)

    Median   3 (5th and 6th values are both 3)

(b) We know that the mean ($\bar{x}$) is 4.

x_i	4	2	7	3	0	3	1	13	4	3
$x_i - \bar{x}$	0	−2	3	−1	−4	−1	−3	9	0	−1
$(x_i - \bar{x})^2$	0	4	9	1	16	1	9	81	0	1

$$\sum_{i=1}^{10} (x_i - \bar{x})^2 = 122$$

$$\text{Variance} = \frac{\sum_{i=1}^{10} (x_i - \bar{x})^2}{n - 1}$$

$$= \frac{122}{9}$$

$$= 13.56$$

$$\text{Standard deviation} = \sqrt{13.56}$$

$$= 3.68$$

(c) Discrete.

Solution to 2.2

(a) Mean $= \dfrac{\Sigma x_i}{n} = \dfrac{760}{8} = 95$

Median is 96.5 (4th and 5th values are 96 and 97 respectively).

x_i	90	99	97	89	108	99	82	96
$x_i - \bar{x}$	−5	4	2	−6	13	4	−13	1
$(x_i - \bar{x})^2$	25	16	4	36	169	16	169	1

$$\Sigma(x_i - \bar{x})^2 = 436$$

$$\text{Variance} = \frac{\Sigma(x_i - \bar{x})^2}{n - 1} = \frac{436}{7}$$

$$= 62.29$$

$$\text{Standard deviation} = \sqrt{62.29}$$

$$= 7.89$$

$$\text{Coefficient of variation} = \frac{\text{SD}}{\text{mean}} \times 100$$

$$= \frac{7.89}{95} \times 100$$

$$= 8.30\%$$

(b) Continuous. (It seems that the experiment may well have been conducted by adding additional one pound weights to the skeins until they broke. Hence the data observed are in whole units. However, a skein which eventually broke under a load of 94 lb could well have broken under a load of 93.6 lb, or of 93.55 lb. The breaking strength is measured on a continuous scale.)

Solution to 2.3

(a) Using the grouped frequency table:

With the 50 batches in the sample, the loss due to out-of-specification batches is:

No. of batches above specification = 3
Cost   = 3 × £1000
Total cost   = £3000

If the process mean had been reduced by 1.0 the sample mean would have been reduced from 245.49 to 244.49. The loss would then have been:

Cost of batches above specification = 1 × £1000 = £1000
Cost of batches below specification = 2 × £300   £600
Total cost   £1600

If the process mean had been reduced by 2.0 the sample mean would have been reduced from 245.49 to 243.49 and the loss would then have been:

Cost of batches below specification = 4 × £300 = £1200
Total cost   = £1200

Any further decrease will increase the number of batches below specification. It would appear that reducing the mean by 2.0 will minimize the loss.

We have, however, fallen into a major trap in decision-making.

We are interested in the population of batches and the sample has been taken to represent the population. Yet we have made the decision based on the sample completely disregarding the population. (The sample itself is of no interest since these batches have already been produced and their values cannot be altered.) What can we say about the population? One comment is that there is little information about how the population behaves at its extremities. It is, however, these extremes which are of major concern in this problem. Crosswell have two options to solve this problem:

(i) They obtain considerably more data. This could well be impossible.
(ii) They can fit a probability model as outlined in Chapter 3.

A realistic solution will therefore have to wait until after Chapter 3 has been undertaken. However, common sense will indicate that by reducing the process mean by 2.0, the lower specification limit is close to where the frequency of batches rapidly increases. It may well be prudent in the first instance to reduce the mean by less than 2.0.

(b) It is possible that this slight lack of symmetry in the histogram is due to sampling error. It is therefore quite likely that the distribution of the population is symmetrical.

(c) This is a situation in which it is possible to have more than one definition of the population. Two possible definitions are:

(i) All present and future batches. To make decisions about this population we have to assume that the first 50 batches are representative of the population.
(ii) All possible batches that theoretically could have been obtained during the period of production of the first 50 batches. The 50 batches are therefore a truly representative sample of this population but to use the information we have to assume the population will remain unchanged.

CHAPTER 3

Solution to 3.1

(a) See Figure S.1.

$$x_1 = 82 \quad Z_1 = (82 - 80)/2 = 1.$$
$$x_2 = 78 \quad Z_2 = (78 - 80)/2 = -1.$$

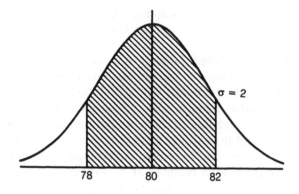

Figure S.1

Using Table A: 'area' to right of 1 is 0.1587.
Therefore also 'area' to left of −1 is 0.1587.
Therefore shaded 'area' is
   $1.0 - (0.1587 + 0.1587) = 1.00 - 0.3174 = 0.6826$
Therefore 68.26% will have resistance of between 78 and 82 ohms.

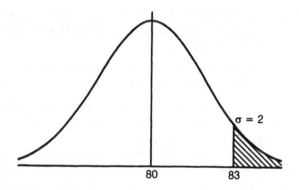

Figure S.2

(b) See Figure S.2.

$$x = 83 \qquad Z = (83 - 80)/2 = 1.5.$$

Using Table A: 'area' to right of 1.5 is 0.0668.
Therefore 6.68% will have resistance of more than 83 ohms.

(c) See Figure S.3.

$$x = 79 \qquad Z = (79 - 80)/2 = -0.5.$$

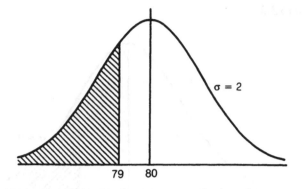

Figure S.3

Using Table A: 'area' to right of 0.5 is 0.3085.
Therefore also 'area' to left of −0.5 is 0.3085.
Therefore P (resistor has resistance of less than 79 ohms) = 0.3085.

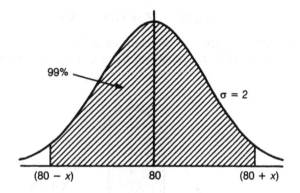

Figure S.4

(d) See Figure S.4. For 99% between $(80 - x)$ and $(80 + x)$ we need 0.5% above $(80 + x)$ and 0.5% below $(80 - x)$. Using Table A (second part), we find that the standardization values of $(80 + x)$ and $(80 - x)$ must be +2.58 and −2.58; i.e.

$$2.58 = \frac{(80 + x) - \text{mean}}{\text{standard deviation}}$$

$$= \frac{(80 + x) - 80}{2}$$

$$= \frac{x}{2}$$

$$\therefore x = 5.16$$

Hence the limits are 74.84 ohms and 85.16 ohms.

Solution to 3.2

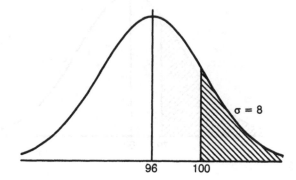

Figure S.5

(a) See Figure S.5.

$$x = 100 \qquad Z = (100 - 96)/8 = 0.5.$$

Using Table A: 'area' to the right of 0.5 is 0.3085.
Therefore 30.85% of skeins will have strength in excess of 100 lb.

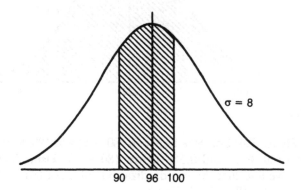

Figure S.6

(b) See Figure S.6.

$$x_1 = 100 \qquad Z_1 = (100 - 96)/8 = 0.5.$$

$$x_2 = 90 \qquad Z_2 = (90 - 96)/8 = -0.75$$

Using Table A: 'area' to the right of 0.5 is 0.3085
'area' to the left of −0.75 is 0.2266.

Therefore P (strength between 90 and 100 lb)

$$= 1 - 0.3085 - 0.2266$$
$$= 0.4649$$

(c) See Figure S.7. Using Table A (in 'reverse' direction), the standardized value which has an 'area' of 0.2 to the right of itself is (approx.) 0.84; i.e.

$$0.84 = \frac{x - \text{mean}}{\text{standard deviation}} = \frac{x - 96}{8}$$

$$\therefore x = 96 + (0.84 \times 8)$$
$$= 96 + 6.72$$
$$= 102.72 \, \text{lb}$$

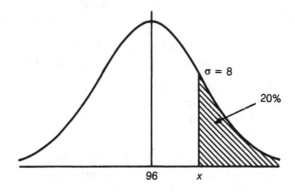

Figure S.7

(d) From (a) percentage of skeins with strength of 100 lb or more is 30.85%; i.e.

P (strength more than 100 lb) $= 0.3085$

$\therefore P$ (strength less than 100 lb) $= 0.6915$

$\therefore P$ (two skeins both have strength less than 100 lb) $= (0.6915)^2$
$$= 0.4782$$

Solution to 3.3

This is a situation in which the Poisson distribution might be applicable. By examination of the manufacturer's records we could establish the type of distribution that would best match the distribution of spotting faults in 15 metre lengths. Since we are unable to do this we will use a Poisson distribution with an appropriate value of μ.

(a) We will use a Poisson distribution with $\mu = 1$, since the fault rate is 1 fault per 15 metres when the process is working normally.

$$\text{Prob. of } r \text{ faults} = (\mu^r/r!)e^{-\mu}$$
$$\text{Prob. of 0 faults} = (1^0/0!)e^{-1}$$
$$= 0.3679$$

(b) As we are now taking a 30 metre length we use a Poisson distribution with $\mu = 2$.

(i) Prob. (0 faults) = $(2^0/0!)e^{-2} = 0.1353$

(ii) Prob. (1 fault) = $(2^1/1!)e^{-2} = 0.2707$
Prob. (2 faults) = $(2^2/2!)e^{-2} = 0.2707$
Prob. (3 faults) = $(2^3/3!)e^{-2} = 0.1804$
Prob. (4 faults) = $(2^4/4!)e^{-2} = 0.0902$
Prob. (less than 5 faults)   = 0.9473
Prob. (more than 4 faults)   = $1 - 0.9473 = 0.0527$

(c) (i) We are again concerned with 30 metre lengths from normal production so we will use the Poisson distribution with $\mu = 2$.

Prob. of less than 5 faults in a 30 metre length = 0.9473.
Prob. that a normal roll will be subjected to full inspection
= $0.9473(1 - 0.9473) + (1 - 0.9473)0.9473 + (1 - 0.9473)^2$
= 0.1026

(ii) The number of faults in 30 metre lengths taken from a roll with a fault rate of one occurrence of spotting per 5 metres of paper will have a Poisson distribution with $\mu = 6$.

Prob. (0 faults) = 0.0025
Prob. (1 fault) = 0.0149
Prob. (2 faults) = 0.0446
Prob. (3 faults) = 0.0892
Prob. (4 faults) = 0.1339

———————
0.2851
———————

Prob. of less than 5 faults in a 30 metre length taken from a roll with one fault per 5 metres = 0.2851

Probability of such a roll avoiding full inspection
$$= (0.2851)^2 = 0.0813$$

Solution to 3.4

(a) The higher the mean weight setting the smaller will be the percentage of packets which have net weights less than 500 grams. Since the packet weights have a normal distribution the percentage of under-weight packets cannot be reduced to zero however high we set the mean. Unfortunately we cannot comply with this regulation unless we weigh every packet individually and reject the underweight packets.

There is a movement away from wording laws in absolute terms such as 'no package should have a weight which is less than the nominal weight', to wording laws in probabilistic terms.

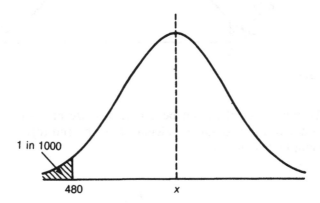

1 in 1000

480   x

Figure S.8

(b) See Figure S.8. For a tail probability of 0.001 the standardized value is 3.10 (Table A).

$$\text{Hence} \quad -3.10 = \frac{480 - x}{10}$$

$$x = 480 + 31.0$$

$$= 511 \text{ grams}$$

The process mean should be set to 511 grams at least in order to ensure that there is less than 1 in 1000 chance of a packet containing less than 480 grams.

(c) See Figure S.9. For a tail probability of 0.025 the standardized value is 1.96.

$$\text{Hence} \quad -1.96 = \frac{500 - x}{10}$$

$$x = 500 + 19.6$$

$$= 519.6$$

The process mean should be set to 519.6 grams at least in order to ensure that there is less than 1 in 40 chance of a packet containing less than 500 grams.

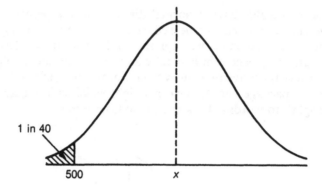

Figure S.9

(d) The mean weight should be set at a figure at least as great as the greater of the two values calculated in (b) and (c). A suitable value would be 520 grams.

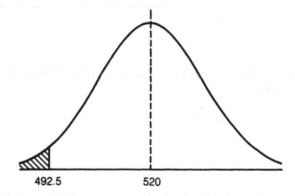

Figure S.10

(e) With the process mean set at 520 grams:

   (i) Rule 1 will be satisfied as the mean is greater than 500.
   (ii) See Figure S.10.

$$\text{Standardized value} = \frac{492.5 - 520}{10}$$

$$= -2.75$$

Thus 99.7% of packages would have weights above the tolerance limit and Rule 2 is satisfied.

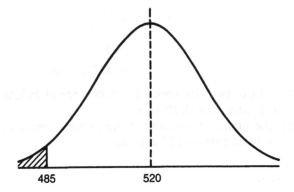

Figure S.11

    (iii) Rule 3 cannot be satisfied because of the infinitely long tail of the normal distribution.

(f) See Figure S.11.

$$\text{Standardized value} = \frac{485 - 520}{10}$$
$$= -3.5$$

Thus only 99.98% of packages would have weights above the absolute tolerance limit and Rule 3 would be violated.

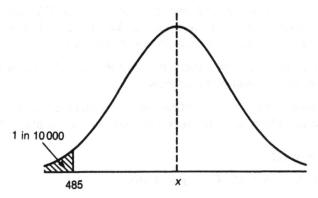

Figure S.12

(g) The mean must be increased to a value greater than 520 grams if Rule 3 is to be satisfied (see Figure S.12).

$$-3.72 = \frac{485 - x}{10}$$
$$x = 485 + 37.2$$
$$= 522.2 \text{ grams}$$

Smith and Co. must increase the process mean weight to 522.2 grams in order to comply with Rule 3.

Thus the minimum setting for the process mean weight which will satisfy all three rules is 522.2 grams.

CHAPTER 4

Solution to 4.1

$n = 6 \quad \bar{x} = 6.27 \quad s = 0.0645$

Null hypothesis: $\mu = 6.20$ True mean of consignment is 6.2.

Alternative hypothesis: $\mu > 6.20$

$$\text{Test statistic} = \frac{|\bar{x} - \mu|}{s/\sqrt{n}}$$
$$= \frac{|6.27 - 6.20|}{0.0645/\sqrt{6}}$$
$$= 2.66$$

Critical value – t-table with $5(n-1)$ degrees of freedom with a one-sided risk gives value of 2.02 and 3.36 at 5% and 1% significance levels respectively. We use a one-sided risk because the alternative hypothesis indicates a change from 6.20 in one direction.

Decision – since the test statistic is above the critical value at a 5% level we accept the alternative hypothesis.

Conclusion – the results confirm, at a 5% significance level, the manufacturer's claim that the mean intrinsic viscosity is above 6.20.

Solution to 4.2

(a) $n = 8 \quad \bar{x} = 49.60 \quad s = 0.968$

Null hypothesis – the true mean is 50.0 ($\mu = 50.0$).

Alternative hypothesis – the true mean is not 50.0 ($\mu \neq 50.0$).

$$\text{Test statistic} = \frac{|\bar{x} - \mu|}{s/\sqrt{n}}$$

$$= \frac{|49.6 - 50.0|}{0.968/\sqrt{8}}$$

$$= 1.17$$

Critical value – t-table with 7 degrees of freedom and a two-sided risk gives a value of 2.36 at a 5% significance level.

Decision – since the test statistic is less than the critical value we cannot reject the null hypothesis.

Conclusion – we cannot conclude that the method is biased at a 5% significance level.

(b) The 95% confidence interval is given by:

$$\bar{x} \pm \frac{ts}{\sqrt{n}} = 49.60 \pm \frac{2.36 \times 0.968}{\sqrt{8}}$$

$$= 49.60 \pm 0.81$$

$$= 48.79 \text{ to } 50.41$$

The maximum bias, at a 95% confidence level, is $(50 - 48.79)1.21$.

(c) If this is an initial evaluation it may well be carried out by a single highly skilled operator on the same day. The population for this situation is 'all possible values that could be obtained by the operator on that day'.

If, however, the method is shortly to become a standard procedure in the laboratory the population of interest is 'all possible values that could be obtained by all possible operators under all possible conditions'. The sample will have to be reasonably representative of this population.

Solution to 4.3

(a) $n = 10$ $\bar{x} = 4.5$ $s = 1.144$

Null hypothesis – the true mean for the modified process is 5.4 ($\mu = 5.4$).

Alternative hypothesis: $\mu < 5.4$

$$\text{Test statistic} = \frac{|\bar{x} - \mu|}{s/\sqrt{n}}$$

$$= \frac{|4.5 - 5.4|}{1.144/\sqrt{10}}$$

$$= 2.49$$

Critical value – t-table one-tailed, 9 degrees of freedom gives 1.83 and 2.82 at 5% and 1% significance levels.

Decision – we reject the null hypothesis and accept the alternative at the 5% significance level.

Conclusion – there is evidence, at the 5% significance level, that the mean impurity has decreased.

(b) t is 2.26 with 9 degrees of freedom and a two-sided risk.

95% confidence interval for the mean is given by:

$$\bar{x} \pm ts/\sqrt{n} = 4.5 \pm 2.26 \times 1.144/\sqrt{10}$$
$$= 4.5 \pm 0.82$$
$$= 3.68 \text{ to } 5.32$$

(c) Sample size $= \left(\dfrac{ts}{c}\right)^2$

$$= \left(\frac{2.26 \times 1.144}{0.15}\right)^2$$

$$= 297.08$$

Rounding to the next highest integer gives a sample size of 298.

CHAPTER 5

Solution to 5.1

$n_A = 10$ $s_A = 2.951$
$n_B = 9$ $s_B = 1.323$

Null hypothesis – the population variances are equal ($\sigma_A^2 = \sigma_B^2$).

Alternative hypothesis – the population variances are unequal ($\sigma_A^2 \neq \sigma_B^2$).

$$\text{Test statistic} = \frac{\text{sample variance of A}}{\text{sample variance of B}}$$

$$= \frac{2.951^2}{1.323^2}$$

$$= 4.98$$

Critical value – with 9 degrees of freedom for the numerator and 8 for the denominator, the F-table gives values of 4.36 and 7.34 at the 5% and 1% levels respectively for a two-sided test.

Decision – since the test statistic is greater than the critical value at a 5% significance level we reject the null hypothesis and accept the alternative.

Conclusion – from this evidence it would appear that method B is more repeatable but it must be emphasized that the population consists of determinations from one operator and one sample of cement and therefore the conclusion is of limited applicability.

Solution to 5.2

$n_A = 6$ $\bar{x}_A = 34.6$ $s_A = 6.046$
$n_W = 6$ $\bar{x}_W = 33.4$ $s_W = 5.476$

(a) Null hypothesis – the population variances are equal $(\sigma_A{}^2 = \sigma_W{}^2)$.

Alternative hypothesis – the population variances are unequal $(\sigma_A{}^2 \neq \sigma_B{}^2)$.

$$\text{Test statistic} = \frac{\text{variance of A}}{\text{variance of W}}$$
$$= (6.046)^2/(5.476)^2$$
$$= 1.22$$

Critical value – F-table with a two-sided test, 5 and 5 degrees of freedom, gives 7.15 and 14.94 at the 5% and 1% levels respectively.

Decision – we cannot reject the null hypothesis at a 5% significance level.

(b) Since there is not conclusive evidence that the variances are unequal we can now combine them to give a better estimate of the population standard deviation.

$$s^2 = \frac{\text{d.f.}_A s_A{}^2 + \text{d.f.}_W s_W{}^2}{\text{d.f.}_A + \text{d.f.}_W}$$
$$= \frac{5 \times 6.046^2 + 5 \times 5.476^2}{5 + 5}$$
$$= 33.270$$
$$s = 5.768$$

Null hypothesis – the population means are equal $(\mu_A = \mu_W)$.

Alternative hypothesis – the population mean for the oil containing additive is greater than the population mean for oil without additive $(\mu_A > \mu_W)$.

$$\text{Test statistic} = \frac{(\bar{x}_A - \bar{x}_W)}{s\sqrt{\left(\dfrac{1}{n_A} + \dfrac{1}{n_W}\right)}}$$

$$= \frac{(34.6 - 33.4)}{5.768\sqrt{\left(\dfrac{1}{6} + \dfrac{1}{6}\right)}}$$

$$= 0.36$$

Critical value – t-table, one-sided risk, with 10 degrees of freedom (5 + 5) gives values of 1.81 and 2.76 at the 5% and 1% level respectively.

Decision – we cannot reject the null hypothesis at a 5% significance level.

Conclusion – there is not conclusive evidence to suggest that the additive decreases petrol consumption.

(c) t, with a one-sided risk and 10 degrees of freedom, has a value of 1.81, $s = 5.768$, $c = 1.0$.

$$n_1 = n_2 = 2\left(\frac{ts}{c}\right)^2$$

$$= 2\left(\frac{1.81 \times 5.768}{1.0}\right)^2$$

$$= 218.0$$

436 cars ($n_1 = n_2 = 218$) are required to be 95% certain of estimating an improvement to within ± 1.0 m.p.g.

(d) The company presumably would like the population to be: All cars being used in the UK at the present time.

Obtaining a random sample would necessitate that each car had an equal chance of being chosen in the sample. Clearly this would be impossible. However, by taking into account factors like engine size, car size and popularity, a reasonably representative sample can be chosen by selecting certain makes of cars.

Solution 5.3

(a) d is the increase in m.p.g.

Car	A	B	C	D	E	F
d	1.6	1.1	1.7	0.3	0.3	2.2

$\bar{x}_d = 1.2$ $s_d = 0.780$

Null hypothesis $\mu_d = 0$

Alternative hypothesis $\mu_d > 0$

$$\text{Test statistic} = \frac{|\bar{x}_d - \mu_d|}{s_d/\sqrt{n}}$$

$$= \frac{|1.2 - 0|}{0.780/\sqrt{6}}$$

$$= 3.77$$

Critical value – t-table, one-sided risk, with 5 degrees of freedom gives values of 2.02 and 3.36 at the 5% and 1% levels respectively.

Decision – we accept the alternative hypothesis at a 1% significance level.

Conclusion – there is substantial evidence of a decrease in petrol consumption.

(b)
$$n = \left(\frac{ts}{c}\right)^2$$

t(one-sided, 5%, 5 d.f.) = 2.02, $s = 0.780$, $c = 1.0$

$$n = \left(\frac{2.02 \times 0.78}{1.0}\right)^2$$

$$= 2.48$$

A sample size of three cars is needed to be 95% certain of estimating an increase to within ± 1.0 m.p.g. Since there are two trials per car, a total of six trials are needed. This is a considerable improvement on the previous design which required 436 cars.

CHAPTER 6

Solution to 6.1

(a) Null hypothesis – the colour-matcher cannot discriminate between the fabrics.

Alternative hypothesis – the colour-matcher can discriminate between the fabrics.

Test statistic = 7.

Critical value – the table of critical values for triangular test gives $8\frac{1}{2}$ at a 5% level and $9\frac{1}{2}$ at a 1% level.

Decision – we cannot reject the null hypothesis. It has not been proved that the colour-matcher can discriminate.

(b) In this situation it is possible that the colour-matcher could make the 'right' decision for the wrong reason. For example he may choose correctly the odd-one-out in the belief that it is lighter when it is in fact darker. A better design would be to present him with one sample from each fabric and ask him to choose the darker.

Solution to 6.2

(a) Null hypothesis $\pi = 10\%$

Alternative hypothesis $\pi > 10\%$

$$\text{Test statistic} = \frac{|p - \pi| - (50/n)}{\sqrt{[\pi(100 - \pi)/n]}}$$

$$(p = (31/200)100 = 15.5\%)$$

$$= \frac{|15.5 - 101 - (50/200)}{\sqrt{[10(90)/200]}}$$

$$= 2.47$$

Critical value – t-table with a one-sided risk and infinite degrees of freedom gives 1.64 at 5% and 2.33 at 1%.

Decision – accept the alternative hypothesis at a 1% significance level.

Conclusion – there is evidence at the 1% significance level that the furnace has deteriorated in performance.

(b) 95% confidence interval for the true percentage is given by:

$$p \pm \{1.96\sqrt{[p(100 - p)/n]} + (50/n)\}$$
$$= 15.5 \pm \{1.96\sqrt{[15.5(84.5)/200]} + (50/200)\}$$
$$= 15.5 \pm 5.3$$
$$= 10.2\% \text{ to } 20.8\%$$

(c) Every process is subject to random perturbations which result in each item being slightly different. In this process every insulator will be different and any slight change in conditions would have resulted in the random perturbations combining to give a different set of 200 insulators. We can therefore consider the day's actual production as a sample of 200 from all the insulators that could have been produced on that day.

(d) Observed values

	Inspector		
Insulator	A	B	Total
Hazed	8	23	31
Not-hazed	72	97	169
Total	80	120	200

(e) Expected frequency $= \dfrac{\text{row total} \times \text{column total}}{\text{grand total}}$

For hazed insulators by Inspector A: $\quad E = \dfrac{31 \times 80}{200} = 12.4$

All the other values can be obtained by subtraction from row and column totals to give:

	Inspector		
Insulator	A	B	Total
Hazed	12.4	18.6	31
Not-hazed	67.6	101.4	169
Total	80	120	200

Null hypothesis – no relationship between insulator classification and inspector ($\pi_A = \pi_B$).

Alternative hypothesis – insulator classification is related to inspector ($\pi_A \neq \pi_B$).

$$\text{Test statistic} = \Sigma \frac{(O - E)^2}{E}$$

$$= \frac{(8 - 12.4)^2}{12.4} + \frac{(23 - 18.6)^2}{18.6} + \frac{(72 - 67.6)^2}{67.6}$$

$$+ \frac{(97 - 101.4)^2}{101.4}$$

$$= 3.079$$

Critical value – chi-squared table with 1 degree of freedom gives values of 3.841 and 6.635 at 5% and 1% significance levels respectively.

Degrees of freedom = (rows − 1) × (column − 1)

$$= 1 \times 1 = 1$$

Decision – there is not enough evidence to reject the null hypothesis at a 5% significance level.

Conclusion – although there is no significant difference between percentages at a 5% level there is a significant difference at a 10% level (critical value 2.69). There is a possibility that the inspectors give different percentages but more evidence will be needed to confirm this.

(f) Prob. (r hazings) = $\dfrac{n!}{r!(n-r)!}(\pi/100)^r[1 - (\pi/100)]^{n-r}$

$$(r = 1, n = 10, \pi = 10\%)$$

$$= \frac{10!}{1!9!} \times 0.1 \times 0.9^9$$

$$= 10 \times 0.1 \times 0.387$$

$$= 0.387$$

Prob. (5 and above) = 1.0 − (0.349 + 0.387 + 0.194 + 0.057 + 0.011)

$$= 0.002$$

Prob. (2 or more hazings = 0.194 + 0.057 + 0.011 + 0.002

$$= 0.264$$

(g) $r = 0$ $n = 10$ $\pi = 20\%$

Prob. (0 hazings) = $\dfrac{10!}{0!\,10!} \times 0.2^0 \times 0.8^{10}$ ($0! = 1$, $0.2^0 = 1$)

$$= 0.8^{10}$$

$$= 0.107$$

Prob. (4 hazings) = 1.0 − (0.107 + 0.268 + 0.302 + 0.201 + 0.034)

$$= 0.088$$

Prob. (less than 2 hazings) = 0.107 + 0.268

$$= 0.375$$

(h) Using the results from parts (f) and (g):

(i) 0.264 (ii) 0.375

Both these probabilities are high and there is a large chance of making a wrong decision. This is not too surprising since ten is a very small sample, when the only information from each sample is whether it is defective or not. A sample size of ten can, however, be adequate for a measured variable.

CHAPTER 7

Solution to 7.1

(a) Rearranging the results in order of magnitude gives:

27.3   30.7   31.7   32.3   32.4   32.5   32.7   34.1

With eight results, Dixon's test statistic uses the formulae:

$$\frac{x_2 - x_1}{x_{n-1} - x_1} \quad \text{and} \quad \frac{x_n - x_{n-1}}{x_n - x_2}$$

With the lowest value this gives:

$$\frac{30.7 - 27.3}{32.7 - 27.3} = \frac{3.4}{5.4} = 0.629$$

With the highest value this gives:

$$\frac{34.1 - 32.7}{34.1 - 30.7} = \frac{1.4}{3.4} = 0.411$$

The critical values for a sample of eight are 0.608 and 0.717 for 5% and 1% significance levels respectively. We can therefore conclude that the value of 27.3 is an outlier at the 5% significance level.

(b) The rejection (or not) of the outlier will depend upon the answers to two questions:

(i) Are the assumptions underlying the outlier test valid in this situation? Dixon's test is dependent upon the distribution being normal. We have no information regarding this assumption but it is highly likely that the information will be available to the car manufacturer from previous tests.

(ii) How are the results to be used? If the data usually follow a normal distribution and if a confidence interval is to be calculated it is preferable that the outlier be rejected. However, the outlier might be the most important result since it could be indicative of a major fault in the car's design.

Solution to 7.2

(a) Combined SD $= \sqrt{\left(\dfrac{\Sigma \, \text{dfs}^2}{\Sigma \, \text{d.f.}}\right)}$

$\quad\quad = \sqrt{[(2 \times 0.192^2 + 2 \times 0.032^2 + 2 \times 0.035^2}$
$\quad\quad\quad\; + \, 2 \times 0.050^2 + 2 \times 0.380^2 + 2 \times 0.069^2)/$
$\quad\quad\quad\; (2 + 2 + 2 + 2 + 2 + 2)]$

$\quad\quad = 0.178$

(b) 95% confidence interval is given by:

$$\bar{x} \pm ts/\sqrt{n}$$

$\bar{x} = 1.93$, t(12 degrees of freedom) $= 2.18$, $s = 0.178$, $n = 3$.

$$1.93 \pm 2.18 \times 0.178/\sqrt{3}$$
$$1.93 \pm 0.224$$
$$1.71 \text{ to } 2.15$$

(c)

	Test statistic	
Laboratory	Smallest value	Largest value
A	0.368	0.632
B	0.167	0.833
C	0.000	1.000
D	0.400	0.600
E	0.871	0.129
F	1.000	0.000

Critical value at a 5% significance level is 0.970.

Therefore the largest value for laboratory C and smallest value for laboratory F could be classed as outliers.

(d) The combined SD is unduly high compared with the individual SDs for laboratories B, C, D and F. This is due to the large influence of laboratory E on the calculation.

The 95% confidence interval for laboratory B is unduly wide, owing to the use of a rather high combined SD.

A visual examination of the results would lead us to believe that the laboratories A and E might have an outlier among their results. Nobody would expect an outlier in laboratories C and F.

In conclusion, none of the three analyses seem fairly to reflect the pattern of results.

(e) (i) There are probably too few determinations to draw a conclusion. However, previous experience indicates that the normal distribution is usually valid for analytical determinations.

(ii) An examination of the standard deviation reveals great differences between them. An outlier test for standard deviations (Cochran's test) would indicate that the laboratories A and E are outliers as far as variability is concerned. The assumption of equal standard deviations has been violated.

(iii) To obtain independent determinations it is necessary that the value of the second (or third) determination should not be related to the first (or second) determination in each laboratory. A repeat determination should therefore result from a repeat of full analytical procedure and not just from part of it. It is also necessary that an operator is unaware of the value of previous determinations on the same sample. Hopefully the assumption of independent observations has not been violated.

(iv) Clearly we cannot obtain a random sample but hopefully our sample is representative of the personnel and procedures that will be used in future determinations.

(v) A continuous variable has been used but the results are rounded to two decimal places. It is this rounding that has resulted in outliers being found in laboratories C and F.

Solution to 7.3

(a) $\bar{x} = 31.5$, $s = 28.01$, $n = 4$, t with 3 degrees of freedom $= 3.18$.

$$\bar{x} \pm ts/\sqrt{n} = 31.5 \pm 3.18 \times 28.01/\sqrt{4}$$
$$= 31.5 \pm 44.5$$
$$= -13.0 \text{ to } 76.0$$

(b) $p = 50\%$, $k = 1.96$, $n = 4$.

$$p \pm \{k\sqrt{[p(100 - p)/n]} + (50/n)\} = 50 \pm \{1.96\sqrt{[50(50)/4]} + (50/4)\}$$
$$= 50 \pm 61.5$$
$$= -11.5\% \text{ to } 111.5\%$$

(c) (i) If the population is 'the time period under review' the sample might be considered as representative of the population. However, only at certain times during this period was the subject in contact with radiation. It would therefore appear that the population is not homogeneous and to draw inferences from the sample in relation to a single population would be dubious.

(ii) After the first result a decision was made to remove the subject from contact with radiation. Therefore the second result cannot be independent of the first.

(iii) Even with only four results there is an indication that the results do not follow a normal distribution.

(iv) The number of 'successes' is only two which is below the minimum of five.

(d) The fact that the samples are not independent invalidates both the confidence interval for the mean and the confidence interval for the percentage. The confidence interval for the mean is also invalid due to non-normality whereas the confidence interval for the percentage does not fill a necessary condition for use of the formula in part (b). The nature of the population under consideration is also unclear.

CHAPTER 8

Solution to 8.1

(a)

Yield				Mean	Range	SD
69.0	71.2	74.2	68.1	70.625	6.1	2.716
72.1	64.3	71.2	71.0	69.65	7.8	3.599
74.0	72.4	67.6	76.7	72.675	9.1	3.820
69.0	70.8	78.0	73.6	72.85	9.0	3.920
67.3	72.5	70.7	69.3	69.95	5.2	2.199
75.8	63.8	72.6	70.5	70.675	12.0	5.075
76.0	69.5	72.6	66.9	71.25	9.1	3.932
74.2	72.9	69.6	76.2	73.225	6.6	2.772
70.4	68.3	65.9	71.5	69.025	5.6	2.470
69.9	70.4	74.1	73.5	71.975	4.2	2.131
68.3	70.1	75.0	78.1	72.875	9.8	4.489
67.2	71.5	73.0	71.8	70.875	5.8	2.534

(b) Mean range = 7.525

Short-term SD = (mean range)/(Hartley's constant)

$$= 7.525/2.06$$

$$= 3.653$$

(c) Combined SD $= \sqrt{\{\Sigma[(\mathrm{df})(\mathrm{SD}^2)]/[\Sigma\,\mathrm{df}]\}}$

$$= \sqrt{\{[(3)(2.716^2)] + [(3)(3.599^2)] + \ldots/[3+3+\ldots]\}}$$

$$= 3.431$$

(d) The three standard deviations are

Long-term SD = 3.251
Short-term SD = 3.653
Combined SD = 3.431

We see that the long-term SD is smaller than the other two, but not significantly smaller. With some sets of data we would find that the long-term SD was significantly greater than the other estimates. This would occur if there was a large, long-term change in mean yield during the time when these 50 batches were produced.

Solution to 8.2

(a) For the four point moving mean chart

Centre line = 71.0

Lower action line $= \bar{X} - 3\sigma/\sqrt{n}$
$= 71.0 - 3(3.5)/\sqrt{4}$
$= 65.75$

Lower warning line $= \bar{X} - 2\sigma/\sqrt{n}$
$= 71.0 - 2(3.5)/\sqrt{4}$
$= 67.5$

Upper warning line $= \bar{X} + 2\sigma/\sqrt{n}$
$= 71.0 + 2(3.5)/\sqrt{4}$
$= 74.5$

Upper action line $= \bar{X} + 3\sigma/\sqrt{n}$
$= 71.0 + 3(3.5)/\sqrt{4}$
$= 76.25$

For the four point moving range chart we multiply the standard deviation by five constants from Table J:

Centre line $= 2.06(3.5)$
$= 7.21$

Lower action line $= 0.20(3.5)$
$= 0.70$

Lower warning line $= 0.59(3.5)$
$= 2.07$

Upper warning line $= 3.98(3.5)$
$= 13.93$

Upper action line $= 5.30(3.5)$

$\qquad\qquad\qquad\quad = 18.55$

These lines, together with suitable axes, are drawn on graph paper to get the blank control charts on which points will be plotted as subsequent batches are produced.

Solution to 8.3

(a)

Batch no.	51	52	53	54	55	56	57	58	59	60
Moving mean	—	—	—	71.7	73.5	73.2	72.0	71.8	69.3	69.9

Batch no.	61	62	63	64	65	66	67	68	69	70
Moving mean	68.2	67.4	68.6	66.6	68.3	67.4	65.6	64.9	64.4	66.0

Batch no.	51	52	53	54	55	56	57	58	59	60
Moving range	—	—	—	4.1	4.7	4.7	7.6	7.6	5.7	7.8

Batch no.	61	62	63	64	65	66	67	68	69	70
Moving range	12.1	12.1	12.1	9.1	4.7	7.9	5.5	5.5	3.3	6.1

The moving means are plotted in Figure S.13 and the moving ranges in Figure S.14.

(b) At batch 67 we get an indication from the four-point moving mean chart that the mean yield has fallen. This conclusion is confirmed by the three subsequent points. Earlier we had some indication of a decrease in yield when points fell between warning and action lines. However, these indications were not decisively confirmed until the yield for batch 67 became available.

On the moving range chart we get no action indications. Whilst plotting the points on the moving range charts you will have noticed the tendency for ranges to repeat. The 12.1, for example, occurs three times. Clearly it would be unwise to use the rule 'Take action if **two** consecutive points fall between action and warning lines', with a four-point moving range chart.

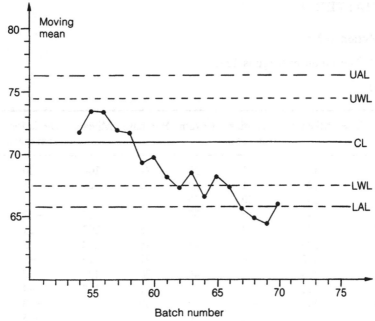

Figure S.13

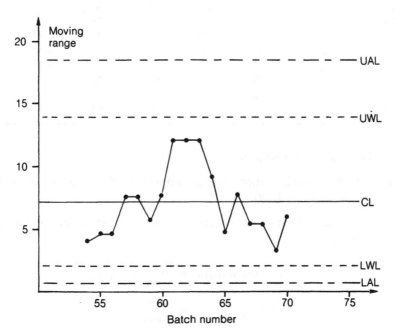

Figure S.14

CHAPTER 9

Solution to 9.1

(a) The mean mileage is 124.

(b)

Purchase	Mileage	Deviation from mean	Cusum	Purchase	Mileage	Deviation from mean	Cusum
1	135	11	11	21	104	−20	74
2	123	−1	10	22	123	−1	73
3	134	10	20	23	115	−9	64
4	141	+17	37	24	103	−21	43
5	127	3	40	25	103	−21	22
6	126	2	42	26	122	−2	20
7	132	8	50	27	115	−9	11
8	120	−4	46	28	131	7	18
9	122	−2	44	29	120	−4	14
10	141	+17	61	30	111	−13	1
11	129	+5	66	31	119	−5	−4
12	121	−3	63	32	126	2	−2
13	148	+24	87	33	134	+10	8
14	137	+13	100	34	126	+2	10
15	132	+8	108	35	128	+4	14
16	136	+12	120	36	110	−14	0
17	120	−4	116	37	114	−10	−10
18	103	−21	95	38	122	−2	−12
19	116	−8	87	39	123	−1	−13
20	131	+7	94	40	137	+13	0

For graph see Figure S.15.

(c) Null hypothesis – there was no change in the mean mileage during observations 1–40.

Alternative hypothesis – there was a change in mean mileage during this period.

$$\text{Test statistic} = \frac{\text{maximum cusum}}{\text{localized SD}}$$

$$= \frac{120}{8.848}$$

$$= 13.56$$

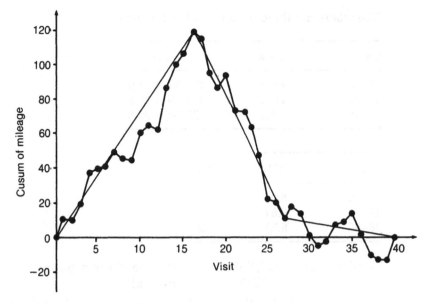

Figure S.15

Critical value – for a span of 40 observations, see table:

8.0 at the 5% significance level
9.3 at the 1% significance level.

Decision – reject the null hypothesis at the 1% significance level.

Conclusion – there has been a change in mean mileage about the time of observation 16.

A further series of significance tests needs to be carried out. These are summarized below:

Section	Observation number for maximum cusum	Maximum cusum	Test statistic	Critical value 5%	Critical value 1%	Decision
1–40	16	120	13.56	8.0	9.3	Reject n.h.
1–16	12	27	3.05	4.7	6.0	Not significant
17–40	27	54	6.10	5.9	7.2	Reject n.h.
17–27	20	13.6	1.53	3.8	5.1	Not significant
28–40	39	13.8	1.56	4.2	5.5	Not significant

Thus there are three sections within the data:

Section	Mean	m.p.g.
1–16	131.5	131.5/4 = 32.9
17–27	114.1	114.1/4 = 28.5
28–40	123.2	123.2/4 = 30.8

(d)

Section	SD	d.f.
1–16	8.09	15
17–27	9.69	10
28–40	8.38	12

Combined SD = $\sqrt{\{[\Sigma(\text{degrees of freedom})(\text{within group SD})^2]/}$
$[\Sigma \text{ degrees of freedom}]\}}$

$= \sqrt{\{[15 \times (8.09)^2 + 10 \times (9.69)^2 + 12 \times (8.38)^2]/}$
$[15 + 10 + 12]\}}$

$= 8.64$

This is approximately equal to the localized SD, 8.848. Note, however, that we must use the localized SD when doing a cusum test. Table I is based on the localized SD.

CHAPTER 10

Solution to 10.1

(a) The plant manager believes reaction time is dependent upon concentration so reaction time is the dependent variable (y) and concentration is the independent variable (x).

(b) SD(x) = 2.216, SD(y) = 34.902.

(c) Correlation coefficient (r) = 0.741.

(d) Null hypothesis – within the population of batches the correlation between reaction time and concentration is zero ($\rho = 0$).

Alternative hypothesis – there is a positive correlation between reaction time and concentration ($\rho > 0$).

Critical value – from Table H with a sample size of 12 and a one-sided test:

0.497 at a 5% significance level
0.658 at a 1% significance level.

Test statistic – the sample correlation coefficient of 0.741.

Decision – reject the null hypothesis at the 1% significance level.

Conclusion – the reaction time is related to concentration in such a way that reaction time increases as concentration increases.

(e) $b = 11.66$

$$a = \bar{y} - b\bar{x}$$
$$= 340 - 11.66 \times 11$$
$$= 211.7$$

The least squares regression equation is

$$y = 211.7 + 11.66x$$

(f) See Figure S.16.

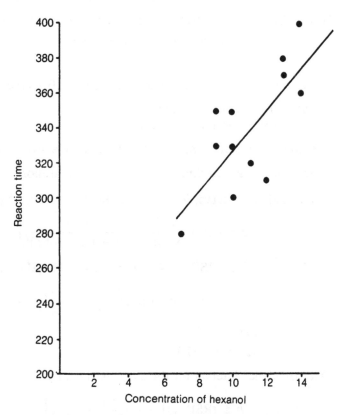

Figure S.16   Relationship between reaction time and concentration

(g) Percentage fit $= 100r^2$

$\qquad = 100(0.741)^2$

$\qquad = 54.9\%$

(h) Residual sum of squares $= (n - 1)(\text{SD of } y)^2(1 - r^2)$

$\qquad = (12 - 1)(34.902)^2(1 - 0.741^2) = 6042.2$

Residual standard deviation $= \sqrt{(\text{residual sum of squares/}}$
$\qquad\qquad\qquad\qquad\qquad \text{degrees of freedom)}$

$\qquad\qquad\qquad\qquad = \sqrt{(6042.2/10)}$

$\qquad\qquad\qquad\qquad = 24.6$

(i) Null hypothesis – there is no relationship between reaction time and concentration ($\beta = 0$).

Alternative hypothesis – there is a positive relationship between reaction time and concentration ($\beta > 0$).

Test statistic $= \sqrt{[(\text{new } \% \text{ fit} - \text{old } \% \text{ fit})(n - k - 1)/(100 - \text{new } \% \text{ fit})]}$

$\qquad = \sqrt{[(54.9 - 0)(12 - 1 - 1)/(100 - 54.9)]}$

$\qquad = 3.49$

Critical values – from the one-sided t-table with 10 degrees of freedom:

1.81 at the 5% significance level
2.76 at the 1% significance level.

Decision – we reject the null hypothesis at the 1% level of significance.

Conclusion – reaction time is related to concentration.

(j) 95% confidence interval for the true intercept is given by:

$$a \pm t\text{RSD} \sqrt{\left(\frac{1}{n} + \frac{(\bar{x})^2}{(n - 1)(\text{SD of } x)^2}\right)}$$

$a = 211.7$, $t = 2.23$, $\text{RSD} = 24.6$, $n = 12$, $\bar{x} = 11$, $(\text{SD of } x)^2 = 4.911$.

$$211.7 \pm 2.23 \times 24.6 \sqrt{\left[\frac{1}{12} + \frac{11^2}{11 \times 4.911}\right]}$$

211.7 ± 83.6

128.1 to 295.3

95% confidence interval for the slope is given by:

$$b \pm t\text{RSD} \sqrt{\left[\frac{1}{(n - 1)(\text{SD of } x)^2}\right]}$$

$$11.66 \pm 2.23 \times 24.6 \sqrt{\left[\frac{1}{11 \times 4.911}\right]}$$

11.66 ± 7.46

4.20 to 19.12

(k) 95% confidence interval for the true mean reaction time at a specified concentration is given by:

$$(a + bX) \pm t\text{RSD} \sqrt{\left[\frac{1}{n} + \frac{(X - \bar{x})^2}{(n - 1)(\text{SD of } x)^2}\right]}$$

(i) When $X = 7$:

$$(211.7 + 11.66 \times 7) \pm 2.23 \times 24.6 \sqrt{\left[\frac{1}{12} + \frac{(7 - 11)^2}{11 \times 4.911}\right]}$$

293.3 ± 33.8

259.5 to 327.1

(ii) When $X = 11$:

$$(211.7 + 11.66 \times 11) \pm 2.23 \times 24.6 \sqrt{\left[\frac{1}{12} + \frac{(11 - 11)^2}{11 \times 4.911}\right]}$$

340.0 ± 15.8

324.2 to 355.8

(l) 95% confidence interval for a single reaction time at a specified concentration is given by:

$$a + bX \pm t\text{RSD} \sqrt{\left[1 + \frac{1}{n} + \frac{(X - \bar{x})^2}{(n - 1)(\text{SD of } x)^2}\right]}$$

(i) When $X = 7$:

$$(211.7 + 11.66 \times 7) \pm 2.23 \times 24.6 \sqrt{\left[1 + \frac{1}{12} + \frac{(7 - 11)^2}{11 \times 4.911}\right]}$$

293.3 ± 64.4

228.9 to 357.7

(ii) When $X = 11$:

$$(211.7 + 11.66 \times 11) \pm 2.23 \times 24.6 \sqrt{\left[1 + \frac{1}{12} + \frac{(11 - 11)^2}{11 \times 4.911}\right]}$$

340.0 ± 57.1

282.9 to 397.1

(m) In his search of past production records the plant manager will have unwittingly carried out many subjective significance tests and will have found a set of data which looks 'significant'. The formal significance tests have just confirmed his informal judgement. This is therefore not in accordance with scientific method since both the theory and the conclusion have come from the same set of data.

Solution to 10.2

(a) $r = 0.989$

(b) $b = 0.3937$
    $a = 11.1$

(c) $\text{RSD} = \sqrt{[(n - 1)s_y^2(1 - r^2)/(n - 2)]}$
    $= \sqrt{[5(7.448)^2(1 - 0.989^2)/4]}$
    $= 1.24$

95% confidence interval for the slope is given by:

$$b \pm t\text{RSD} \sqrt{\left[\frac{1}{(n - 1)(\text{SD of } x)^2}\right]}$$

$$0.394 \pm 2.78 \times 1.24 \sqrt{\left[\frac{1}{5 \times (18.708)^2}\right]}$$

$$0.394 \pm 0.082$$

(d) 95% confidence interval for the true value of y at a given value of x is:

$$(a + bx) \pm t\text{RSD} \sqrt{\left[\frac{1}{n} + \frac{(x - \bar{x})^2}{(n - 1)(\text{SD of } x)^2}\right]}$$

$$(11.1 + 0.3937x) \pm 2.78 \times 1.24 \sqrt{\left[\frac{1}{6} + \frac{(x - 175)^2}{5 \times (18.708)^2}\right]}$$

Wind-up speed	95% confidence for true value of birefringence
150	70.2 ± 2.49
160	74.1 ± 1.87
175	80.0 ± 1.41
190	85.9 ± 1.87
200	89.8 ± 2.49

For plots see Figure S.17.

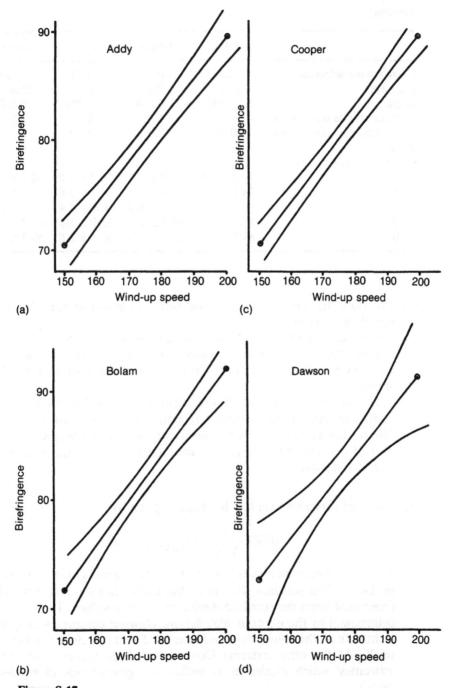

Figure S.17

Summary

	Addy	Bolam	Cooper	Dawson
Correlation coefficient	0.989	0.987	0.995	0.935
Intercept	11.1	9.3	10.5	13.5
Slope	0.394	0.404	0.397	0.380
Residual standard deviation	1.24	1.18	1.24	1.20
95% confidence interval for slope	±0.082	±0.093	±0.056	±0.199
95% CI for true line at:				
150	70.2 ± 2.5	69.9 ± 2.7	70.2 ± 2.0	70.5 ± 5.2
160	74.1 ± 1.9	73.9 ± 1.9	74.1 ± 1.6	74.3 ± 3.3
175	80.0 ± 1.4	80.0 ± 1.3	80.0 ± 1.4	80.0 ± 1.4
190	85.9 ± 1.9	86.1 ± 1.9	85.9 ± 1.6	85.7 ± 3.3
200	89.8 ± 2.5	90.1 ± 2.7	89.8 ± 2.0	89.5 ± 5.2

(e) No! Both the slope and the residual standard deviation are subject to sampling variation.

If we examine the confidence intervals for the slope coefficients we can see that these are wide compared with the range of the four experiments. They have been fortunate to obtain four such similar slopes.

The same can be said of the residual standard deviation. The confidence interval for a standard deviation can be found using Table F. Use of this table, for Addy's experiment, as an example, gives values of 0.74 to 3.56 for a 95% confidence interval for the true population standard deviation.

(f) The confidence interval for the slope is given by:

$$t\text{RSD} \sqrt{\left[\frac{1}{(n-1)(\text{SD of } x)^2}\right]}$$

Since the number of results was the same and there is little variability in the residual standard deviation, the confidence interval is mainly dependent upon the standard deviation of the x values. This value is determined by the experimental design. Dawson designed an experiment with all the values clustered round 175 and therefore s_x was very small. At the other extreme Cooper had three values at the two 'extremes' which maximizes s_x within the given range of wind-up speeds.

(g) The correlation coefficients together with the standard deviation of the wind-up speeds are given below:

	Dawson	Bolam	Addy	Cooper
Correlation coefficient	0.935	0.987	0.989	0.995
Standard deviation of wind-up speed	3.0	6.5	7.5	10.9

This table indicates a strong relationship between the range of wind-up speeds in the design, as indicated by the standard deviation, and the correlation coefficient. Interpretation of a correlation coefficient must always take into account the spread of the values. The designs chosen by Bolam and Cooper, with clusters of wind-up speeds, and large 'gaps', make the correlation coefficient difficult to interpret.

(h) Again this is dependent upon the design. Bolam has four central values with only two values at the 'extremes', while Cooper has no values in the centre.

(i) The confidence interval for Dawson's experiment is similar to the other three experiments within the range of wind-up speeds (165 to 185) used by him. Extrapolating outside this range produces wide confidence intervals and will be little use in determining 'start-up' conditions for the new product.

(j) Yes! The danger is that the linearity assumptions may not be valid. With Addy's design we would obtain a strong indication of curvature because of the spread of points. With Cooper's design having only two values this is not possible. He could be markedly wrong as shown in Figure S.18.

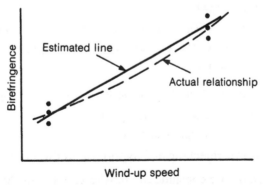

Figure S.18 Is the relationship linear or curved?

CHAPTER 11

Solution to 11.1

(a) As the first independent variable (z) enters the equation the percentage fit becomes 50.7%. We test the statistical significance of the draw ratio as follows:

Null hypothesis – tensile strength of the yarn is not related to the draw ratio (z).

Alternative hypothesis – tensile strength of the yarn is related to the draw ratio (z).

Test statistic $= \sqrt{[(\text{new } \% \text{ fit} - \text{old } \% \text{ fit})(n - k - 1)/(100 - \text{new } \% \text{ fit})]}$
$= \sqrt{[(50.7 - 0)(12 - 1 - 1)/(100 - 50.7)]}$
$= 3.21$

Critical values – from the two-sided t-table with 10 degrees of freedom:

2.23 at the 5% significance level
3.17 at the 1% significance level.

Decision – we reject the null hypothesis at the 1% level of significance.

Conclusion – we conclude that the tensile strength of the yarn is related to the draw ratio used in the drawing process.

As the second independent variable (x) enters the equation the percentage fit increases from 50.7% to 69.3%. We test the statistical significance of this increase as follows:

Null hypothesis – tensile strength is not related to the spinning temperature (x).

Alternative hypothesis – tensile strength is related to the spinning temperature (x).

Test statistic $= \sqrt{[(\text{new } \% \text{ fit} - \text{old } \% \text{ fit})(n - k - 1)/(100 - \text{new } \% \text{ fit})]}$
$= \sqrt{[(69.3 - 50.7)(12 - 2 - 1)/(100 - 69.3)]}$
$= 2.34$

Critical values – from the two-sided t-test with 9 degrees of freedom:

2.26 at the 5% significance level
3.25 at the 1% significance level.

Decision – we reject the null hypothesis at the 5% level of significance.

Conclusion – we conclude the tensile strength of the yarn is related to the spinning temperature (x) as well as the draw ratio (z).

As the third independent variable (w) enters the equation the percentage fit increases from 69.3% to 73.8%. This small increase gives a test statistic equal to 1.17 which is less than the critical value for 5% significance (2.31, with 8 degrees of freedom). We therefore conclude that the tensile strength of the yarn is not related to the drying time (w).

(b) Before we draw any conclusions about the relationships between the independent variables on the one hand and the response on the other, we would be wise to examine the correlation matrix, paying particular attention to the intercorrelation of the three independent variables. Unfortunately the regression program did not print out a correlation matrix.

What can we deduce from the regression equations and their percentage fits? We will consider the equations in the order in which they were fitted:

(i) The percentage fit of the first equation is 50.7%. Dividing by 100 and taking the square root we obtain 0.712, so the correlation between draw ratio (z) and tensile strength (y) must be either +0.712 or −0.712. As the coefficient of z in the first equation is positive the correlation between y and z must be positive. We conclude therefore that $r_{zy} = 0.712$.

(ii) As the second independent variable (x) enters the equation the percentage fit increases from 50.7% to 69.3%. Comparing the first and second equations we note that the coefficient of z does not change as x enters the equation. We will conclude, therefore, that the correlation between x and z is equal to zero. (This conclusion is beyond dispute provided that there are no rounding errors in the coefficients.) It is quite possible that the research chemist chose the values of draw ratio (z) and spinning temperature (z) in his 12 trials so that the two variables would not be correlated with each other. Because there is zero correlation between the two independent variables in the second equation there is a simple relationship between the percentage fit and the correlations with the response:

$$\% \text{ fit} = 100(r_{zy}^2 + r_{xy}^2)$$
$$69.3 = 100(0.712^2 + r_{xy}^2)$$
$$r_{xy} = +0.431 \text{ or } -0.431$$

As the coefficient of x in the second equation is negative (and $x_{xz} = 0$) we can conclude that the correlation between the spin temperature (x) and tensile strength (y) is equal to −0.431.

(iii) As the third independent variable enters the equation there is a small increase in percentage fit which is not significant as we have demonstrated in part (a). We note that the coefficients of x and z change considerably as drying time (w) enters the equation. This indicates that r_{xw} and r_{zw} are rather high, but we are unable to say exactly what their values are. Nor are we able to estimate the correlation between drying time (w) and tensile strength. Our findings are summarized in the correlation matrix below:

	w	x	z	y
w	1.000	High	High	?
x	High	1.000	0.000	−0.431
z	High	0.000	1.000	0.712
y	?	−0.431	0.712	1.000

Solution to 11.2

(1) response
(2) independent
(3) independent
(4) cross-product
(5) interaction
(6) interaction
(7) response
(8) value (or level)
(9) x
(10) z
(11) z
(12) t-test
(13) percentage fit
(14) percentage fit
(15) degrees of freedom
(16) x
(17) percentage fit
(18) zero
(19) x
(20) z
(21) r_{xz}
(22) r_{yz}
(23) residual sum of squares
(24) residuals
(25) percentage fit
(26) degrees of freedom
(27) residual standard deviation
(28) response
(29) variation
(30) fixed

Solution to 11.3

(a) Two of the six equations have tensile strength (y) as the dependent variable:

$$y = 1.100 + 0.0100z \qquad 85.0\% \text{ fit}$$

$$y = 2.867 + 0.0133x \qquad 53.3\% \text{ fit}$$

The percentage fit is calculated by squaring the appropriate correlation coefficient. Clearly the first of the two equations offers the better explanation of the variability in tensile strength. (This assertion is based solely on a statistical criterion, of course.)

(b) Percentage fits of the six equations are calculated from:

% fit $= 100\,r^2$

% fit $= 100(0.9220)^2 = 85.0\%$	for equation (1)
% fit $= 100(0.9220)^2 = 85.0\%$	for equation (2)
% fit $= 100(0.7303)^2 = 53.3\%$	for equation (3)
% fit $= 100(0.7303)^2 = 53.3\%$	for equation (4)
% fit $= 100(0.8416)^2 = 70.8\%$	for equation (5)
% fit $= 100(0.8416)^2 = 70.8\%$	for equation (6)

Note that equations (1) and (2) have the same percentage fit even though the two equations represent distinctly different lines on a graph. The same can be said of equations (3) and (4) or of equations (5) and (6).

(c) The partial correlation coefficient for y and x with z as the fixed variable can be calculated from:

$$[r_{xy} - r_{xz}r_{yz}]/\sqrt{[(1 - r_{xz}^2)(1 - r_{yz}^2)]}$$

$$= [0.7303 - (0.8416)(0.9220)]/\sqrt{[(1 - 0.8416^2)(1 - 0.9220^2)]}$$

$$= \frac{-0.04566}{0.20912} = -0.2183$$

Note that this partial correlation coefficient is negative though the simple correlation between y and x (0.7303) is positive. When the effect of the spinning temperature (z) has been taken into account it appears that an increase in drying time (x) will result in a decrease in tensile strength, rather than an increase as suggested by the positive simple correlation between x and y.

(d) Increase in % fit $= (100 - \text{old \% fit})(\text{partial correlation})^2$

$$= (100 - 85.0)(0.2183)^2$$

$$= 0.71\%$$

This minute increase in percentage fit may appear to be in conflict with the large correlation (0.7303) between x and y. The explanation lies in the huge correlation between the first independent variable (z) and the second independent variable (x).

(e) Percentage fit $= 100(r_{zy}^2 + r_{xy}^2 - 2r_{zy}r_{xy}r_{xz})/(1 - r_{xz}^2)$

$= 100[0.9220^2 + 0.7303^2 - 2(0.9220)(0.7303)$
$\qquad (0.8416)]/(1 - 0.8416^2)$

$= 100(0.8501 + 0.5333 - 1.1334)/(0.2917)$

$= 100(0.25)/(0.2917)$

$= 85.70\%$

This result is in agreement with our earlier finding, the percentage fit could be expected to increase by 0.71% from the base line of 85.00%.

(f) The missing lines are:

E   3.8   250   3.6   0.2
E   45   250   55   −10

(g) For the residuals in Table 11.6:

Mean = 0.00   Standard deviation = 0.1225

For the residuals in Table 11.7:

Mean = 0.00   Standard deviation = 9.3541

The correlation of the two sets of residuals

$$= \frac{-0.2500}{(0.1225)(9.3541)}$$

$$= -0.2182$$

Note that this result is equal to the partial correlation coefficient calculated in (c).

CHAPTER 12

Solution to 12.1

(a) Since we have three variables, and each variable has two levels, and each of the eight possible treatment combinations is used in one (and only one) trial the experiment is a 2^3 factorial experiment.

(b) You would be able to estimate:

   (i) three main effects;
  (ii) three two-variable interactions;
 (iii) one three-variable interaction.

(c) Adding 6 to the response values will not alter the values of the estimates or the sums of squares.

(d)

	Temperature (°C)			
	80		90	
	pH		pH	
Speed of agitation	6	8	6	8
Slow	6	2	4	2
Fast	4	8	0	6

The table above is not the only possible three-way table. You may have produced a different arrangement and you may have called the variables A, B and C.

(e)

| | Temperature (°C) | | | pH | | | pH | |
Speed	80	90	Speed	6	8	Temperature	6	8
Slow	4	3	Slow	5	2	80	5	5
Fast	6	3	Fast	2	7	90	2	4

(f) Speed main effect = $4.5 - 3.5 = 1.0$
Temperature main effect = $3.0 - 5.0 = -2.0$
pH main effect = $4.5 - 3.5 = 1.0$
Speed × temp. interaction = $3.5 - 4.5 = -1.0$
Speed × pH interaction = $6.0 - 2.0 = 4.0$
Temp. × pH interaction = $4.5 - 3.5 = 1.0$
Speed × temp. × pH interaction

$$= \tfrac{1}{2}\{[\tfrac{1}{2}(6 + 4) - \tfrac{1}{2}(2 + 0)] - [\tfrac{1}{2}(6 + 8) - \tfrac{1}{2}(4 + 2)]\}$$
$$= 0.0$$

(g) The largest effect estimate is the speed × pH interaction. We will test this first:

Null hypothesis – there is no interaction between speed and pH.

Alternative hypothesis – there is an interaction between speed and pH.

$$\text{Test statistic} = \frac{\text{effect estimate}}{RSD/\sqrt{(p2^{n-2})}}$$

$$= \frac{4.0}{1.15/\sqrt{2}}$$

$$= 4.92$$

Critical values – from the two-sided t-table with 3 degrees of freedom:

3.18 at the 5% significance level
5.84 at the 1% significance level.

Decision – we reject the null hypothesis at the 5% level of significance.

Conclusion – we conclude that there is an interaction between speed of agitation and pH.

Further significance testing reveals that no other effect estimates are statistically significant.

Solution to 12.2

(a)

Trial	s	t	p	$s \times t$	$s \times p$	$t \times p$	$s \times t \times p$
1	1	80	6	80	6	480	480
2	1	90	8	90	8	720	720
3	0	90	6	0	0	540	0
4	1	80	8	80	8	640	640
5	0	80	8	0	0	640	0
6	0	80	6	0	0	480	0
7	1	90	6	90	6	540	540
8	0	90	8	0	0	720	0

(b) If we examine the values of s and $(s \times t)$ all trials with $s = 0$ must also have $(s \times t) = 0$. All trials with $s = 1$ have $(s \times t) = 80$ or 90. Clearly this must result in a high correlation between s and $(s \times t)$. The use of 0, 1 for slow and fast speeds is the worst choice possible for obtaining high correlations between a main effect and its interactions.

(c) Using a coding of -1 and $+1$ for s and t gives:

Trial	s	t	$(s \times t)$
1	1	−1	−1
2	1	1	1
3	−1	1	−1
4	1	−1	−1
5	−1	−1	1
6	−1	−1	1
7	1	1	1
8	−1	1	−1

There is now a zero correlation between s and $(s \times t)$ compared with 0.99 with the uncoded data.

(d) Using the formula:

Test statistic $= \sqrt{[(\text{new \% fit} - \text{old \% fit})(n - k - 1)/(100 - \text{new \% fit})]}$

		Critical value	
% fit	Test statistic	5%	1%
16.67	1.09	2.45	3.71
28.43	0.91	2.57	4.03
77.08	2.91	2.78	4.60
94.60	3.12	3.18	5.84
98.77	2.60	4.30	9.92
100%	Infinite	12.71	63.66

(e) Clearly there are no variables or interactions above the critical value at the first stage and therefore the conclusion would be drawn that the variables and their interactions were not significant. Even with the full analysis as given in part (d), it is impossible to reach any conclusion apart from noting that the sudden jumps in percentage fit are a sign of a strange set of data.

(f) The final regression equation is:

$$y = 76.0 - 10s - 0.8t - 10p - 0.2(s \times t) + 4.0(s \times p) + 0.1(t \times p)$$

When $s = 1$, $t = 90$, $p = 8$:

$$y = 76.0 - (10 \times 1) - (0.8 \times 90) - (10 \times 8) - 0.2(1 \times 90)$$
$$+ 4(1 \times 8) + 0.1(90 \times 8)$$
$$= 76 - 10 - 72 - 80 - 18 + 32 + 72$$
$$= 0$$

(g) The three-variable interaction has an effect estimate of zero in Problem 1. Clearly all the variation can be explained without this interaction and it has no part to play in the multiple regression analysis.

CHAPTER 13

Solution to 13.1

(a) There are four variables at two levels which would give 16 trials in a full experiment. This is therefore a $\frac{1}{4}$ (2^4) design.

(b)

Trial	A	B	C	D	AB	AC	AD	BC	BD	CD	ABC	ABD	ACD	BCD	ABCD
1	−	−	−	+	+	+	−	+	−	−	−	+	+	+	−
2	+	−	+	−	−	+	−	−	+	−	−	+	−	+	+
3	−	+	−	−	−	+	+	−	−	+	+	+	−	+	−
4	+	+	+	+	+	+	+	+	+	+	+	+	+	+	+

(c) The defining contrasts are:

   AC   ABD   BCD

(d) The alias groups are:

   (A, C, BD, ABCD)
   (B, AD, CD, ABC)
   (D, AB, BC, ACD)

Solution to 13.2

(a) With the information you have been given in Chapter 13 the easiest way for you to answer this problem is to find a half replicate that will give the chemist the estimates he requires. Such a half replicate does exist. The design matrix for a 2^4 factorial experiment is given in Chapter 13. You might find it useful.

(b) Each of the design matrices below represents a half replicate of a 2^4 factorial design. Either of these half replicates will yield the estimates needed by the chemist.

$$
\begin{array}{cccc}
A & B & C & D \\
+ & - & - & - \\
- & + & - & - \\
- & - & + & - \\
+ & + & + & - \\
+ & - & - & + \\
- & + & - & + \\
- & - & + & + \\
+ & + & + & +
\end{array}
\qquad
\begin{array}{cccc}
A & B & C & D \\
- & - & - & - \\
+ & + & - & - \\
+ & - & + & - \\
- & + & + & - \\
- & - & - & + \\
+ & + & - & + \\
+ & - & + & + \\
- & + & + & +
\end{array}
$$

These two half replicates together constitute the full 2^4 experiment. For each of the half replicates the defining contrast is ABC and the alias pairs are (A, BC), (B, AC), (C, AB), $(D, ABCD)$, (AD, BCD), (BD, ACD), (CD, ABD).

Solution to 13.3

(1) n (2) 2
(3) 4   (4) 2
(5) interaction   (6) t-test
(7) residual standard deviation   (8) 8
(9) 4   (10) 8
(11) $y = a + bx + cz$ (12) $y = a + bx + cz + dxz$
(13) cross-product   (14) correlated
(15) scale (or standardize)   (16) zero
(17) replicate   (18) defining contrast
(19) alias pairs   (20) interaction XW
(21) interaction ZW (22) main effect X
(23) interaction XZ (24) interaction
(25) confounded

CHAPTER 14

Solution to 14.1

(a) Coding of design
    Use -1 for F1, low temperature, enzyme
    and $+1$ for F4, high temperature, bioenzyme

Cell	Malt (A)	Temperature (B)	Yeast (C)
1	-1	-1	$+1$
2	-1	$+1$	-1
3	$+1$	-1	-1
4	$+1$	-1	$+1$
5	$+1$	-1	-1
6	-1	$+1$	-1
7	-1	$+1$	-1
8	$+1$	-1	$+1$
9	-1	-1	$+1$
10	-1	-1	$+1$

Correlation matrix

	A	B	C
A	1.0	-0.53	0.0
B	-0.53	1.0	-0.65
C	0.0	-0.65	1.0

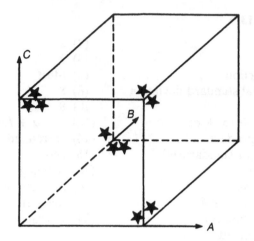

Figure S.19

(b) The design can be represented as stars at the corners of a cube with each star denoting the experimental conditions for a particular cell (Figure S.19). There are four stars without any crosses and it is desirable that two cells are chosen at one of these corners. It is also desirable that the imbalance in the number of cells at different levels for A and B be rectified. However, our main concern must be to reduce the correlation of (AB) and (BC). Since both of these are negative it is imperative that the products of AB and AC are positive. Thus we can have A, B and C either all negative or all positive. All negative values will worsen the imbalance, so all positive values are chosen. The extra cells are:

Cell	A	B	C
11	+1	+1	+1
12	+1	+1	+1

The experimental conditions for the new cells are

Cell	Malt	Temperature	Yeast
11	F4	High	Bioenzyme
12	F4	High	Bioenzyme

Note: These are the best two cells when only main effects are being investigated. The inclusion of two-factor interactions, which is essential for a realistic investigation, would lead to different cells.

(c) *Correlation matrix*

	A	B	C
A	1.0	−0.17	0.17
B	−0.17	1.0	−0.31
C	0.17	−0.31	1.0

Solution to 14.2

(a) Coding of levels

Temperature	270	−1
	275	0
	280	+1
Pressure	600	−1
	700	0
	800	+1
Concentration	0.15	−1
	0.20	0
	0.25	+1

The levels of quadratic and interactions are obtained by multiplying the appropriate levels from the main effects.

Design matrix

Cell	Temp.	Press.	Conc.	(Temp.)^2	(Press. × conc.)
1	0	0	−1	0	0
2	0	0	0	0	0
3	0	0	+1	0	0
4	0	−1	+1	0	−1
5	0	+1	+1	0	+1
6	−1	−1	+1	+1	−1
7	+1	−1	+1	+1	−1

(b) Correlation matrix

	Temp.	Press.	Conc.	(Temp.)^2	(Press. × conc.)
Temp.	1.0	0.0	0.0	0.0	0.0
Press.		1.0	−0.24	−0.65	1.0
Conc.			1.0	+0.37	−0.24
(Temp.)^2				1.0	−0.65
Press. × conc.					1.0

(c) The matrix reveals high correlations between three effects, the highest being press. and (press. × conc.) interaction. To reduce this correlation it is necessary that these effects have opposite signs. However, if press. has the same sign for both cells it will result in increasing the correlation between press. and conc. The choice of cells is given below.

Cell	Temp.	Press.	Conc.	(Temp.)^2	(Press. × conc.)
8		+1	−1		−1
9		−1	−1		+1

Negative values of concentration also reduce the imbalance towards positive cells in the first seven cells.

It remains to choose values of temp. which reduce the negative correlation between (temp.)^2 and (press. × conc.) interaction. For cell 8, (temp.)^2 must take its lowest level of zero while for cell 9 it must have a level of +1. The level for temp. in cell 9 can therefore be either −1 or +1, it being irrelevant which is chosen. The full design for the extra two cells is given below.

Cell	Temp.	Press.	Conc.	(Temp.)^2	(Press. × conc.)
8	0	+1	−1	0	−1
9	+1	−1	−1	+1	+1

(d) Correlation matrix

	Temp.	Press.	Conc.	(Temp.)^2	(Press. × conc.)
Temp.	1.0	−0.19	−0.26	0.27	0.30
Press.		1.0	−0.24	−0.70	0.28
Conc.			1.0	0.08	−0.24
(Temp.)^2				1.0	−0.10
Press. × conc.					1.0

Thus it can be seen that two of the high correlations have been reduced but the correlation between (temp.)^2 and press. still remains high. It is impossible to reduce all three correlations with only two extra cells.

CHAPTER 15

Solution to 15.1

(a) Total sum of squares

$$= \text{(overall SD)}^2 \text{(degrees of freedom)}$$
$$= (0.27775)^2 (14)$$
$$= 1.0800$$

(b) Within samples sum of squares

$$= \Sigma\{(\text{sample SD})^2(\text{degrees of freedom})\}$$
$$= (0.100)^2(4) + (0.18708)^2(4) + (0.22361)^2(4)$$
$$= 0.3800$$

(c) Between samples sum of squares

$$= (\text{SD of sample means})^2(\text{number of samples} - 1)$$
$$\times (\text{number of observations on each sample})$$
$$= (0.26458)^2(3 - 1)(5)$$
$$= 0.7000$$

(d)

Source of variation	Sum of squares	Degrees of freedom	Mean square
Between samples	0.7000	2	0.3500
Within samples	0.3800	12	0.0317
Total	1.0800	14	—

(e) Null hypothesis – there is no difference in impurity at the three depths in the tank.

Alternative hypothesis – there is variation in impurity from depth to depth.

$$\text{Test statistic} = \frac{\text{between samples mean square}}{\text{within samples mean square}}$$
$$= \frac{0.350}{0.0317}$$
$$= 11.04$$

Critical values – from the one-sided F-table with 2 and 12 degrees of freedom:

3.89 at the 5% significance level
6.93 at the 1% significance level.

Decision – we reject the null hypothesis at the 1% level of significance.

Conclusion – the concentration of impurity is different at different depths within the tank.

(f) Testing standard deviation $= \sqrt{(\text{within samples mean square})}$
$$= \sqrt{0.0317}$$
$$= 0.178$$

(g) There is a strong indication in the data that the concentration of impurity is greater at greater depths, as we can see in the table below. The F-test has proved, beyond reasonable doubt, that the apparent increase in impurity is not simply due to testing error.

Depth (m)	0.5	2.5	4.5
Mean impurity	3.2	3.3	3.7

(h) In order to explore the relationship between impurity concentration and depth we could use regression analysis. We will do this in Problem 2.

Solution to 15.2

(a) The total sum of squares is equal to 1.08. This was calculated in Problem 15.1. It is the same variation in impurity that we are exploring in both problems.

(b) Residual sum of squares

$$= (1 - r^2)(n - 1)(\text{SD of } y)^2$$
$$= (1 - 0.5787)(15 - 1)(0.27775)^2$$
$$= 0.455$$

(c) Regression sum of squares

$$= r^2(n - 1)(\text{SD of } y)^2$$
$$= 0.5787(15 - 1)(0.27775)^2$$
$$= 0.625$$

(d)

Source of variation	Sum of squares	Degrees of freedom	Mean square
Due to regression on x	0.6250	1	0.6250
Residual	0.4550	13	0.0350
Total	1.0800	14	

(e) Null hypothesis – impurity concentration (y) is not related to depth (x).

Alternative hypothesis – impurity concentration (y) is related to depth (x).

$$\text{Test statistic} = \frac{\text{regression mean square}}{\text{residual mean square}}$$
$$= \frac{0.6250}{0.0350}$$
$$= 17.86$$

Critical values – from the one-sided F-table with 1 and 13 degrees of freedom:

4.67 at the 5% significance level
9.07 at the 1% significance level.

Decision – we reject the null hypothesis at the 1% level of significance.

Conclusion – we conclude that the concentration of impurity is related to the depth from which the sample is taken.

We have explored the possibility that there is a linear relationship between impurity and depth. Perhaps the relationship is curved. With samples taken from three depths (0.5, 2.5 and 4.5 m) we could fit a quadratic curve, and we will do so in Problem 3.

Solution to 15.3

(a) Percentage fit

$$= \frac{\text{due to regression on } X \text{ and } X^2 \text{ sum of squares}}{\text{total sum of squares}} \times 100$$

Due to regression on X and X^2 sum of squares

$$= (0.64815)(1.08)$$

$$= 0.7000$$

Residual sum of squares

$$= \text{total sum of squares} - \text{regression sum of squares}$$

$$= 1.0800 - 0.7000$$

$$= 0.3800$$

Source of variation	Sum of squares	Degrees of freedom	Mean square
Due to regression on X	0.6250	1	—
Due to introduction of X^2	0.0750	1	0.0750
Due to regression on X and X^2	0.7000	2	—
Residual	0.3800	12	0.0317
Total	1.0800	14	—

(b) Null hypothesis – the relationship between concentration of impurity (y) and depth (x) is a linear relationship.

Alternative hypothesis – the relationship between concentration of impurity (y) and depth (x) is a quadratic relationship.

$$\text{Test statistic} = \frac{\text{due to introduction of } X^2 \text{ mean square}}{\text{residual mean square}}$$

$$= \frac{0.0750}{0.0317}$$

$$= 2.37$$

Critical values – from the one-sided F-table with 1 and 12 degrees of freedom:

4.75 at the 5% significance level
9.33 at the 1% significance level.

Decision – we cannot reject the null hypothesis.

Conclusion – we conclude that the relationship between concentration of impurity (y) and depth (x) is a linear relationship.

(c) We see that the bottom three rows of the analysis of variance table in Problem 3 are exactly the same as the three rows in the table in Problem 1 except for the names given to the sources of variation.

CHAPTER 16

Solution to 16.1

(a)

X	Z	W	A	B	C	T
−	−	+	+	+	+	267
+	−	−	+	+	+	240
−	+	−	+	+	+	223
+	+	+	+	+	+	241
−	−	+	−	+	−	227
+	−	−	−	+	−	239
−	+	−	−	+	−	226
+	+	+	−	+	−	208
−	−	+	+	−	−	227
+	−	−	+	−	−	253
−	+	−	+	−	−	212
+	+	+	+	−	−	226
−	−	+	−	−	+	246
+	−	−	−	−	+	255
−	+	−	−	−	+	218
+	+	+	−	−	+	256

(b) Effect estimates:

$$X = (-267 + 240 - 223 + 241 - 227 + 239 - 226 + 208 - 227 + 253$$
$$- 212 + 226 - 246 + 255 - 218 + 256)/8$$

$X = 9.0$
$Z = -18.0$
$W = 4.00$
$A = 1.75$
$B = -2.75$
$C = 16.00$

(c)

AX	AZ	AW	BX	BZ	BW	CX	CZ	CW	T
−	−	+	−	−	+	−	−	+	267
+	−	−	+	−	−	+	−	−	240
−	+	−	−	+	−	−	+	−	223
+	+	+	+	+	+	+	+	+	241
+	+	−	−	−	+	+	+	−	227
−	+	+	+	−	−	−	+	+	239
+	−	+	−	+	−	+	−	+	226
−	−	−	+	+	+	−	−	−	208
−	−	+	+	+	−	+	+	−	227
+	−	−	−	+	+	−	+	+	253
−	+	−	+	−	+	+	−	+	212
+	+	+	−	−	−	−	−	−	226
+	+	−	+	+	−	−	−	+	246
−	+	+	−	+	+	+	−	−	255
+	−	+	+	−	+	−	+	−	218
−	−	−	−	−	−	+	+	+	256

(d) Effect estimates

$AX = -1.25$
$AZ = -3.25$
$AW = 4.25$
$BX = -12.75$
$BZ = -0.75$
$BW = -0.25$
$CX = 0.50$
$CZ = 0.50$
$CW = 14.50$

(e)

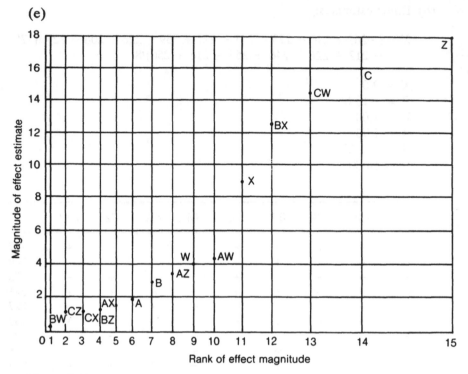

Figure S.20

(f) On the half normal plot, 10 of the points lie roughly on a straight line passing through the origin. It is reasonable to conclude that the other 5 points represent significant effects. We conclude that main effects Z, C and X, together with interactions CW and BX, are the only significant effects.

(g) Let us first interpret the two significant interactions. The interaction CW is illustrated in Figure S.21 and interaction BX in Figure S.22. In both diagrams each point is the mean of 8 results and confidence intervals were calculated using a standard deviation of 5.0 estimated from the half normal plot.

We see in Figure S.21 that the variation in tensile strength due to the use of different storage locations will be greatly reduced if we use 70% natural rubber rather than 80%. We see in Figure S.22 that the variation in tensile strength due to differences between deliveries is very much the same whether we use a mill speed of 50 rpm or 60 rpm. However, there is a strong indication that this variation would be

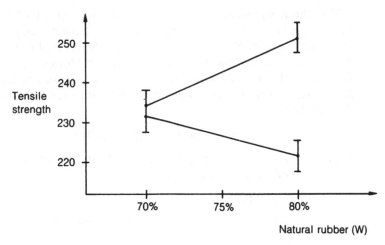

Figure S.21

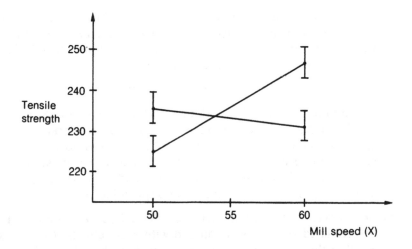

Figure S.22

reduced by using a mill speed of 55 rpm. It is highly desirable, of course, that a plausible scientific explanation be found to support these findings. It is also desirable to confirm the existence of the interactions by further experimentation. Despite the absence of either explanation or confirmation, I recommend the use of 70% natural rubber and a mill speed of 55 rpm, in order to minimize the influence of carbon black variation on the tensile strength.

The three significant main effects are mill speed (X), type of accelerator (Z) and type of storage (C). We have already decided to use a mill speed of 55 rpm and we have set the level of natural rubber at 70% to minimize the effect of variation in storage (Figure S.21). Thus the only task remaining is to select the type of accelerator. To predict the tensile strength we can expect with each type we can use the equation

$$T = \mu + (Z/2)z + (C/2)c + (X/2)x + (CW/2)cw + (BX/2)bx$$

in which T is tensile strength, Z is the effect estimate for main effect Z, z is the coded value of Z $(-1$ or $+1)$ etc.

Substituting the effect estimates we get:

$$T = 235.25 - 9z + 8c + 4.5x + 7.25cw + 6.375bx$$

Substituting $w = -1$, for 70% natural rubber, we get

$$T = 235.25 - 9z + 0.75c + 4.5x + 6.375bx$$

The very small coefficient of c confirms our finding that the use of 70% natural rubber would nullify the influence of variation in storage. Thus we could ignore the $0.75c$ term. For our chosen mill speed of 55 rpm we substitute $x = 0$ into the equation to get:

$$T = 235.25 - 9z$$

Substituting $z = -1$ and $z = +1$ into this final equation gives a predicted tensile strength of 226.3 kg/cm^2 with the HLX accelerator and 244.3 kg/cm^2 with the MD accelerator. If neither figure is close to the desired level of tensile strength for this rubber, the R & D chemist would need to consider adjustments to the mill speed (X) or the percentage natural rubber (W), or taking other appropriate action.

(h) The effects of interactions AB, AC, BC, XZ, XW and ZW could not be estimated as these were aliased with main effects. I do not know whether these interactions are less important than those which can be assessed. Taguchi suggests that an interaction between a pair of control variables, or an interaction between a pair of noise variables, is less important than an interaction between a control variable and a noise variable.

(i) This experiment is a quarter replicate of a 2^6 factorial experiment.

(j) With a 2^6 factorial experiment there are 63 columns in the design matrix. If you extend your design matrix to include three-factor interactions, four-factor interactions, etc., you will find amongst the 63 columns only three which contain all + signs. They are ABC, XZW

and *ABCXZW*. Multiplying each effect by these three, as we did in Chapter 13, gives us the alias groups which are:

X	*Z*	*W*	*A*	*B*	*C*
ABCX	*ABCZ*	*ABCW*	*BC*	*AC*	*AB*
ZW	*XW*	*XZ*	*AXZW*	*BXZW*	*CXZW*
ABCZW	*ABCXW*	*ABCXZ*	*BCXZW*	*ACXZW*	*ABXZW*

AX	*AW*	*AZ*	*BX*	*BW*	*BZ*
BCX	*BCW*	*BCZ*	*ACX*	*ACW*	*ACZ*
AZW	*AXZ*	*AXW*	*BZW*	*BXZ*	*BXW*
BCZW	*BCXZ*	*BCXW*	*ACZW*	*ACXZ*	*ACXW*

CX	*CW*	*CZ*
ABX	*ABW*	*ABZ*
CZW	*CXZ*	*CXW*
ABZW	*ABXZ*	*ABXW*

We can see in the 15 alias groups that interactions between pairs of control factors, (*ZW*, *XW*, and *XZ*) are aliased with main effects. Also, interactions between pairs of noise factors are aliased with main effects. However, each interaction between a control factor and a noise factor (e.g. *AX*, *AW*, etc.) is aliased only with three-factor and higher interactions.

Solution to 16.2

(a)

Control factors				Response	
X	*Z*	*W*	Mean	SD	*S/N*
−	−	+	241.75	19.0679	22.061
+	−	−	246.75	8.4212	29.338
−	+	−	219.75	6.1305	31.089
+	+	+	232.75	20.5487	21.082

(b) Using mean tensile strength as response:

Main effect $X = (-241.75 + 246.75 - 219.75 + 232.75)/2$
$$= 9.0$$

Main effect $Z = (-241.75 - 246.75 + 219.75 + 232.75)/2$
$$= -18.0$$

Main effect $W = (241.75 - 246.75 - 219.75 + 232.75)/2$
$$= 4.0$$

(c) Using SD of tensile strength as response:

Main effect $X = (-19.0679 + 8.4212 - 6.1305 + 20.5487)/2$
$= 1.89$

Main effect $Z = (-19.0679 - 8.4212 + 6.1305 + 20.5487)/2$
$= -0.40$

Main effect $W = (19.0679 - 8.4212 - 6.1305 + 20.5487)/2$
$= 12.53$

(d) Using the signal-to-noise ratio as response:

Main effect $X = (-22.061 + 29.338 - 31.089 + 21.082)/2$
$= -1.36$

Main effect $Z = (-22.061 - 29.338 + 31.089 + 21.082)/2$
$= 0.39$

Main effect $W = (22.061 - 29.338 - 31.089 + 21.082)/2$
$= -8.64$

(e) The analysis we have carried out in parts (b), (c) and (d) has not given us an estimate of the residual standard deviation. Thus we cannot check the statistical significance of the effect estimates. However, in each group of three estimates we see that there is one which is much larger than the other two.

From our analysis of the standard deviations in part (c) we conclude that the percentage natural rubber (W) should be set to the low level (70%) in order to reduce the *variability* in tensile strength. From our analysis of the means in part (b) we conclude that the type of accelerator (Z) should be chosen to achieve the required *level* of tensile strength. These conclusions are entirely consistent with the conclusions we drew in question 1, when we focused on the interaction between control factors and noise factors. However, I am sure you will agree that the earlier analysis had a precision that this one lacks.

Turning to the analysis in part (d) of the signal-to-noise ratios, we see that the three effect estimates are very similar in proportion to those from part (c), but with opposite signs. Taguchi's objective is to maximize the signal-to-noise ratio, in order to minimize the variation in tensile strength. Thus we are driven to using the low level of percentage natural rubber (W), as we concluded from our analysis of the standard deviations. With this set of data it would not matter which of the three formulae we used to calculate the signal-to-noise ratio:

$$S/N \text{ ratio} = 10 \log [(\text{Mean}/\text{SD})^2]$$

$$S/N \text{ ratio} = 10 \log [(1/\text{SD})^2]$$
$$S/N \text{ ratio} = 10 \log [(\text{Mean/SD})^2 - (1/r)]$$

where r = no. of results in each row of Table 2.

All three formulae, and others, appear in different texts on Taguchi methods.

References and further reading

Some of the books and papers listed below are referred to in the chapters others are included for the benefit of readers who wish to study further.

Barnett, V. and Lewis, T. (1979) *Outliers in Statistical Data*, Wiley, Chichester.
Box, G. E. P., Hunter, W. G. and Hunter, J. S. (1978) *Statistics for Experiments*, Wiley, New York.
British Standards 2846 *Guide to Statistical Interpretation of Data*:
　　Part 1 (1975) *Routine Analysis of Quantitative Data*;
　　Part 2 (1981) *Estimation of the Mean: Confidence Interval*;
　　Part 3 (1975) *Determination of a Statistical Tolerance Interval*;
　　Part 4 (1976) *Techniques of Estimation and Tests Relating to Means and Variances*;
　　Part 5 (1977) *Power of Test Relating to Means and Variances*;
　　Part 6 (1976) *Comparison of Two Means in the Case of Paired Observations*;
Caulcutt, R. (1989) *Data Analysis in the Chemical Industry* (Vol. 1), *Basic Techniques*, Ellis Horwood, Chichester.
Caulcutt, R. and Boddy, R. (1983) *Statistics for Analytical Chemists*, Chapman and Hall, London.
Champ, C. W. and Woodall, W. H. (1987) Exact results for Shewhart control charts with supplementary runs rules. *Technometrics*, **28** (4), 393–9.
Chatfield, C. (1978) *Statistics for Technology*, Chapman and Hall, London.
Chatterjee, S. and Price, B. (1977) *Regression Analysis by Example*, Wiley, Chichester.
Cox, D. R. (1958) *Planning of Experiments*, Wiley, Chichester.
Cox, D. R. and Snell, E. J. (1981) *Applied Statistics: Principles and Examples*, Chapman and Hall, London.
Daniel, C. (1976) *Applications of Statistics to Industrial Experimentation*, Wiley, New York.
Daniel, C. and Wood, F. S. (1971) *Fitting Equations to Data*, Wiley, New York.
Davies, O. L. (1978) *Design and Analysis of Industrial Experiments*, Longman, London.
Davies, O. L. and Goldsmith, P. L. (1972) *Statistical Methods in Research and Production*, Longman, London.
Deming, W. E. (1986) *Out of the Crisis*, MIT Press, Cambridge, Mass.
Diamond, W. J. (1981) *Practical Experiments Designs*, Van Nostrand Reinhold, New York.

Draper, N. and Smith, H. (1966) *Applied Regression Analysis*, Wiley, New York.
Ishikawa, K. (1985) *What is Total Quality Control?* Prentice-Hall.
Johnson, N. and Leone, F. C. (1964) *Statistics and Experimental Design* (Vols 1 and 2), Wiley, New York.
Montgomery, D. C. (1976) *Design and Analysis of Experiments*, Wiley, New York.
Moroney, M. J. (1966) *Facts from Figures*, Penguin.
Neave, H. R. (1990) *The Deming Dimension*, SPC Press, Knoxville, Tennessee.
Oakland, J. S. (1989) *Total Quality Management*, Heinemann, Oxford.
Oakland, J. S. and Followell, R. F. (1990) *Statistical Process Control*, Heinemann, Oxford.
Pearson, E. S. and Hartley, H. P. (eds) (1966) *Biometrical Tables for Statisticians*, Biometrica Trust, London.
Ross, P. J. (1988) *Taguchi Techniques for Quality Engineering*, McGraw-Hill, New York.
Taguchi, G. (1987) *System of Experimental Design*, Volumes 1 and 2, UNIPUB.
Wheeler, D. J. (1987) *Understanding Industrial Experimentation*, SPC Press, Knoxville, Tennessee.
Wilson, A. L. (1979) Approach for achieving comparable analytical results. *The Analyst*, **104**, 273–89.

Statistical tables

TABLE A   NORMAL DISTRIBUTION

z = standardized value
Prob. = probability of the standardized value being exceeded

z	Prob.	z	Prob.	z	Prob.	z	Prob.
0.00	0.5000	0.31	0.3783	0.62	0.2676	0.93	0.1762
0.01	0.4960	0.32	0.3745	0.63	0.2643	0.94	0.1736
0.02	0.4920	0.33	0.3707	0.64	0.2611	0.95	0.1711
0.03	0.4880	0.34	0.3669	0.65	0.2578	0.96	0.1685
0.04	0.4840	0.35	0.3632	0.66	0.2546	0.97	0.1660
0.05	0.4801	0.36	0.3594	0.67	0.2515	0.98	0.1635
0.06	0.4761	0.37	0.3557	0.68	0.2483	0.99	0.1611
0.07	0.4721	0.38	0.3520	0.69	0.2451	1.00	0.1587
0.08	0.4681	0.39	0.3483	0.70	0.2420	1.01	0.1562
0.09	0.4641	0.40	0.3446	0.71	0.2389	1.02	0.1539
0.10	0.4602	0.41	0.3409	0.72	0.2358	1.03	0.1515
0.11	0.4562	0.42	0.3372	0.73	0.2327	1.04	0.1492
0.12	0.4522	0.43	0.3336	0.74	0.2296	1.05	0.1469
0.13	0.4483	0.44	0.3300	0.75	0.2266	1.06	0.1446
0.14	0.4443	0.45	0.3264	0.76	0.2236	1.07	0.1423
0.15	0.4404	0.46	0.3228	0.77	0.2206	1.08	0.1401
0.16	0.4364	0.47	0.3192	0.78	0.2177	1.09	0.1379
0.17	0.4325	0.48	0.3156	0.79	0.2148	1.10	0.1357
0.18	0.4286	0.49	0.3121	0.80	0.2119	1.11	0.1335
0.19	0.4247	0.50	0.3085	0.81	0.2090	1.12	0.1314
0.20	0.4207	0.51	0.3050	0.82	0.2061	1.13	0.1292
0.21	0.4168	0.52	0.3015	0.83	0.2033	1.14	0.1271
0.22	0.4129	0.53	0.2981	0.84	0.2005	1.15	0.1251
0.23	0.4090	0.54	0.2946	0.85	0.1977	1.16	0.1230
0.24	0.4052	0.55	0.2912	0.86	0.1949	1.17	0.1210
0.25	0.4013	0.56	0.2877	0.87	0.1922	1.18	0.1190
0.26	0.3974	0.57	0.2843	0.88	0.1894	1.19	0.1170
0.27	0.3936	0.58	0.2810	0.89	0.1867	1.20	0.1151
0.28	0.3897	0.59	0.2776	0.90	0.1841	1.21	0.1131
0.29	0.3859	0.60	0.2743	0.91	0.1814	1.22	0.1112
0.30	0.3821	0.61	0.2709	0.92	0.1788	1.23	0.1093

z	Prob.	z	Prob.	z	Prob.	z	Prob.
1.24	0.1075	1.54	0.0618	1.84	0.0329	2.70	0.00346
1.25	0.1056	1.55	0.0606	1.85	0.0322	2.75	0.00297
1.26	0.1038	1.56	0.0594	1.86	0.0314	2.80	0.00255
1.27	0.1020	1.57	0.0582	1.87	0.0307	2.85	0.00219
1.28	0.1003	1.58	0.0571	1.88	0.0301	2.90	0.00187
1.29	0.0985	1.59	0.0559	1.89	0.0294	2.95	0.00159
1.30	0.0968	1.60	0.0548	1.90	0.0287	3.00	0.00135
1.31	0.0951	1.61	0.0537	1.91	0.0281	3.05	0.00114
1.32	0.0934	1.62	0.0526	1.92	0.0274	3.10	0.00097
1.33	0.0918	1.63	0.0516	1.93	0.0268	3.15	0.00082
1.34	0.0901	1.64	0.0505	1.94	0.0262	3.20	0.00069
1.35	0.0885	1.65	0.0495	1.95	0.0256	3.25	0.00058
1.36	0.0869	1.66	0.0485	1.96	0.0250	3.30	0.00048
1.37	0.0853	1.67	0.0475	1.97	0.0244	3.35	0.00040
1.38	0.0838	1.68	0.0465	1.98	0.0239	3.40	0.00034
1.39	0.0823	1.69	0.0455	1.99	0.0233	3.45	0.00028
1.40	0.0808	1.70	0.0446	2.00	0.0228	3.50	0.00023
1.41	0.0793	1.71	0.0436	2.05	0.0202	3.55	0.00019
1.42	0.0778	1.72	0.0427	2.10	0.0179	3.60	0.00016
1.43	0.0764	1.73	0.0418	2.15	0.0158	3.65	0.00013
1.44	0.0749	1.74	0.0409	2.20	0.0139	3.70	0.00011
1.45	0.0735	1.75	0.0401	2.25	0.0122	3.75	0.00009
1.46	0.0721	1.76	0.0392	2.30	0.0107	3.80	0.00007
1.47	0.0708	1.77	0.0384	2.35	0.0094	3.85	0.00006
1.48	0.0694	1.78	0.0375	2.40	0.0082	3.90	0.00005
1.49	0.0681	1.79	0.0367	2.45	0.0071	3.95	0.00004
1.50	0.0668	1.80	0.0359	2.50	0.0062	4.00	0.00003
1.51	0.0665	1.81	0.0351	2.55	0.0054		
1.52	0.0643	1.82	0.0344	2.60	0.0047		
1.53	0.0630	1.83	0.0336	2.65	0.0040		

PERCENTAGE POINTS

Significance level					
Two-sided			One-sided		
10% (0.10)	5% (0.05)	1% (0.01)	10% (0.10)	5% (0.05)	1% (0.01)
1.64	1.96	2.58	1.28	1.64	2.33

TABLE B CRITICAL VALUES FOR THE *t*-TEST

	Significance level					
	Two-sided test			One-sided test		
Degrees of freedom	10% (0.10)	5% (0.05)	1% (0.01)	10% (0.10)	5% (0.05)	1% (0.01)
1	6.31	12.71	63.66	3.08	6.31	31.82
2	2.92	4.30	9.92	1.89	2.92	6.97
3	2.35	3.18	5.84	1.64	2.35	4.54
4	2.13	2.78	4.60	1.53	2.13	3.75
5	2.02	2.57	4.03	1.48	2.02	3.36
6	1.94	2.45	3.71	1.44	1.94	3.14
7	1.89	2.36	3.50	1.42	1.89	3.00
8	1.86	2.31	3.36	1.40	1.86	2.90
9	1.83	2.26	3.25	1.38	1.83	2.82
10	1.81	2.23	3.17	1.37	1.81	2.76
11	1.80	2.20	3.11	1.36	1.80	2.72
12	1.78	2.18	3.06	1.36	1.78	2.68
13	1.77	2.16	3.01	1.35	1.77	2.65
14	1.76	2.15	2.98	1.35	1.76	2.62
15	1.75	2.13	2.95	1.34	1.75	2.60
16	1.75	2.12	2.92	1.34	1.75	2.58
17	1.74	2.11	2.90	1.33	1.74	2.57
18	1.73	2.10	2.88	1.33	1.73	2.55
19	1.73	2.09	2.86	1.33	1.73	2.54
20	1.72	2.08	2.85	1.32	1.72	2.53
25	1.71	2.06	2.78	1.32	1.71	2.49
30	1.70	2.04	2.75	1.31	1.70	2.46
40	1.68	2.02	2.70	1.30	1.68	2.42
60	1.67	2.00	2.66	1.30	1.67	2.39
120	1.66	1.98	2.62	1.29	1.66	2.36
Infinite	1.64	1.96	2.58	1.28	1.64	2.33

TABLE C CRITICAL VALUES FOR THE *F*-TEST

One-sided at 5% significance level

Degrees of freedom for smaller variance	Degrees of freedom for larger variance														
	1	2	3	4	5	6	7	8	9	10	12	15	20	60	Infinity
1	161.4	199.5	215.7	224.6	230.2	234.0	236.8	238.9	240.5	241.9	243.9	246.0	248.0	252.2	254.3
2	18.51	19.00	19.16	19.25	19.30	19.33	19.35	19.37	19.38	19.40	19.41	19.43	19.45	19.48	19.50
3	10.13	9.55	9.28	9.12	9.01	8.94	8.89	8.85	8.81	8.79	8.74	8.70	8.66	8.57	8.53
4	7.71	6.94	6.59	6.39	6.26	6.16	6.09	6.04	6.00	5.96	5.91	5.86	5.80	5.69	5.63
5	6.61	5.79	5.41	5.19	5.05	4.95	4.88	4.82	4.77	4.74	4.68	4.62	4.56	4.43	4.36
6	5.99	5.14	4.76	4.53	4.39	4.28	4.21	4.15	4.10	4.06	4.00	3.94	3.87	3.74	3.67
7	5.59	4.74	4.35	4.12	3.97	3.87	3.79	3.73	3.68	3.64	3.57	3.51	3.44	3.30	3.23
8	5.32	4.46	4.07	3.84	3.69	3.58	3.50	3.44	3.39	3.35	3.28	3.22	3.15	3.01	2.93
9	5.12	4.26	3.86	3.63	3.48	3.37	3.29	3.23	3.18	3.14	3.07	3.01	2.94	2.79	2.71
10	4.96	4.10	3.71	3.48	3.33	3.22	3.14	3.07	3.02	2.98	2.91	2.85	2.77	2.62	2.54
12	4.75	3.89	3.49	3.26	3.11	3.00	2.91	2.85	2.80	2.75	2.69	2.62	2.54	2.38	2.30
15	4.54	3.68	3.29	3.06	2.90	2.79	2.71	2.64	2.59	2.54	2.48	2.40	2.33	2.16	2.07
20	4.32	3.49	3.10	2.87	2.71	2.60	2.49	2.45	2.39	2.35	2.28	2.20	2.12	1.95	1.84
60	4.00	3.15	2.76	2.53	2.37	2.25	2.17	2.10	2.04	1.99	1.92	1.84	1.75	1.53	1.39
Infinity	3.84	3.00	2.60	2.37	2.21	2.10	2.01	1.94	1.88	1.83	1.75	1.67	1.57	1.32	1.00

One-sided at 1% significance level

Degrees of freedom for smaller variance	Degrees of freedom for larger variance														
	1	2	3	4	5	6	7	8	9	10	12	15	20	60	Infinity
1	4052	5000	5403	5625	5764	5859	5928	5982	6022	6056	6106	6157	6209	6313	6366
2	98.50	99.00	99.17	99.25	99.30	99.33	99.36	99.37	99.39	99.40	99.42	99.43	99.45	99.48	99.50
3	34.12	30.82	29.46	28.71	28.24	27.91	27.67	27.49	27.35	27.23	27.05	26.87	26.69	26.32	26.13
4	21.20	18.00	16.69	15.98	15.52	15.21	14.98	14.80	14.66	14.55	14.37	14.20	14.02	13.65	13.46
5	16.26	13.27	12.06	11.39	10.97	10.67	10.46	10.29	10.16	10.05	9.89	9.72	9.55	9.20	9.02
6	13.75	10.92	9.78	9.15	8.75	8.47	8.26	8.10	7.98	7.87	7.72	7.56	7.40	7.06	6.88
7	12.25	9.55	8.45	7.85	7.46	7.19	6.99	6.84	6.72	6.62	6.47	6.31	6.16	5.82	5.65
8	11.26	8.65	7.59	7.01	6.63	6.37	6.18	6.03	5.91	5.81	5.67	5.52	5.36	5.03	4.86
9	10.56	8.02	6.99	6.42	6.06	5.80	5.61	5.47	5.35	5.26	5.11	4.96	4.81	4.48	4.31
10	10.04	7.56	6.55	5.99	5.64	5.39	5.20	5.06	4.94	4.85	4.71	4.56	4.41	4.08	3.91
12	9.33	6.93	5.95	5.41	5.06	4.82	4.64	4.50	4.39	4.30	4.16	4.01	3.86	3.54	3.36
15	8.68	6.36	5.42	4.89	4.56	4.32	4.14	4.00	3.89	3.80	3.67	3.52	3.37	3.05	2.87
20	8.10	5.85	4.94	4.43	4.10	3.87	3.70	3.56	3.46	3.37	3.23	3.09	2.94	2.61	2.42
60	7.08	4.98	4.13	3.65	3.34	3.12	2.95	2.82	2.72	2.63	2.50	2.35	2.20	1.84	1.60
Infinity	6.63	4.61	3.78	3.32	3.02	2.80	2.64	2.51	2.41	2.32	2.18	2.04	1.88	1.47	1.00

Two-sided at 5% significance level

Degrees of freedom for smaller variance	Degrees of freedom for larger variance														
	1	2	3	4	5	6	7	8	9	10	12	15	20	60	Infinity
1	647.8	799.5	864.2	899.6	921.8	937.1	948.2	956.7	963.3	968.6	976.7	984.9	993.1	1010.0	1061.8
2	38.51	39.00	39.17	39.25	39.30	39.33	39.36	39.37	39.39	39.40	39.41	39.43	39.45	39.48	39.50
3	17.44	16.04	15.44	15.10	14.88	14.73	14.62	14.54	14.47	14.42	14.34	14.25	14.17	13.99	13.90
4	12.22	10.65	9.98	9.60	9.36	9.20	9.07	8.98	8.90	8.84	8.75	8.66	8.56	8.36	8.26
5	10.01	8.43	7.76	7.39	7.15	6.98	6.85	6.76	6.68	6.62	6.52	6.43	6.33	6.12	6.02
6	8.81	7.26	6.60	6.23	5.99	5.82	5.70	5.60	5.52	5.46	5.37	5.27	5.17	4.96	4.85
7	8.07	6.54	5.89	5.52	5.29	5.12	4.99	4.90	4.82	4.76	4.67	4.57	4.47	4.25	4.14
8	7.57	6.06	5.42	5.05	4.82	4.65	4.53	4.43	4.36	4.30	4.20	4.10	4.00	3.78	3.67
9	7.21	5.71	5.08	4.72	4.48	4.32	4.20	4.10	4.03	3.96	3.87	3.77	3.67	3.45	3.33
10	6.94	5.46	4.83	4.47	4.24	4.07	3.95	3.85	3.78	3.72	3.62	3.52	3.42	3.20	3.08
12	6.55	5.10	4.47	4.12	3.89	3.73	3.61	3.51	3.44	3.37	3.28	3.18	3.07	2.85	2.72
15	6.20	4.77	4.15	3.80	3.58	3.41	3.29	3.20	3.12	3.06	2.96	2.86	2.76	2.52	2.40
20	5.87	4.46	3.86	3.51	3.29	3.13	3.01	2.91	2.84	2.77	2.68	2.57	2.46	2.22	2.09
60	5.29	3.93	3.34	3.01	2.79	2.63	2.51	2.41	2.33	2.27	2.17	2.06	1.94	1.67	1.48
Infinity	5.02	3.69	3.12	2.79	2.57	2.41	2.29	2.19	2.11	2.05	1.94	1.83	1.71	1.39	1.00

Two-sided at 1% significance level

Degrees of freedom for smaller variance	Degrees of freedom for larger variance														
	1	2	3	4	5	6	7	8	9	10	12	15	20	60	Infinity
1	16211	20000	21615	22500	23056	23437	23715	23925	24091	24224	24426	24630	24836	25253	25465
2	198.5	199.0	199.2	199.2	199.3	199.3	199.4	199.4	199.4	199.4	199.4	199.4	199.4	199.5	199.5
3	55.55	49.80	47.47	46.19	45.39	44.84	44.43	44.13	43.88	43.69	43.29	43.08	42.78	42.15	41.83
4	31.33	26.28	24.26	23.15	22.46	21.97	21.62	21.35	21.14	20.97	20.70	20.44	20.17	19.61	19.32
5	22.78	18.31	16.53	15.56	14.94	14.51	14.20	13.96	13.77	13.62	13.38	13.15	12.90	12.40	12.14
6	18.63	14.54	12.92	12.03	11.46	11.07	10.79	10.57	10.39	10.25	10.03	9.81	9.59	9.12	8.88
7	16.24	12.40	10.88	10.05	9.52	9.16	8.89	8.68	8.51	8.38	8.18	7.97	7.75	7.31	7.08
8	14.69	11.04	9.60	8.81	8.30	7.95	7.69	7.50	7.34	7.21	7.01	6.81	6.61	6.18	5.95
9	13.61	10.11	8.72	7.96	7.47	7.13	6.88	6.69	6.54	6.42	6.23	6.03	5.83	5.41	5.19
10	12.83	9.43	8.08	7.34	6.87	6.54	6.30	6.12	5.97	5.85	5.66	5.47	5.27	4.86	4.64
12	11.75	8.51	7.23	6.52	6.07	5.76	5.52	5.35	5.20	5.09	4.91	4.72	4.53	4.12	3.90
15	10.80	7.70	6.48	5.80	5.37	5.07	4.85	4.67	4.54	4.42	4.25	4.07	3.88	3.48	3.26
20	9.94	6.99	5.82	5.17	4.76	4.47	4.26	4.09	3.96	3.85	3.68	3.50	3.32	2.92	2.69
60	8.49	5.79	4.73	4.14	3.76	3.49	3.29	3.13	3.01	2.90	2.74	2.57	2.39	1.96	1.69
Infinity	7.88	5.30	4.28	3.72	3.35	3.09	2.90	2.74	2.62	2.52	2.36	2.19	2.00	1.53	1.00

TABLE D   CRITICAL VALUES FOR THE TRIANGULAR TEST

Number of sets of three samples	Significance level		
	5% (0.05)	1% (0.01)	0.1% (0.001)
5	$3\frac{1}{2}$	$4\frac{1}{2}$	—
6	$4\frac{1}{2}$	$5\frac{1}{2}$	—
7	$4\frac{1}{2}$	$5\frac{1}{2}$	$6\frac{1}{2}$
8	$5\frac{1}{2}$	$6\frac{1}{2}$	$7\frac{1}{2}$
9	$5\frac{1}{2}$	$6\frac{1}{2}$	$7\frac{1}{2}$
10	$6\frac{1}{2}$	$7\frac{1}{2}$	$8\frac{1}{2}$
11	$6\frac{1}{2}$	$7\frac{1}{2}$	$9\frac{1}{2}$
12	$7\frac{1}{2}$	$8\frac{1}{2}$	$9\frac{1}{2}$
13	$7\frac{1}{2}$	$8\frac{1}{2}$	$10\frac{1}{2}$
14	$8\frac{1}{2}$	$9\frac{1}{2}$	$10\frac{1}{2}$
15	$8\frac{1}{2}$	$9\frac{1}{2}$	$11\frac{1}{2}$
16	$8\frac{1}{2}$	$10\frac{1}{2}$	$11\frac{1}{2}$
17	$9\frac{1}{2}$	$10\frac{1}{2}$	$12\frac{1}{2}$
18	$9\frac{1}{2}$	$11\frac{1}{2}$	$12\frac{1}{2}$
19	$10\frac{1}{2}$	$11\frac{1}{2}$	$13\frac{1}{2}$
20	$10\frac{1}{2}$	$12\frac{1}{2}$	$13\frac{1}{2}$
25	$12\frac{1}{2}$	$14\frac{1}{2}$	$16\frac{1}{2}$
30	$14\frac{1}{2}$	$16\frac{1}{2}$	$18\frac{1}{2}$
35	$16\frac{1}{2}$	$18\frac{1}{2}$	$21\frac{1}{2}$
40	$18\frac{1}{2}$	$20\frac{1}{2}$	$23\frac{1}{2}$
45	$20\frac{1}{2}$	$23\frac{1}{2}$	$25\frac{1}{2}$
50	$22\frac{1}{2}$	$25\frac{1}{2}$	$27\frac{1}{2}$
60	$26\frac{1}{2}$	$29\frac{1}{2}$	$32\frac{1}{2}$
70	$30\frac{1}{2}$	$33\frac{1}{2}$	$36\frac{1}{2}$
80	$34\frac{1}{2}$	$37\frac{1}{2}$	$40\frac{1}{2}$
90	$37\frac{1}{2}$	$41\frac{1}{2}$	$44\frac{1}{2}$
100	$41\frac{1}{2}$	$45\frac{1}{2}$	$48\frac{1}{2}$

TABLE E   CRITICAL VALUES FOR THE CHI-SQUARED TEST

Degrees of freedom	Significance level	
	5% (0.05)	1% (0.01)
1	3.841	6.635
2	5.991	9.210
3	7.816	11.35
4	9.488	13.28
5	11.07	15.08
6	12.59	16.81
7	14.07	18.49
8	15.51	20.09
9	16.92	21.67
10	18.31	23.21
11	19.68	24.72
12	21.03	26.22
13	22.36	27.69
14	23.68	29.14
15	25.00	30.58
16	26.30	32.00
17	27.59	33.41
18	28.87	34.81
19	30.14	36.19
20	31.41	37.57

TABLE F CONFIDENCE INTERVAL FOR A POPULATION STANDARD DEVIATION

Degrees of freedom	90%		95%	
	L_1	L_2	L_1	L_2
1	0.51	15.9	0.45	31.9
2	0.58	4.42	0.52	6.28
3	0.62	2.92	0.57	3.73
4	0.65	2.37	0.60	2.87
5	0.67	2.09	0.62	2.45
6	0.69	1.92	0.64	2.20
7	0.71	1.80	0.66	2.04
8	0.72	1.71	0.68	1.92
9	0.73	1.65	0.69	1.83
10	0.74	1.59	0.70	1.75
12	0.76	1.52	0.72	1.65
15	0.77	1.44	0.74	1.55
20	0.80	1.36	0.77	1.44
24	0.81	1.32	0.78	1.39
30	0.83	1.27	0.80	1.34
40	0.85	1.23	0.82	1.28
60	0.87	1.18	0.85	1.22
Infinity	1.00	1.00	1.00	1.00

Level of confidence

When estimating a population standard deviation (σ) by means of a sample standard deviation (s) a confidence interval for σ is given by:

Lower limit = $L_1 s$

Upper limit = $L_2 s$

TABLE G   CRITICAL VALUES FOR DIXON'S TEST

Sample size	Significance level	
	5%	1%
3	0.970	0.994
4	0.829	0.926
5	0.710	0.821
6	0.628	0.740
7	0.569	0.680
8	0.608	0.717
9	0.564	0.672
10	0.530	0.635
11	0.502	0.605
12	0.479	0.579
13	0.611	0.697
14	0.586	0.670
15	0.565	0.647
16	0.546	0.627
17	0.529	0.610
18	0.514	0.594
19	0.501	0.580
20	0.489	0.567
21	0.478	0.555
22	0.468	0.544
23	0.459	0.535
24	0.451	0.526
25	0.443	0.517
26	0.436	0.510
27	0.429	0.502
28	0.423	0.495
29	0.417	0.489
30	0.412	0.483
31	0.407	0.477
32	0.402	0.472
33	0.397	0.467
34	0.393	0.462
35	0.388	0.458
36	0.384	0.454
37	0.381	0.450
38	0.377	0.446
39	0.374	0.442
40	0.371	0.438

TABLE H   CRITICAL VALUES OF THE PRODUCT MOMENT CORRELATION

Degrees of freedom	Two-sided test		One-sided test	
	5% (0.05)	1% (0.01)	5% (0.05)	1% (0.01)
2	0.950	0.990	0.900	0.980
3	0.878	0.959	0.805	0.934
4	0.811	0.917	0.729	0.882
5	0.754	0.875	0.669	0.833
6	0.707	0.834	0.621	0.789
7	0.666	0.798	0.582	0.750
8	0.632	0.765	0.549	0.715
9	0.602	0.735	0.521	0.685
10	0.576	0.708	0.497	0.658
11	0.553	0.684	0.476	0.634
12	0.532	0.661	0.457	0.612
13	0.514	0.641	0.441	0.592
14	0.497	0.623	0.426	0.574
15	0.482	0.606	0.412	0.558
20	0.423	0.537	0.360	0.492
30	0.349	0.449	0.296	0.409
40	0.304	0.393	0.257	0.358
60	0.250	0.325	0.211	0.295

TABLE I CRITICAL VALUES FOR GOLDSMITH'S CUSUM TEST

Length of span	Significance level		
	10% (0.10)	5% (0.05)	1% (0.01)
5	2.4	2.7	3.3
6	2.7	3.0	3.6
7	2.9	3.2	4.0
8	3.2	3.5	4.3
9	3.4	3.7	4.6
10	3.6	3.9	4.9
12	3.9	4.3	5.3
15	4.2	4.8	5.8
20	5.2	5.0	6.6
25	5.6	6.0	7.3
30	6.2	6.7	8.0
40	7.2	7.8	9.3
50	8.0	8.6	10.4
60	8.8	9.5	11.3
70	9.5	10.3	12.2
80	10.1	10.8	12.9
90	10.5	11.3	13.7
100	11.0	11.8	14.3

TABLE J   CONTROL LINES FOR RANGE CHARTS

Sample size	Multiplier				
	LAL	LWL	CL	UWL	UAL
2	0.00	0.04	1.13	3.17	4.65
3	0.06	0.30	1.69	3.68	5.05
4	0.20	0.59	2.06	3.98	5.30
5	0.37	0.85	2.33	4.20	5.45
6	0.54	1.06	2.53	4.36	5.60
7	0.69	1.25	2.70	4.49	5.70
8	0.83	1.41	2.85	4.61	5.80
9	0.96	1.55	2.97	4.70	5.90
10	1.08	1.67	3.08	4.79	5.95
11	1.20	1.78	3.17	4.86	6.05
12	1.30	1.88	3.26	4.92	6.10

Control line = multiplier × standard deviation.
The multiplier is taken from the appropriate column of the above table.
The standard deviation is a measure of process variability based on an appropriate set of data.

Index